OECD-FAO Agricultural Outlook 2012-2021

OECD
BETTER POLICIES FOR BETTER LIVES

FOOD AND AGRICULTURE ORGANIZATION
OF THE UNITED NATIONS

Please cite this publication as:
OECD/FAO (2012), *OECD-FAO Agricultural Outlook 2012-2021*, OECD Publishing and FAO.
http://dx.doi.org/10.1787/agr_outlook-2012-en

ISBN 978-92-64-17302-6 (print)
ISBN 978-92-64-17307-1 (PDF)

Annual: OECD-FAO Agricultural Outlook
ISSN 1563-0447 (print)
ISSN 1999-1142 (online)

Photo credits:
Cover illustrations
© Unclesam - Fotolia
© Istockphoto/Aydin Mutlu
© Akalong Suitsuit – Fotolia
© HONG SIEW MEE – Fotolia
© HLPhoto – Fotolia

Corrigenda to OECD publications may be found on line at: *www.oecd.org/publishing/corrigenda*.

Foreword

The Agricultural Outlook *is prepared jointly by the Organisation for Economic Co-operation and Development (OECD) and the Food and Agriculture Organization (FAO) of the United Nations. The main purpose of the report is the attempt to build consensus on global prospects for the agriculture, fisheries and food sectors, and on emerging issues which affect them. Accordingly, the projections and assessments provided in the report are the result of close co-operation with national experts in OECD countries as well as some key non-OECD countries and agro-industry organisations, reflecting the combined knowledge and expertise of this wide group of collaborators. A jointly developed modelling system, based on the OECD's Aglink and FAO's Cosimo models facilitates consistency and analysis of the projections. The fully documented outlook database, including historical data and projections, is available through the OECD-FAO joint internet site* www.agri-outlook.org.

This annual report provides market projections for biofuels, cereals, oilseeds, sugar, meats, dairy products and fish and seafood over the 2012-21 period. The market assessments are contingent on a set of underlying assumptions regarding macroeconomic factors and the continuation of domestic agricultural and trade policies. They also assume normal weather conditions and long-term productivity trends. As such, the Outlook *presents a plausible view on the evolution of global agricultural markets over the next decade and provides a baseline for further analysis of alternative economic or policy assumptions.*

The setting for this Outlook *is one of lowering of agricultural commodity prices from near record levels in response to a large rebound in supplies of major crops and a weaker macroeconomic environment. Food price inflation has eased but remains a concern in developing countries. Looking ahead, the short run global economic outlook has weakened with inflation over the medium term in the major emerging economies expected to be above the OECD area average. Other key assumptions are for a slower population growth and much higher energy prices in the coming decade.*

A critical question addressed in this report is whether the supply response to projected higher commodity prices will be sufficient to meet the future demand for food, feed, fuel and fibre. Rising input costs, increasing resource constraints, growing environmental pressures and the uncertainties of climate change will all have an impact on agricultural output. The key issue facing global agriculture is how to increase productivity in a more sustainable way.

Acknowledgements

This *Agricultural Outlook* is jointly prepared by the OECD and FAO Secretariats.

At the OECD, the *Outlook* report was authored by the Agro-Food Trade and Markets Division of the Trade and Agriculture Directorate: Wayne Jones (Division Head), Pavel Vavra (*Outlook* and baseline co-ordinator), Linda Fulponi, Céline Giner, Ignacio Pérez Domínguez, Garry Smith, Grégoire Tallard and Shinichi Taya. The OECD Secretariat is grateful for the contributions provided by Marcel Adenäuer, Institute for Food and Resource Economics, University Bonn and Pierre Charlebois, Agriculture and Agri-Food Canada. Additional drafting, research and statistical assistance were provided by Armelle Elasri, Alexis Fournier, Gaëlle Gouarin and Claude Nenert, with additional input from Coralie David (DAF/INV) and Ane Kathrine Christensen (ECO/MPD). Meetings organisation and document preparation were provided by Christine Cameron and editing by Michèle Patterson. Technical assistance in the preparation of the *Outlook database* was provided by Frano Ilicic. Many other colleagues in the OECD Secretariat and member country delegations furnished useful comments on earlier drafts of the report.

At the FAO, the team of economists and commodity officers from the Trade and Markets Division contributing to this edition consisted of David Hallam (Division Director), Boubaker BenBelhassen (Principal Officer), Merritt Cluff (Team Leader), Holger Matthey, Abdolreza Abbassian, ElMamoun Amrouk, Pedro Arias, Concepcion Calpe, Darja Regorsek and Peter Thoenes. Arno Becker from AFC joined the team as biofuel consultant. Stefania Vannuccini and Audun Lem contributed from the Fisheries and Aquaculture Department, with technical support from Pierre Charlebois. Keith Wiebe, Leslie Lipper and Seth Meyer from the Agricultural Development Economics Division contributed to the thematic and biofuel chapters. Research assistance and database preparation were provided by Claudio Cerquiglini, Barbara Ferraioli, Berardina Forzinetti, Marco Milo, Fiorella Picchioni and Barbara Senfter. Meeting organisation, document preparation and team administration were provided by Emily Carroll. Several other colleagues from FAO and member country institutions improved this report through valuable details and comments.

Finally, the valuable information and feedback provided by the International Dairy Federation, International Fertilizer Industry Association, International Fishmeal and Fish Oil Organisation, International Grains Council, International Meat Secretariat, International Seed Federation, International Sugar Organisation and CropLife International is gratefully acknowledged.

Table of Contents

Tables

Figures

This book has...

StatLinks

A service that delivers Excel® files from the printed page!

Look for the *StatLinks* at the bottom right-hand corner of the tables or graphs in this book. To download the matching Excel® spreadsheet, just type the link into your Internet browser, starting with the ***http://dx.doi.org*** prefix.

If you're reading the PDF e-book edition, and your PC is connected to the Internet, simply click on the link. You'll find *StatLinks* appearing in more OECD books.

Acronyms and Abbreviations

ACP	African, Caribbean and Pacific countries
AI	Avian influenza
AIS	Agriculture Innovation System
AMAD	Agricultural Market Access Database
AMIS	Agricultural Market Information System
ARS	Argentinean peso
ASEAN	Association of South East Asian Nations
AUD	Australian dollars
AUSFTA	Australia and United States Free Trade Agreement
BN	Billion
BNGY	Billion gallons per year
Bnl	Billion litres
BNLY	Billion litres per year
BRIC	Emerging economies of Brazil, Russian Federation, India and China
BRIIC	Emerging economies of Brazil, Russian Federation, India, Indonesia and China
BRL	Real (Brazil)
BSE	Bovine Spongiform Encephalopathy
Bt	Billion tonnes
BTL	Biomass to liquid
CAD	Canadian dollar
CAFTA	Central American Free Trade Agreement
CAP	Common Agricultural Policy (EU)
CCC	Commodity Credit Corporation
CET	Common External Tariff
CGIAR	Consultative Group on international Agricultural Research
CIS	Commonwealth of Independent States
CN	Combined Nomenclature
CNY	Yuan (China)
COOL	Country of Origin Labelling
CMO	Common Market Organisation for sugar (EU)
CO_2	Carbon dioxide
CPI	Consumer Price Index
CRP	Conservation Reserve Program of the United States
Cts/lb	Cents per pound
Cwe	Carcass weight equivalent
DBES	Date-based export scheme
DDA	Doha Development Agenda
DDG	Dried Distiller's Grains

Dw	Dressed weight
EBA	Everything-But-Arms Initiative (EU)
ECOWAP	West Africa Regional Agricultural Policy
ECOWAS	Economic Community of West African States
EEP	Export Enhancement Program (US)
EISA Act	Energy Independence and Security Act of 2007 (US)
EPA	US Environmental Protection Agency
EPAs	Economic Partnership Agreements (between EU and ACP countries)
ERS	Economic Research Service of the US Department for Agriculture
Est	Estimate
E85	Blends of biofuel in transport fuel that represent 85% of the fuel volume
EU	European Union
EU15	Fifteen member states of the European Union
EU10	Ten new member states of the European Union from May 2004
EU27	Twenty seven member states of the European Union (including Bulgaria and Romania from 2007)
EUR	Euro (Europe)
FAO	Food and Agriculture Organization of the United Nations
FCE Act	Food, Conservation and Energy Act of 2008 US Farm Bill
FDP	Fresh dairy products
FDP	Fertiliser Deep Placement
FFV	Flex fuel Vehicles
FMD	Foot and Mouth Disease
FOB	Free on board (export price)
FR	Federal Reserve (US central bank)
FSRI ACT	Farm Security and Rural Investment Act (US) of 2002
FTA	Free Trade Agreement
G10	Group of ten countries (see Glossary)
G20	Group of 20 developing countries (see Glossary)
GAEZ	Global Agro-Ecological Zones
GAL	Gallons
GATT	General Agreement on Tariffs and Trade
GDP	Gross domestic product
GDPD	Gross domestic product deflator
GEO-GLAM	Group on Earth Observations Global Agricultural Monitoring Initiative
GHG	Green House Gases
GLB	Gross land balances
GMO	Genetically modified organism
Ha	Hectares
HFCS	High fructose corn syrup
Hl	Hectolitre
HS	Harmonised commodity description and coding system
IBRD	International Bank for Reconstruction and Development
ICARDA	International Center for agricultural Research in the Dry Areas
IDA	International Development Association
IEA	International Energy Agency
IFA	International Fertiliser Industry Association

IFAD	International Fund for Agricultural Development
IFDC	International Fertiliser Development Center
IFPRI	International Food Policy Research Institute
IICA	Inter-American Institute for Cooperation Agriculture
iLUC	Indirect land-use change
IMF	International Monetary Fund
INR	Indian rupees
IPCC	Intergovernmental Panel on Climate Change
IPM	Integrated Pest Management
IPR	Intellectual Property Rights
Kg	Kilogrammes
KORUS	Korean-US Free Trade Agreement
KRW	Korean won
Kt	Thousand tonnes
LAC	Latin America and the Caribbean
Lb	Pound
LDCs	Least Developed Countries
LICONSA	Leche Industralizada
Lw	Live weight
MCI	Multiple cropping index
MERCOSUR	Common Market of South America
MFN	Most Favoured Nation
Mha	Million hectares
Mn	Million
MPS	Market Price Support
Mt	Million tonnes
MTBE	Methyl tertiary butyl ether
MXN	Mexican peso
N	Nitrogen
NP	Nitrogen, phosphate
NPK	Nitrogen, phosphate, potassium
NAFTA	North American Free Trade Agreement
NLB	Net land balances
NZD	New Zealand dollar
OECD	Organisation for Economic Cooperation and Development
OIE	World Organisation for Animal Health
P	Phosphorus
p.a.	Per annum
PCE	Private consumption expenditure
PIK	Payment in kind programme (US)
PPP	Purchasing power parity
PR	Phosphate Rocks
PROCAMPO	Mexican Farmers Direct Support Programme
PRRS	Porcine reproductive and respiratory syndrome
PSE	Producer Support Estimate
Pw	Product weight
R&D	Research and development

RED	Renewable Energy Directive in the EU
RFS2	Renewable Fuels Standard in the US, which is part of the Energy Policy Act
RIN	Renewable Identification Numbers prices
Rse	Raw sugar equivalent
Rtc	Ready to cook
RUB	Russian ruble
RUK	Russian Federation, Ukraine and Kazakhstan
Rwt	Retail weight
SAI	Sustainable Agricultural Initiative
SCP	European Food and Sustainable Consumption and Production round table
SFP	Single Farm Payment scheme (EU)
SI	Supplemental Irrigation
SMP	Skim milk powder
SOFA	FAO State of Food and Agriculture
SPS	Sanitary and phytosanitary measures
SRES	Special Report on Emissions Scenarios
STRV	Short tons raw value
T	Tonnes
TBT	Technical Barriers to Trade
T/ha	Tonnes/hectare
THB	Thai baht
TFP	Total factor productivity
TRQ	Tariff rate quota
UDP	Urea Deep placement
UHT	Ultra-heat treatment is the partial sterilisation of food by heating it for a short time
UK	United Kingdom
UN	The United Nations
UNCTAD	United Nations Conference on Trade and Development
UN HLTF	Un High Level Task Force on the Food security crisis
UNEP	FAO United Nations Environment Program
UNICEF	The United Nations Children's Fund
URAA	Uruguay Round Agreement on Agriculture
US	United States
USD	United States dollar
USDA	United States Department of Agriculture
v-CJD	New Creutzfeldt-Jakob Disease
VAT	Value added tax
VHP	Very high polarization sugar
WAEMU	West African Economic and Monetary Union
WFP	World Food Programme
WMP	Whole milk powder
Wse	White sugar equivalent
WTO	World Trade Organisation
ZAR	South African rand
Zn	Zinc

The Outlook in Brief

Recent *OECD-FAO Agricultural Outlook* reports have focused on high and volatile agricultural commodity prices, stressing that prices would come down as markets respond but would remain on a higher plateau underpinned by continuing strong demand and rising costs for some inputs. As anticipated, prices have started to ease but remain at relatively high levels. Food price inflation at the retail level has fallen significantly from its peak in 2008 and its contribution to overall inflation has moderated. Nevertheless, food price inflation remains high in many developing countries and is still outpacing overall inflation in the majority of countries examined.

Price volatility remains a concern, with weather-related yield variability the main threat as long as stocks remain low. With a rebound in crop production, stocks have improved somewhat and markets in 2012 appear less turbulent. The key issue facing global agriculture is how to increase productivity in a more sustainable way to meet the rising demand for food, feed, fuel and fibre.

Nominal prices of the commodities covered in this *Outlook* are expected to trend upwards over the next ten years. Prices in real terms (adjusted for inflation) will remain flat or decline from current levels, but are projected to average 10%-30% above those of the previous decade.

Global agriculture is increasingly linked to energy markets. Oil price projections contained in the macroeconomic assumptions are on average about USD 25 above those used last year (ranging from USD 110 to USD 140 per barrel over the outlook period). These higher oil prices are a fundamental factor behind the higher agricultural commodity price projections, affecting not only oil-related costs of production but also increasing the demand for biofuels and the agricultural feedstocks used in their production.

Despite strong prices, slower production growth is anticipated. Growth in global agricultural production has been above 2% p.a. over the past several decades, but is projected to slow to 1.7% p.a. over the next decade. Growing resource constraints, environmental pressures, and higher costs for some inputs are anticipated to inhibit supply response in virtually all regions. In this context, this *Outlook* suggests that more attention be paid to increasing sustainable agricultural productivity growth.

Based on their greater potential to increase land devoted to agriculture and to improve productivity, developing countries will provide the main source of global production growth to 2021. Annual production growth in developing countries is projected to average 1.9% p.a. compared to 1.2% p.a. in developed countries. An additional 680 million people are expected to inhabit the planet by 2021 with the fastest population growth rates in Africa and India. Rising incomes and urbanisation will lead to changes in diets that shift consumption to more processed foods, fats and animal protein. This will favour higher value meats and dairy products, and drive the indirect demand for coarse grains and oilseeds for livestock feed.

Emerging economies will capture an increasing share of the expanding world trade in agriculture. Most prominent are countries like Brazil, China, Indonesia, Thailand, the Russian Federation and the Ukraine that have made significant investments to boost agricultural production capacity. By 2021, developing countries will account for the majority of exports of rice, oilseeds, vegetable and palm oil, protein meals, sugar, beef, poultry meat, fish and fish products.

Commodity highlights

Global production of bio-ethanol and bio-diesel is projected to almost double by 2021, heavily concentrated in Brazil, the United States, and the European Union. Biofuels are based mainly on agricultural feedstocks and are expected to consume a growing share of the global production of sugarcane (34%), vegetable oil (16%), and coarse grains (14%) by 2021.

In response to government mandates, biofuel trade between the United States and Brazil is expected to increase. This *Outlook* anticipates that the United States would import sugarcane-based ethanol mainly from Brazil to help meet domestic demand created by its mandate for advanced biofuels, while Brazil would import lower priced maize-based ethanol principally from the United States to satisfy the demand from its large fleet of flex-fuel vehicles. US low-blend ethanol demand is expected to be constrained by the blend wall from 2016 onwards.

Cereal stock-to-use ratios will remain below historical averages, posing the risk of future price volatility. The Russian Federation, Ukraine and Kazakhstan are expected to become much more important sources of wheat exports by 2021, but high production variability in this region may have implications for global trade and world price volatility. Larger exports of rice are projected from Least Developed Countries in Asia, while rice imports are to increase in Africa.

Oilseeds production and exports continue to be dominated by the traditional players, but emerging exporters like the Ukraine and Paraguay are expected to increasingly contribute to global export growth. China, the dominant importer, will account for more than half of total world imports. Brazil's oilseed production growth is expected to slow from 4.9% to less than 2% p.a. over the outlook period.

Food and ethanol demand for sugar crops will be sustained over the medium term, maintaining high sugar prices. Production cycles will continue to characterise sugar markets in Asia, leading to occasional large trade fluctuations and price volatility. Because of Brazil's dominant position in the sugar market, the allocation of its sugarcane crop between ethanol and sugar production remains a key market driver.

Increased demand for meats will mostly stem from large economies in Asia, crude oil exporting countries and Latin America, where income gains are expected to be significant. Poultry meat will lead this anticipated growth as the cheapest and most accessible source of meat protein, overtaking pigmeat as the largest meat sector by the end of the outlook period.

Fish production is one of the fastest growing sources of animal protein. World fisheries and aquaculture production are expected to grow by 15% over the projection period. However, with a 33% growth in aquaculture production, it will surpass capture fisheries as the primary source of fish for human consumption by 2018.

A modest increase in consumption of dairy products is expected in developed countries with the exception of cheese and fresh dairy products, while in developing regions consumption of all products is expected to increase about 30% by 2021. Developing countries are projected to overtake developed countries in milk production by 2013, with large increases in China and India.

A time for change – longer term perspectives

Agricultural production needs to increase by 60% over the next 40 years to meet the rising demand for food. This translates into an additional 1 Bnt of cereals and 200 Mt of meat a year by 2050 compared with 2005/07 levels. Additional production will also be necessary to provide feedstock for expanding biofuel production.

Globally, the scope for area expansion is limited. Total arable land is projected to increase by only 69 Mha (less than 5%) by 2050. Additional production will need to come from increased productivity in the same way as it has for the past 50 years.

Increasing productivity will be central to containing food prices in a context of rising resource constraints and will be a key factor in reducing global food insecurity. Productivity gains in the medium-term may come primarily from reducing the productivity gap in developing countries, but a stylised scenario suggests that a significant share of the increased output of crops, used as feedstocks, could be expected to go into biofuel production.

At the same time, there is a growing need to improve the sustainable use of available land, water, marine ecosystems, fish stocks, forests, and biodiversity. Some 25% of all agricultural land is highly degraded. Critical water scarcity in agriculture is a fact for many countries. Many fish stocks are over-exploited, or in risk of being over-exploited. There is a growing consensus that climate change and extreme weather events will increase.

Encouraging better agronomic practices, creating the right commercial, technical and regulatory environment, and strengthening agricultural innovation systems (*e.g.* research, education, extension, infrastructure), including measures addressing the specific needs of smallholders, are essential policy challenges identified in this report. Measures to reduce food loss and waste are also key to meeting rising demand and improving productivity in the supply chain.

OECD-FAO Agricultural Outlook 2012

Chapter 1

Overview of the OECD-FAO Agricultural Outlook 2012

Introduction

The *Agricultural Outlook* is produced collaboratively by the Organisation for Economic Co-operation and Development (OECD) and the Food and Agriculture Organization (FAO) of the United Nations. It brings together the commodity, policy and country expertise of both organisations and input from collaborating member countries to provide an annual assessment of medium-run projections of national, regional and global agricultural commodity markets. The projections of the *Agricultural Outlook* constitute a plausible scenario of what may happen under a certain set of assumptions, such as normal weather conditions, a specific macroeconomic environment over the coming ten years, population trends, as well as current agricultural and trade policy settings of countries around the world. The projections cover production, consumption, stocks, trade and prices for the different agricultural products for the period 2012 to 2021. The evolution of markets over the outlook period is typically described using the annual growth rate or percentage changes for the final year of 2021 relative to a three-year base period of 2009-11. Given the uncertainties which envelope agricultural markets, the final section of the Overview discusses important assumptions affecting the future evolution of agricultural markets and the sensitivity of the baseline projection to these conditioning factors. One aspect of uncertainties concerning future agricultural supplies is the role to be played by agricultural productivity growth. This is addressed comprehensively in the special feature on achieving sustainable agricultural productivity growth in the second chapter of this report.

The setting – A large rebound in supplies to calm markets

After considerable turbulence in recent years, a large rebound in supplies of major crops in response to high prices has helped to restore market balances. Many commodity prices have started to decline in 2011/12 with the easing of tight fundamentals and a weaker macroeconomic environment. Farmers have been able to achieve strong cash returns from near record prices over the past two or three seasons and are expected to use the accumulated proceeds to re-invest in expanded production. As a result, a strong increase of global production is projected in the near term for most agricultural crops, although prospects vary by agricultural product. This trend to lower crop prices will also relieve pressure on prices of feedstuffs and, consequently, on high prices of livestock products at the start of the outlook period. While high market volatility has abated in the short term with the improvement in stock cover from larger production, it remains a medium-term risk for international markets. A short crop or a restrictive trade policy action in a major producing and exporting region could quickly lead to a reversal in market fundamentals and further price surges.

Recent market tightness for biofuels is set to ease at the start of the outlook period. High prices for bioethanol and biodiesel are currently underpinned by high crude oil and energy prices, but both are expected to decline in 2012/13 with increasing supplies. World wheat and coarse grain prices are projected to decline over 2012/13 with the expected increase in global supplies, although they will still remain at relatively high levels and stay vulnerable to developments in other markets. International rice prices fall in 2012/13, amid ample supplies in Asian exporting countries and weakening import demand, as several importers achieve higher domestic production.

World oilseed prices have started to firm in 2012 as production in 2011/12 was less than needed to satisfy demand, further tightening the market situation. Higher prices may bring forth increased global production and larger export supplies in the United States that will put downward pressure on world oilseed prices in 2012/13. Larger supplies of oilseeds for crushing will in turn put a cap on any rally in protein meal prices in 2012/13, while continuing fast growth in demand leads to further gains in vegetable oil prices in the same period.

Despite a sharply lower output of sugar in Brazil, which has suffered the first decline in production in six years of continuous expansion, larger output in other countries in response to high prices resulted in record production of sugar in 2011/12, and the emergence of a substantial global surplus. World sugar prices are expected to average lower in 2012/13, while still remaining at relatively high levels, and to display less of their characteristic volatility as stock cover improves.

High feed costs over several past seasons and a slowdown in demand with weaker economic conditions have combined to reduce producer returns in the livestock sector, encouraging them to reduce animal inventories and slowing total meat production in the years leading up to the outlook period. Higher producer prices, with feed costs easing in the short term, can be expected to improve meat margins and set the stage for some expansion in production of red meats and poultry in 2012/13.

The production of fish from capture fisheries and from aquaculture are both expected to witness an increase in supplies in 2012. World demand for fishmeal and fish oil is expected to grow faster than supply, boosting their prices. Prices for capture, aquaculture and traded fish products are expected to continue their steady increase and rise further in 2012.

For the dairy sector, strong supply response, stimulated by high returns, excellent fodder and good pasture conditions in leading producing countries in Oceania and parts of Latin America, result in rising global milk production in 2012. Increased production of dairy products and slower growth in consumption is projected to lead to a slowdown in the rate of increase of dairy product prices in 2012.

In addition to the market factors cited above that are specific to each commodity, a number of other developments are playing an important role in conditioning all agricultural markets in both the near term, as well as the longer-run developments in global agriculture. Among the important influences are the generally weaker world macroeconomic environment with the eurozone crisis, high oil and energy prices that continue to fluctuate around a rising trend with increased uncertainty and a depreciated US dollar. In addition, government intervention remains widespread in agricultural markets and continues to influence production, consumption and trade. Box 1.1 discusses the main assumptions underlying the agricultural projections and the uncertainties section of this chapter attempts to highlight the contribution to agricultural market prospects of some of these key assumptions.

Food price inflation slowing across the world

While this *Outlook* does not project food prices, it is instructive to examine recent trends. Rising retail food prices are a concern in most countries, though particularly for low-income countries, as well as for low-income households more generally. Food inflation slowed in the majority of countries across the globe over the past year. It declined in approximately two-thirds of the developing countries and the emerging economies of Brazil, the Russian Federation, India, Indonesia and China (the so-called BRIICs) as well as in over a third of the OECD countries.[1] The widespread slowing of food price inflation coincides with the slowing in the growth of the FAO food commodity price index which increased less than 1% over the year ending January 2012 compared to 3.6% in the previous year.

Though the slowing of food price inflation does not mean that food prices, in absolute terms, have come down, its decline should be viewed as welcome news particularly when coupled with household income increases experienced over the past year in many developing and emerging countries. While food price inflation outpaced overall inflation in the majority of countries examined, its slowing has helped to slow overall inflation.

Examining average annual food price inflation rates for OECD and developing country aggregates over the past ten years indicates that they have been both higher and more variable in the developing world than in the OECD area (Figure 1.1).[2] This reflects the greater weight of basic foodstuffs in the consumer food basket in developing countries and high rates of inflation in prices of inputs such as labour in these countries, despite wage rates remaining relatively low. However, these aggregates mask substantial variation amongst countries (Table 1.1).[3]

Figure 1.1. **Food price inflation rates are higher in developing countries**

Average annual food price inflation rates in per cent: 2001-2011

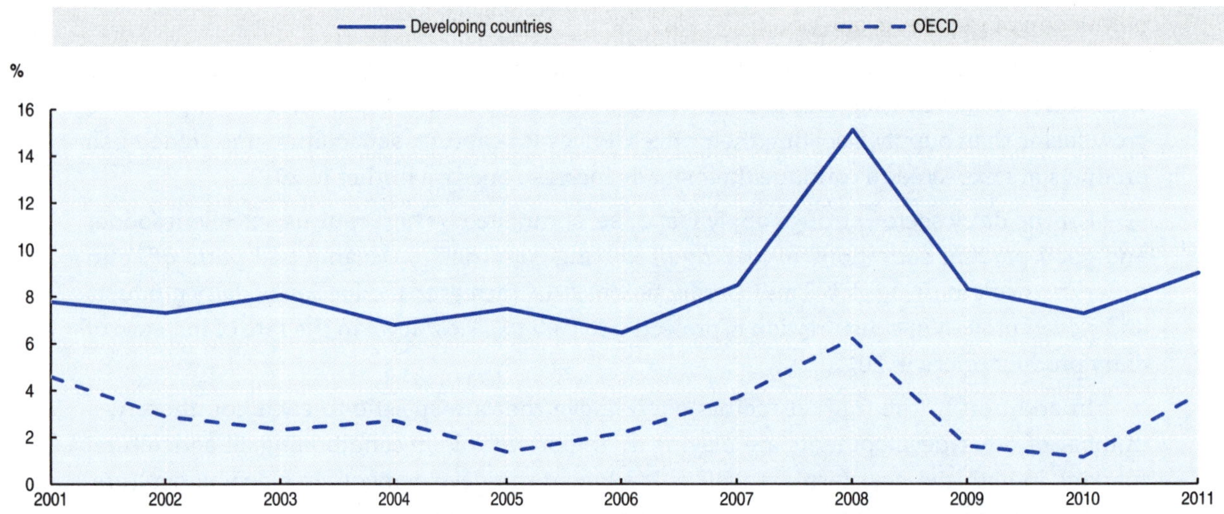

Source: *Main economic indicators*, OECD, OECD Secretariat calculations based on national sources and ILO data.

StatLink http://dx.doi.org/10.1787/888932638773

In the year ending January 2012, food price inflation increased sharply in South Africa, but slowed in Brazil, Indonesia, India and the Russian Federation while remaining fairly stable in China. On the African continent, food inflation was characterised by marked differences: decelerating in much of West and North Africa, while accelerating significantly in East and Southern Africa. It also moderated in many of the large Asian countries with a strong deceleration in Pakistan, India and Bangladesh, often falling by 40% or more. Declines were also observed in large numbers of South and Central American countries, but again there were exceptions, such as Guatemala, Ecuador and Chile, where it rose significantly (Figure 1.2). Detailed information on a country basis is provided in Table B.1 of the Statistical Annex.

Table 1.1. **Food Inflation rates in the BRIICS, January 2000 to January 2011**

Annual per cent change

	1-2001	1-2002	1-2003	1-2004	1-2005	1-2006	1-2007	1-2008	1-2009	1-2010	1-2011	1-2012
OECD total	**3.6**	**5.3**	**0.9**	**3.2**	**1.7**	**2**	**2.5**	**5.3**	**5.4**	**–0.5**	**2.6**	**4.3**
Brazil	3	9.9	21	6.1	3.8	1.3	2	11.5	10.3	3.6	10.4	6.9
China	–0.4	–1.9	2.4	8	4	3.6	5	18.2	4.2	3.7	10.3	10.5
India	–5.8	2.4	4	4.2	2.5	4.8	10.9	5.9	12.7	20.6	7.5	0.3
Indonesia	0.8	15.4	4.5	1.7	8.2	15.2	11.2	11.4	13.5	4.9	16.2	3.3
Russian Federation	19.2	17.6	10.9	9.2	12.9	10.8	7	18.4	16.8	6.3	14.2	2.1
South Africa	5	12.3	17.4	1.3	1.2	3.6	7.7	13.2	16	2	3.1	10.5

Source: *Main economic indicators*, OECD, OECD Secretariat calculations.

StatLink http://dx.doi.org/10.1787/888932640464

OECD countries

Food price inflation also varied across OECD countries (Figure 1.3). Indeed, food price inflation reached 4.6% for the aggregate of OECD countries in 2011, with a number of countries registering inflation rates of 5% or more this past year including the United States, Mexico, Chile, Turkey, Denmark, Iceland and the Czech Republic. However, food price inflation rates fell in about 40% of OECD countries with considerable disparities between countries. Even among countries of the European Union, with similar agricultural and economic policies, food inflation rates differed greatly.

Figure 1.2. **Food price inflation shows considerable variation in BRIICS and developing countries**

Food price inflation per cent change

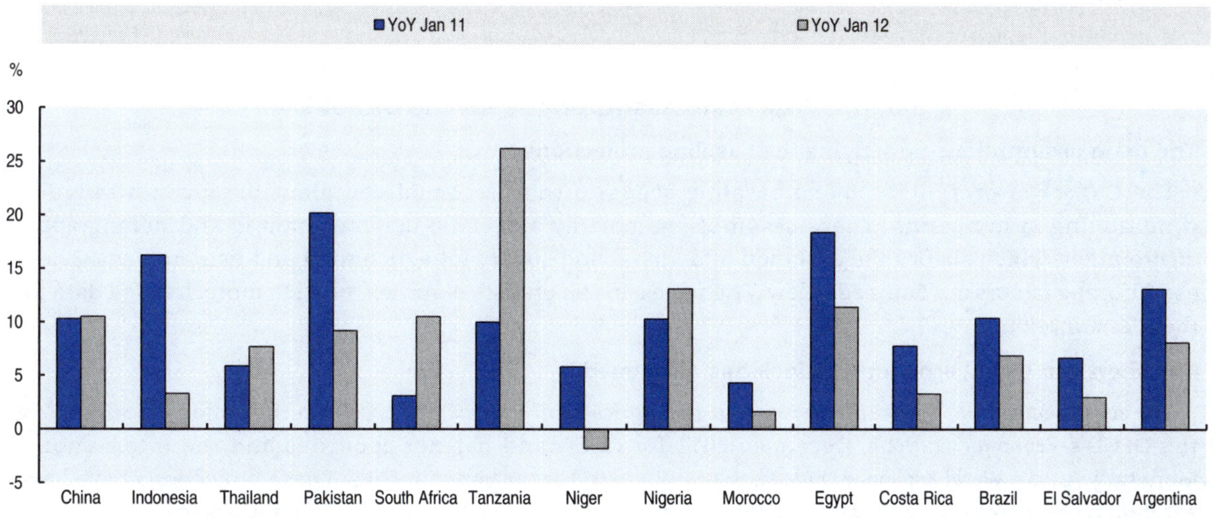

Source: OECD Secretariat calculations based on national sources.

StatLink http://dx.doi.org/10.1787/888932638792

In addition to food price inflation, food expenditure shares are also important in determining the impact of food prices on household welfare. The share of household budgets allocated to food expenditures has declined in most countries over recent decades even though it remains above 25% for many countries in the world (in Table B.1 of the Statistical Annex). For instance, a number of African and South Asian countries have experienced significant decreases in food expenditure shares, often from 50%, or more, to approximately

Figure 1.3. **Annual food price inflation rates: selected OECD countries**

Food price inflation per cent change

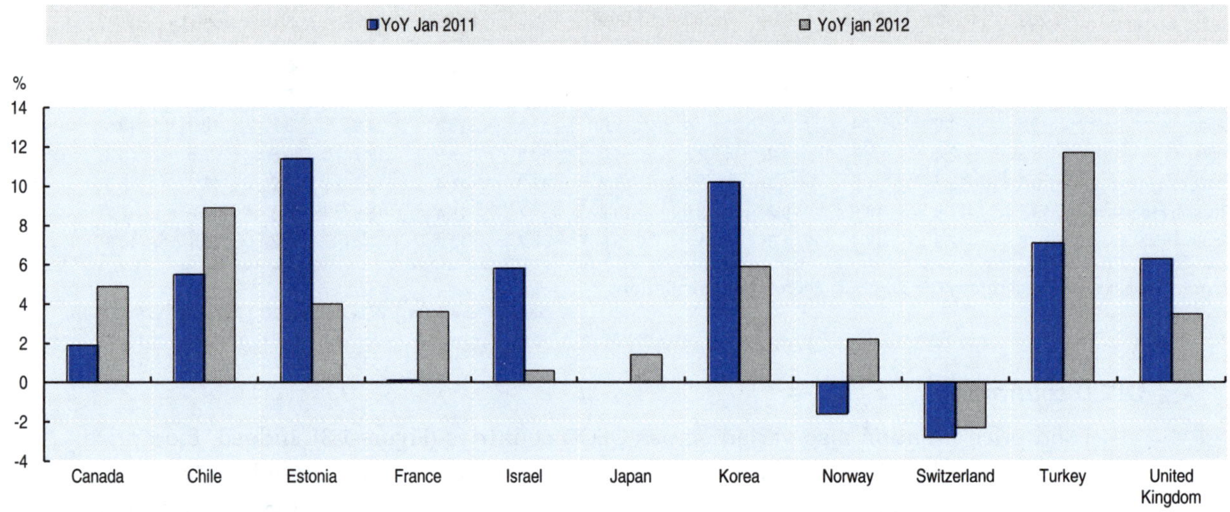

Source: Main economic indicators, OECD.

StatLink http://dx.doi.org/10.1787/888932638811

30-35%.[4] Food price increases in countries where a large share of income is devoted to food expenditures imply reallocation of expenditure to food from other goods and services, such as education, health and transport, which are needed for economic growth. Thus, food price inflation remains an important variable closely watched in both OECD and non-OECD countries.

Box 1.1. **The main assumptions for the Outlook**

The main assumptions underlying the baseline projections

The *Outlook* is presented as one baseline scenario that is considered plausible given a range of conditioning assumptions. These assumptions portray a specific macroeconomic and demographic environment which shapes the evolution of demand and supply for agricultural and fish products. These conditioning factors are outlined below. The tables, in the Statistical Annex, provide more detailed data for these assumptions.

The short-run global economic outlook has weakened

The macroeconomic assumptions used in the *Agricultural Outlook* are based on the "middle scenario" of the OECD's *Economic Outlook*, December 2011 for OECD and partner countries and the International Monetary Fund's *World Economic Outlook*, September 2011 for other countries. These projections have been extended where necessary to cover the full ten-year period of this *Outlook*. This scenario is based on several assumptions including: the continuation of accommodative monetary policy; eventual containment of the sovereign debt and banking sector problems in the Euro area; and a gradual return of confidence amongst economic agents. The short-run economic outlook has deteriorated as fiscal consolidation measures in many OECD countries put a drag on growth. However, some immediate near-term risks have receded. Nonetheless, a two speed growth dichotomy continues to exist between a slow, halting transition back towards relatively weak medium run income and employment growth in much of the OECD area's developed economies, and much stronger income and employment growth in the developing and emerging economies.

Box 1.1. **The main assumptions for the Outlook** *(cont.)*

Relative to expectations from the second half of 2011, the global economy is slowing as many concerns remain about public debt in the Euro area. The OECD's outlook for economic growth in the 17 member Euro area is "unusually uncertain" with large downside risks from lack of effective policy action which could lead to widespread recession. The downside risks lie in the possibility of sovereign default and cross-border effects on creditors. Added to this financial crisis, policy makers must address a social crisis with stubbornly high unemployment levels and with rather gloomy prospects for growth to relieve the unemployment situation. Weak demand in much of the OECD area is also starting to slow growth in the large emerging countries and across the developing world (Figure 1.4).

Figure 1.4. **GDP growth: uncertainties concerning the speed of the recovery**

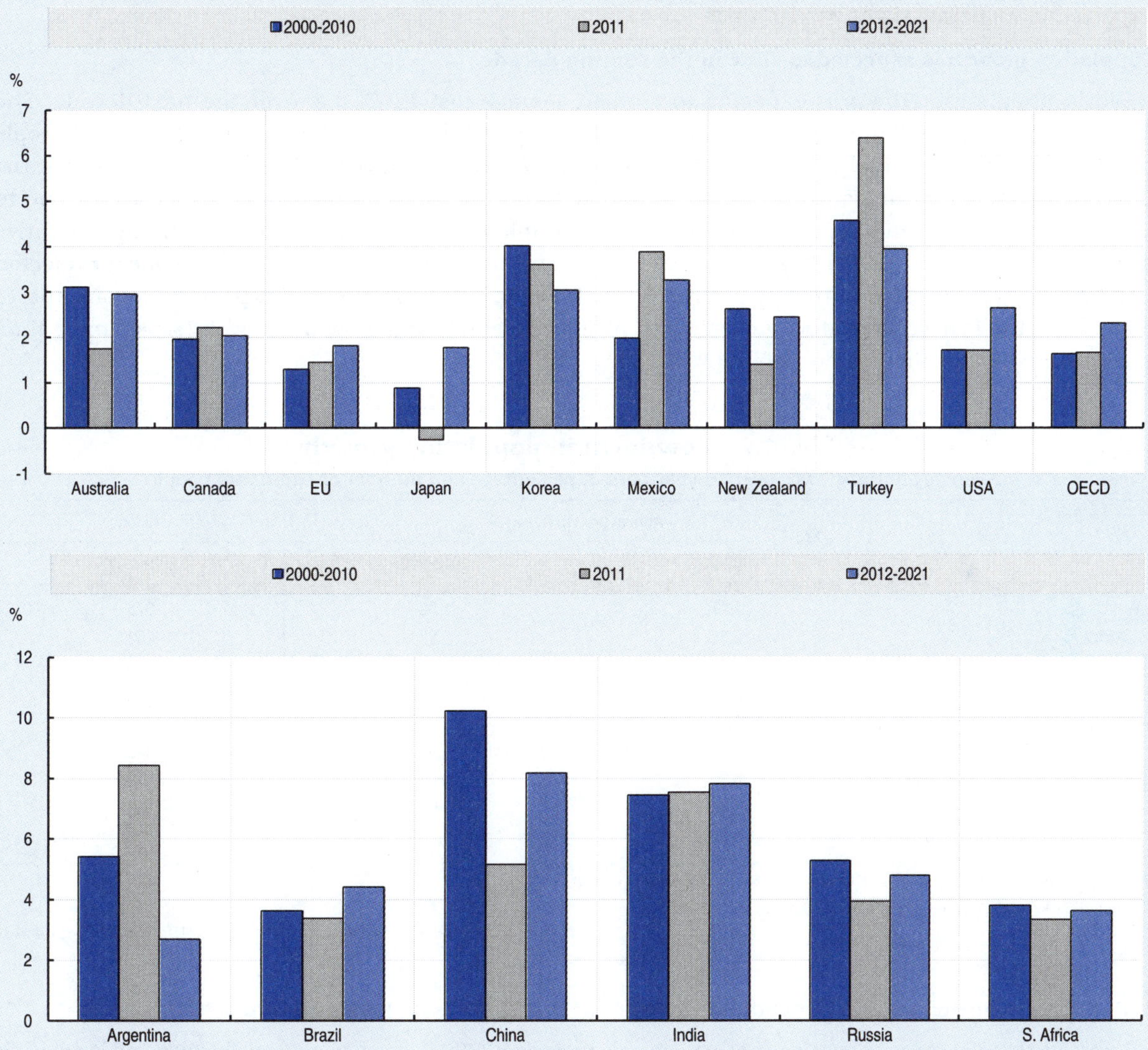

Note: Average annual growth is the least-squares growth rate.

Source: OECD and FAO Secretariats.

StatLink http://dx.doi.org/10.1787/888932638830

Box 1.1. **The main assumptions for the Outlook** *(cont.)*

Under this scenario, growth prospects for the OECD area in the short term are assumed to be weak, with an overall growth rate of 1.6% in 2012. Economic growth prospects for OECD countries in the medium term are also expected to improve to average above 2% p.a. Among OECD countries, Turkey is expected to have the strongest growth during the next ten years, averaging more than 4% p.a. Among the larger economies, the United States is expected to average 2.6% p.a. during the next ten years, while Australia, Korea and Mexico are expected to grow at around 3% p.a.

Economic growth prospects for countries outside the OECD area in the developing world are generally more positive. Assuming the downside risks in the OECD area do not materialise, Brazil, the Russian Federation, India and China are anticipated to continue growing over the next ten years at an impressive pace of around 8% p.a. These countries remain the high growth leaders, with prospects for substantial market expansion. After expanding strongly in the last decade at around 6.6% p.a., growth in the LDCs is expected to moderate slightly to 5.8% p.a.

Population growth is expected to slow in the coming decade

World population growth is expected to slow to average just 1.02% p.a. over the next decade. The slowdown in the growth rate is manifested in all regions. Nonetheless, about 680 million additional people needing food, fibre and fuel will inhabit the planet by 2021. Europe, in general, including members of the European Union, continues to exhibit a low population growth rate which is projected to decline to 0.11% p.a. by 2021. Japan's population is expected to shrink during the next decade, exhibiting a negative growth rate of –0.18% p.a. Turkey, Mexico, Australia and the United States have the highest projected population growth rates within the OECD area. Developing countries are expected to continue to experience the fastest population growth, with Africa as a whole still growing at over 2% p.a., more than double than in any other region (Table 1.2).

Table 1.2. **Slowdown in population growth**

Average annual growth over 10-year period, percentage and number of additional people

	Annual growth rate in %		Millions
	2002-2011	2012-2021	2021-2012
World	1.17	1.02	678
Africa	2.38	2.29	233
Latin America and Caribbean	1.19	0.97	55
North America	0.92	0.82	27
Europe	0.19	0.11	7
Asia and Pacific	1.14	0.91	345
China	0.52	0.30	43
India	1.46	1.20	160
Oceania Developed	1.56	1.16	3

Note: Average annual growth is the least-squares growth rate.
Source: UN World Population Prospects (2010 Revision).

StatLink http://dx.doi.org/10.1787/888932640483

Inflation is expected to remain subdued throughout the OECD, with levels close to 2%

Even with accommodating monetary policy in many OECD countries, weak demand and high unemployment that characterises many of the developed economies, is expected to keep inflation in check. As measured by the Private Consumer Expenditure (PCE) deflator, inflation in the OECD area over the next ten years is assumed to average 2% p.a., slightly below the 2000 to 2011 average.

Box 1.1. **The main assumptions for the Outlook** *(cont.)*

Deflationary pressures in Japan are expected to diminish in the medium term, and the general price level is expected to rise after 2014. In emerging countries, falls in non-agricultural commodity prices and the slower growth of the global economy have mitigated inflationary pressures. Nonetheless, inflation in many of these rapidly growing economies is expected to be above the OECD average. For example, in India and South Africa, inflation is projected to average about 4.8% p.a. during the next ten years. Although not formally included as part of the *Outlook*, low real interest rates are likely to remain an important factor to encourage investment and further consolidation in agriculture and food sectors in the coming ten years.

Inflation also affects exchange rates which play a critical role in the baseline projections as they can strongly influence relative competitiveness for exporters and affordability for importers and hence agricultural trade across regions.

The exchange rates in the baseline relative to the US dollar are held constant at their 2011 levels in real terms for the projection period. The nominal exchange rate for the period 2012-21 is driven by the inflation differentials *vis-à-vis* the United States. Inflation differentials between the United States and certain dynamic economies (the Russian Federation, Argentina, South Africa and Brazil) will drive down the value of their currencies, but no big adjustments are projected, as inflation is expected to remain under control during the projection period.

Energy prices to be much higher

A major upward revision has been made to the assumptions regarding oil and energy prices over the medium term. The world oil price (Brent), rose significantly in 2011.Consequently several agencies providing long term energy price forecasts have revised their expectations upwards. The assumption used for this *Outlook* projection is from the OECD *Economic Outlook*, No. 90 (December 2011). In this projection, the crude oil price, in nominal terms, is expected to increase slowly over the outlook period from USD 111 per barrel in 2011 to USD 142 per barrel by 2021, an average annual growth rate of 2.9%. This oil price projection is well above those included in past *Agricultural Outlooks*, and to a considerable extent is responsible for the higher crop prices anticipated in this *Outlook*, when compared to last year's edition.

Policy considerations

Policy has long been recognised as exerting an important influence in agricultural and fisheries markets. In many cases, the policy reforms of the past decade or so have changed the shape of markets. The introduction of more decoupled payments and progress towards the elimination of direct price supports mean that policy measures have less direct influence on production decisions in many countries. However, import protection and domestic support policies for agriculture still loom large in many developed economies, while the recent application of export taxes or bans in some emerging and developing countries has also had important impacts that reflect the growing importance of these countries in world agricultural trade. This *Outlook* assumes that policies will continue to be applied in line with existing legislation. A conclusion to the Doha Development Agenda of multilateral trade negotiations, that include trade in agricultural products, is not anticipated in this baseline. The Russian Federation is assumed to ratify the protocol of accession signed on 16 December 2011 and to become a full member of the World Trade Organisation (WTO) in 2012.

Summary of main market trends and developments

Agricultural prices to remain on a higher plateau

This *Outlook* maintains the perspective of recent years that agricultural commodity prices will remain on a high plateau throughout the next decade, underpinned by the assumption that oil prices will continue to rise in both nominal and real terms. The eventual strengthening of global economic growth and stronger demand for agricultural products, even with high and rising oil and energy prices, along with growing biofuel demand and slowing production growth, are expected to help keep the prices of agricultural products at relatively high levels over the outlook period. Higher input costs (fertiliser, chemicals) reflecting increasing oil prices will tend to slow yield and productivity growth, and together with resource pressures on water and land availability for agricultural expansion will cut the pace of production increases and lead to less accumulation of stocks.

While world prices for many agricultural crops are projected to remain high, they will nevertheless decline from 2011 levels in the near term as global production continues to respond to past high prices, stocks rebuild and demand initially grows less rapidly with weaker macroeconomic conditions. Prices of many meat products are high at the start of the outlook period, reflecting reduced livestock inventories and producer margins which have been squeezed over several years by high grain and protein meal prices. Beyond the near term, stronger demand growth and rising production costs will underpin high commodity prices over the remainder of the outlook period

The general evolution of prices for selected commodities over the projection period is shown in both nominal and real terms in Figures 1.5, 1.6, and 1.7.

The prices of all the commodities covered in this *Outlook* are projected to average higher in nominal terms than in the previous decade. In the baseline projection, the prices of meats, fish and dairy products rise relative to the costs of feedstuffs of grains and oilseed meals, coarse grains prices rise slightly more than oilseed prices and biofuel prices for bio-ethanol and biodiesel rise more than their feedstock prices of grains, oilseeds, vegetable oils and sugar crops.

When these prices are expressed in real terms (*i.e.* adjusted for inflation) all commodity prices apart from food grains of wheat and rice will be higher than their average level in the previous decade of 2002-11.

Prices are expected to be high enough to encourage producers of crop and livestock products to increase areas harvested and animal inventories and to achieve increased productivity through further investment in the use of improved seed varieties, inputs and high quality feedstuffs as well as to adopt productivity enhancing technologies in the face of rising energy prices. The projected prices, in real terms, for traditional agricultural commodities, are highest for livestock products. While this price development is projected to also hold true for coarse grains, oilseeds, protein meals, vegetable oils, cheese and the two milk powders, the difference with the past decade will be less pronounced.

One reason is that many of these products did not experience a surge in prices in 2007/08 as occurred for cereals and oilseeds. For the non-traditional agricultural commodities of renewable biofuels, projected prices in real terms are expected to be substantially higher, on average, to 2021, particularly for ethanol (up 75%) in comparison with both the last decade and the base period of 2009-11.

The smaller rise in feed costs relative to projected meat prices will improve livestock sector net margins and, with increased demand, will provide the incentives for a moderate pace of expansion in livestock inventories and for increased production of meat and dairy products over the outlook period. Rising *per capita* consumption of fish products is expected to lead to increasing prices of fish from both capture and aquaculture sources over the next ten years. Aquaculture prices are expected to see more rapid growth than capture fisheries, driven by higher cost of inputs.

While the risks of high price volatility are expected to abate somewhat in the near term with more comfortable commodity supply expectations and rising stocks, any

Figure 1.5. **All agricultural commodity prices to average higher in nominal terms in 2012-21 relative to the previous decade**

Per cent change of average nominal prices in 2012-21 relative to different base periods

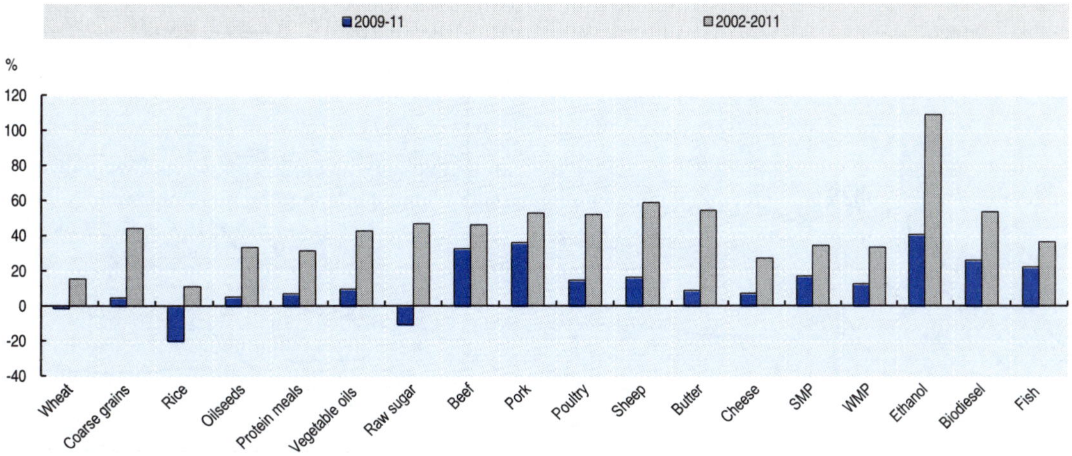

Source: OECD and FAO Secretariats.

StatLink http://dx.doi.org/10.1787/888932638849

Figure 1.6. **In real terms livestock product prices to show some of the highest average gains in 2012-21 relative to the previous decade**

Per cent change of average real prices relative to different base periods

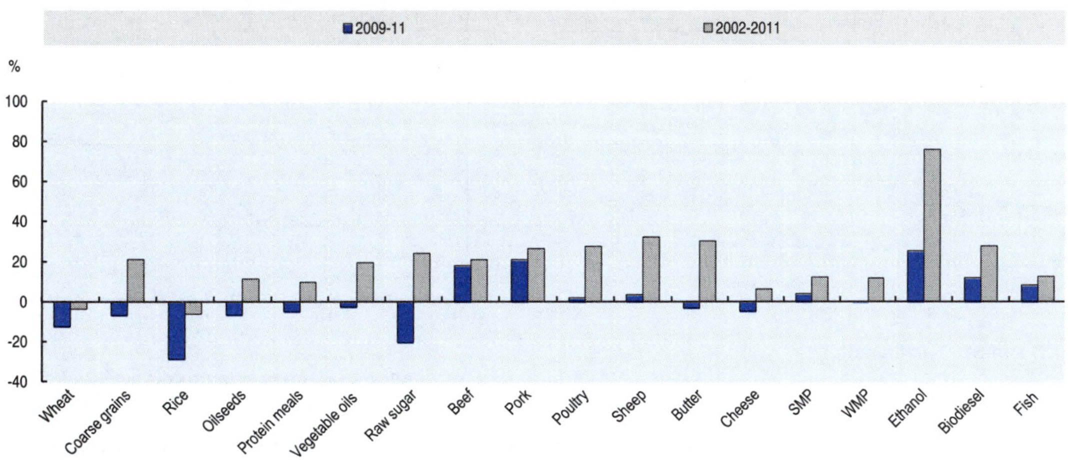

Source: OECD and FAO Secretariats.

StatLink http://dx.doi.org/10.1787/888932638868

unforeseen production shortfalls or trade restricting policy measures in major producing and trading countries could quickly change the situation and lead to further bouts of high volatility and price surges. Recent sharp price swings in world prices of basic food commodities have exposed a number of weaknesses in market information systems and in the co-ordination of actions and policy responses. Weaknesses include lack of reliable and up-to-date information on crop production, utilisation, stocks, and export availability. The resurgence of high food prices in 2010 helped place global food security among the nine key pillars of the Multi-Year Action Plan for Development recognised by the G20 Seoul Summit in November 2010. This led to a *Policy Report on Price Volatility in Food and Agricultural Markets: Policy Responses*, coordinated by FAO and OECD, with contributions by FAO, IFAD,

Figure 1.7. **Agricultural commodity prices to increase in nominal terms**

Price trends in nominal terms for agricultural commodities to 2021

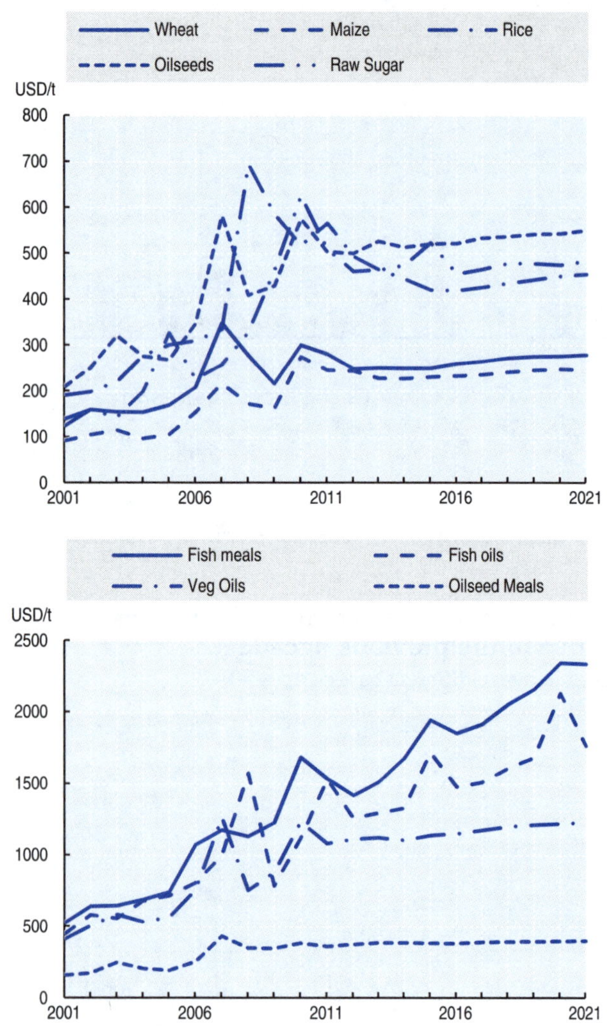

Source: OECD and FAO Secretariats.

StatLink http://dx.doi.org/10.1787/888932638887

IMF, OECD, UNCTAD, WFP, the World Bank, the WTO, IFPRI and the UN HLTF, on ways to manage price volatility and mitigate its impacts. This *Policy Report* included ten recommendations, one of which called for the creation of an Agricultural Market Information System (AMIS). This was endorsed by the G20 Agriculture Ministers and AMIS was officially launched in September 2011 in Rome. The AMIS is described in Box 1.2.

Early in 2012, Mexico, as G20 President, invited international organisations to examine practical actions that could be undertaken to sustainably improve agricultural productivity growth, in particular on small family farms. A report entitled "*Sustainable Agricultural Productivity Growth and Bridging The Gap For Small Family Farms*", responding to this request was prepared, coordinated by FAO and OECD, as a collaborative undertaking by Bioversity, CGIAR Consortium, FAO, IFAD, IFPRI, IICA, OECD, UNCTAD, UN High Level Task Force on the Food Security Crisis, WFP, World Bank and WTO. The final report contains a number of forward looking policy recommendations that define the steps governments might take to

Box 1.2. **Agricultural Market Information System (AMIS): History, objectives and structure**[1]

AMIS is an open global agricultural market information system that will closely monitor major developments in world markets and report on abnormal market conditions, while working in parallel to improve the quality of data, market analysis and forecasts at national and international levels. It will further provide a forum to promote enhanced co-ordination of policy responses. AMIS is intended to build on and complement existing systems in order to improve global food market information. AMIS is not a new international organisation but a platform through which major actors can work together to strengthen synergies and collaboration in order to improve data reliability, timeliness and frequency. AMIS will also build the capacity of developing countries in market outlook analysis and promote policy dialogue.

Initially, AMIS focuses on four leading commodities: maize, rice, soyabeans and wheat. Countries participating in AMIS are the G20 Members and seven invited countries (Egypt, Kazakhstan, Nigeria, the Philippines, Thailand, Ukraine and Viet Nam). These include the world's leading producing, consuming and exporting countries of the commodities concerned and account for a large share of the world food market. The active participation of these countries will ensure that key information on factors that affect the markets for basic food commodities will be available and analysed quickly, thus providing a public good for the international community.

AMIS is located in FAO and supported by nine international organisations (FAO, IFAD, OECD, UNCTAD, WFP, the World Bank, the WTO, IFPRI and the UN HLTF) with capacity to collect, analyse and disseminate information on a regular basis regarding the current and future food market situation and food policies. The International Grains Council (IGC) is an observer co-operating in the area of market information. The Group on Earth Observations Global Agricultural Monitoring Initiative (GEO-GLAM) is assisting AMIS in improving crop production forecasts.

The organisations involved will ensure that the information outputs of AMIS are objective and factual. The AMIS Secretariat is responsible for carrying out global food market outlook analysis based on information provided by the participating countries. It is currently developing appropriate tools, methodologies and comprehensive indicators, reflecting food market developments in a meaningful way. The Secretariat is also responsible for assessing the quality of data provided and for the provision and dissemination of high quality food market outlook information products in a timely manner.

AMIS is composed of two groups:

- *The Global Food Market Information Group*, which provides information on supply and demand as well as prices and policy developments that could impact the market situation for the commodities covered by AMIS, will bring together food market experts from the participating countries. This Group is expected to identify gaps in information collection and, through specific projects, build capacity to collect market outlook information and improve the quality of the data in terms of timeliness, coherence and completeness.
- *The Rapid Response Forum*, which focuses on enhancing policy dialogue when the market situation and outlook indicates a potential market crisis. As such, the Forum will encourage the co-ordination of policies and the development of common strategies. It is made up of senior policy makers from the participating countries who meet when the food market situation warrants. Its objective is to promote discussions on options in order to enhance policy co-ordination.

1. For more details about the AMIS, please check their website on *www.amis-outlook.org/amis-events/en/*.

help close the agriculture productivity gap in a sustainable way, including the importance of innovation and investment in fostering productivity growth and the need for a coherent policy environment.

Global agricultural production to expand, but at a slower pace

A critical question for this *Outlook* is *whether*, *how* and *where* agricultural production will respond to current and projected high commodity prices. Over the past 50 years, trend growth in global agricultural production, as measured by FAO's net production index, has been 2.3% p.a. Past cycles of high prices have been calmed with high production response, and in the presence of inelastic demand, real prices had trended downward for decades.

However, as noted in previous editions of the *Agricultural Outlook*, real commodity prices, as measured by FAO's Food Price Index, reached a trough around year 2000, and have persistently risen since then to double their values of a decade ago.[5] With signs of slowing productivity in some regions, high and rising input costs for fertiliser and energy related inputs, and with growing constraints due to water and land degradation, what will be the future supply response?

Global agricultural output grew by 2.6% p.a. over the last decade, led by growth in Brazil, China, India and the Russian Federation (Figure 1.8). This *Outlook* anticipates a slowing of output growth to 1.7% p.a., which still outpaces population growth and with growth in output per person estimated at 0.7% p.a. While the short-term supply response to price signals is often higher in developed countries, where variable inputs such as fertiliser are more widely used, output in developing countries is expected to grow by 1.9% over the next ten years, resulting in a continual increase in their share of global production.

The fundamentals of slowing growth by commodity are described in the commodity chapters of this report. Chapter 2 discusses the need for greater and more sustainable productivity growth and the ways that this can be achieved.

Figure 1.8. **Average annual growth in net agricultural output to slow down**

Percentage change in least squares growth 2002-11 and 2012-21

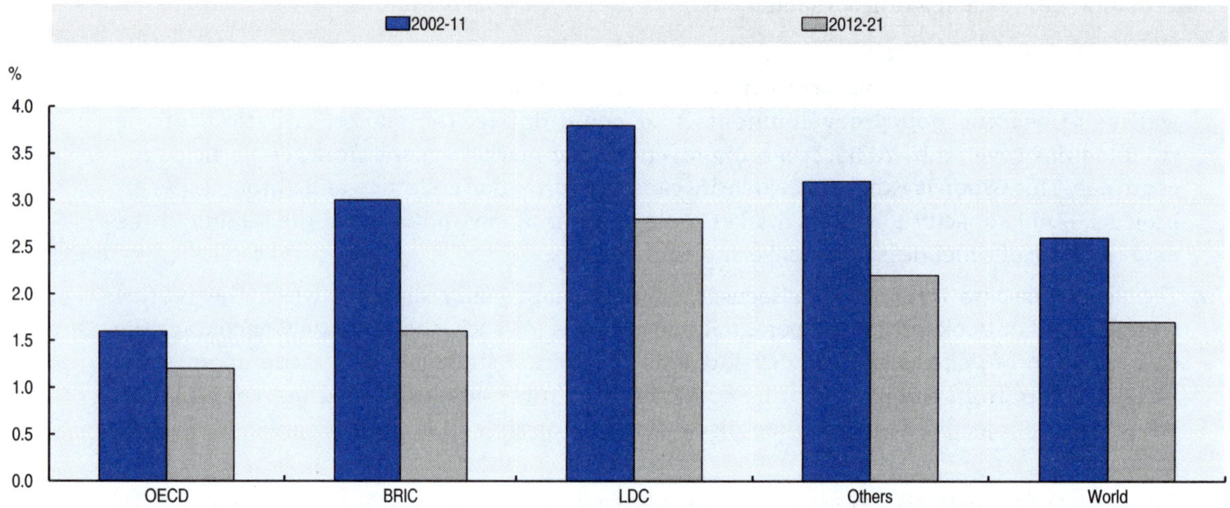

Note: Net output is valued at 2004-06 prices for production of primary commodities in this *Outlook*, as production less seed and feed.

Source: OECD and FAO Secretariats.

StatLink http://dx.doi.org/10.1787/888932638906

Production shares of developing countries to continue to expand

The growth in commodity production in the developing and emerging economies has exceeded that in the developed world during recent decades as countries have invested more in their agricultural sectors. This trend is projected to continue in the coming ten years for all of the 20 agricultural products covered in this *Outlook*. In the case of meats (beef, pigmeat, and poultry meat), dairy products (butter, cheese and milk powders), vegetable oils and sugar, the projected annual growth of production in these developing countries exceeds that in the OECD area by a wide margin, often in excess of 50%. As a result, the OECD share of global production of all these commodities declines over the course of the outlook period. Only in the case of selected dairy products of cheese, milk powders, biofuels, and fish oil does OECD area production continue to dominate the world market output, having lost this status for beef, poultry meat and coarse grains production in the current decade and for the other products in an earlier period.

The production of bio-ethanol and biodiesel from agricultural feedstocks are projected to show the highest growth rates, at 4.8-5% p.a., of all agricultural products over the outlook period. World ethanol and biodiesel production are both projected to almost double to reach some 180 bnl and 41 bnl, respectively, by 2021. Production and use of biofuels in the developed countries are driven mainly by government policies including mandates.

The leading world ethanol producer is the United States which is projected to account for 44% of the increase in world production of ethanol to 2021, with half of the increase projected to come from maize and the other half from cellulosic material. Brazil with its highly integrated sugarcane-based ethanol production system will be responsible for 29% of the increase and the European Union, using a mixture of feedstocks, for 12% of the additional output. Apart from these three countries/trading blocs, China, India and Thailand are expected to expand production and to increase their contributions to world ethanol production by 2-3% each over the outlook period (Figure 1.9).

Figure 1.9. **Larger country shares of biofuels production by 2021**

Biodiesel and ethanol production in 2021, main countries, billion litres

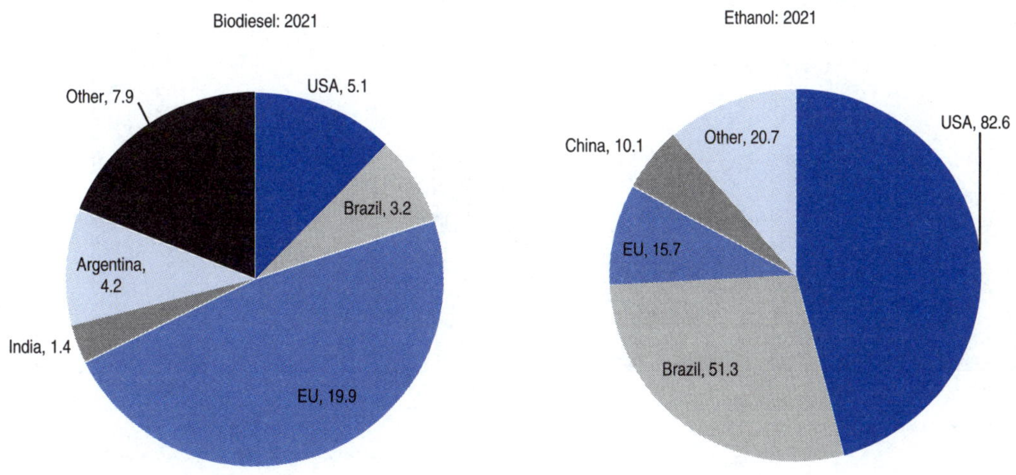

Source: OECD and FAO Secretariats.

StatLink http://dx.doi.org/10.1787/888932638925

In the case of biodiesel, the European Union is by far the major producer accounting for a projected 47% of the increase in world production in 2021. Other producers with smaller shares of global output in 2021 are the United States (11%), Argentina (9%), Brazil (6%), and India and Indonesia (5%), the latter based on palm oil as the main feedstock.

Global cereals production is expected to grow by 1.1% p.a., on average, to 2021 and down from 2.5% p.a. in the previous decade. A slowdown in both yield growth and area expansion are expected to be responsible for the deceleration in the pace of cereal output increase. World production of coarse grains and rice are projected to grow slightly more rapidly at 1.4% and 1.2% p.a., respectively, than global wheat production (0.9% p.a.). The growth in global production of cereals is expected to be marginally less rapid than the projected growth in annual consumption to 2021, implying a general tightening of the cereal market supply situation.

Despite slower projected growth in cereal output, the production of wheat is set to expand strongly in traditional producing regions within the developed countries that will account for 59% of the additional output to 2021 (Figure 1.10). For coarse grains, the United States continues to dominate the global industry, especially for maize, but with strong growth in output anticipated in China, the European Union, Brazil, India, Argentina, Mexico and Canada. Asian countries will continue to dominate rice production. World oilseed production is also projected to slow following strong growth in the previous decade as additional land was drawn into production in response to high prices. Nonetheless, global production of oilseeds is projected to increase by around 20% by 2021, when compared to the base period, with additional oilseed acreage contributing to about half the increase. Global vegetable oil production, which has been growing rapidly in response to strong demand, is projected to increase by more than 28% over the outlook period with seven countries (Argentina, Brazil, China, the United States, India, Indonesia and Malaysia) accounting for 75% of the expansion.

Figure 1.10. **Change in the production of crops**

Percentage change 2021 relative to average 2009-11

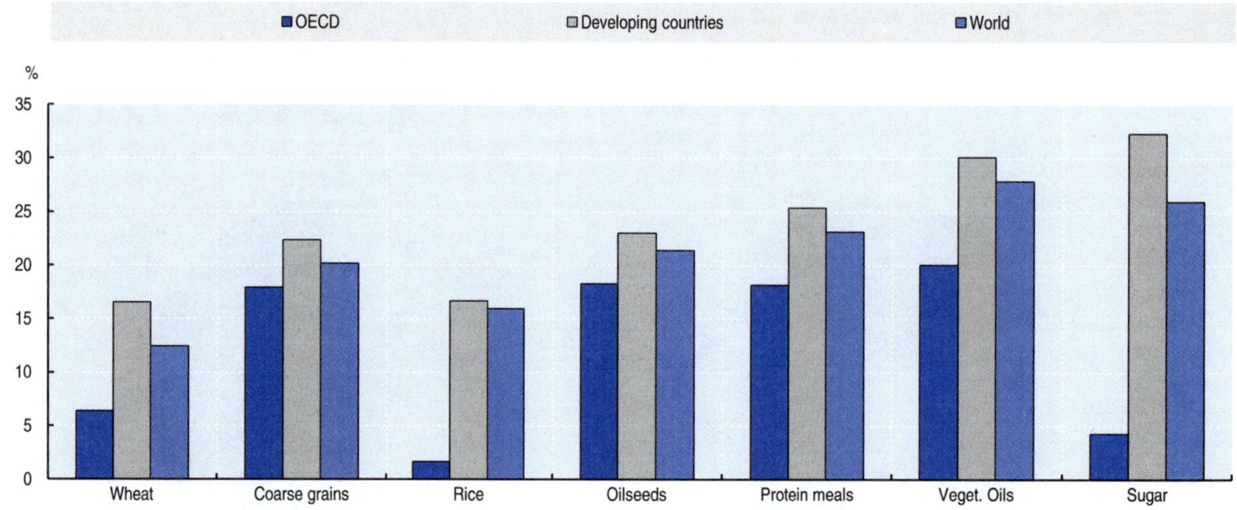

Source: OECD and FAO Secretariats.

StatLink http://dx.doi.org/10.1787/888932638944

Sugar is one of the few products where world production is expected to post superior growth rates in the coming decade than was the case for the previous ten years. Production is expected to grow slightly faster at 1.9% p.a., on average, to 2021, compared to an estimated growth of 1.7% p.a. in the last decade. This slight acceleration reflects continuing strong output growth in the developing countries which are projected to account for 93% of the additional global production to 2021. Higher meat prices relative to the costs of feedstuffs are projected to improve livestock sector returns and increase incentives for an expansion in animal inventories and increased meat production over the projection period (Figure 1.11). Meat production is expected to be led by the poultry sector that boosts output by 29% followed by sheepmeat with 21%, and with beef and pigmeat increasing slightly more than 16%. Meat production in the developing countries is expected to grow at more than double the pace of that in the developed countries to meet strong demand driven by rising incomes and populations. World milk production is expected to grow by 2% p.a., to 2021, down from 2.1% p.a., in the previous decade. This slight slowdown reflects slower growing global milk animal inventories which are not completely compensated by the expected higher growth in milk yields over the outlook period. Milk and dairy products covered in this *Outlook* remain amongst the agricultural commodities for which production exhibits the highest growth rates, particularly for the developing countries.

Capture fisheries has long been the dominant source of fish products, but this role is rapidly being supplanted by aquaculture. Going forward, production from capture fisheries is expected to be stagnant, in particular for developed countries in the OECD area as there is little room to increase harvests from current levels. In contrast, aquaculture is projected to continue to record strong growth over the next ten years and to become the largest source of supply of fish products for human consumption by 2018.

Figure 1.11. **Change in production of livestock and fish products**

Percentage change 2021 relative to average 2009-11

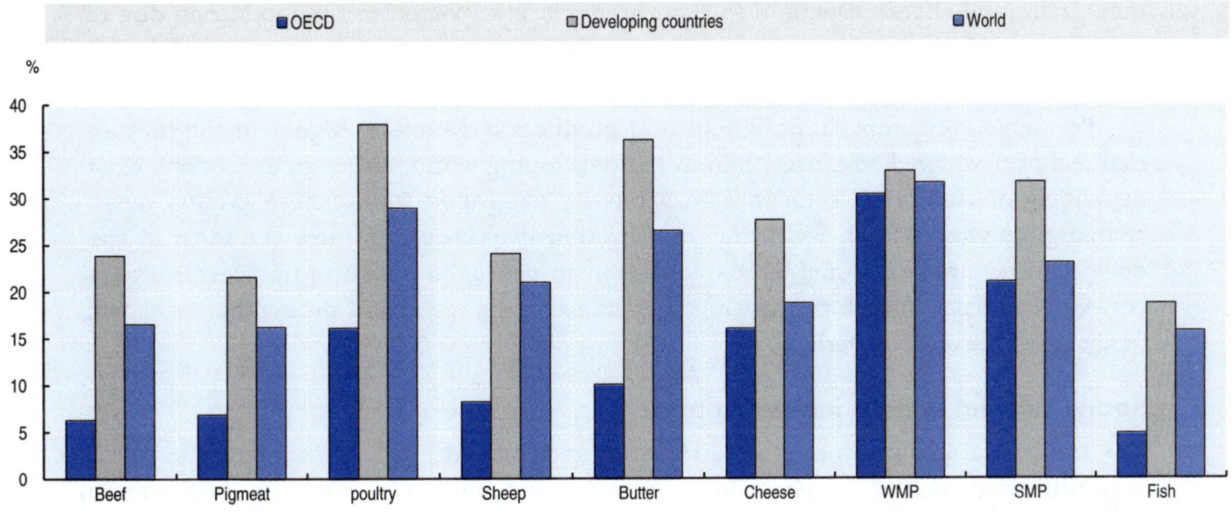

Source: OECD and FAO Secretariats.

StatLink http://dx.doi.org/10.1787/888932638963

Stronger consumption growth set to continue

A feature of recent *Outlooks* is the assessment that consumption growth for agricultural products has been remarkably resilient in the face of high and volatile prices and is expected to strengthen overtime, leading to tighter markets and holding prices of many crops, fish, livestock and renewable biofuels at historically high levels. The main drivers of increasing consumption are rising *per capita* incomes, population growth, and the growing size and affluence of large middle classes in the emerging economies, which along with continuing urbanisation lead to a shift in consumption patterns. Consumption is changing from staple foods towards more fats and oils and more animal protein. This trend increases the demand for vegetable oils, meats, sugar and dairy products including indirect demand for coarse grains and oilseeds in animal rations (Figures 1.12 and 1.13).

The more rapid growth in consumption will continue to take place in the developing and emerging countries. The projections suggest that consumption will increase for all products and in all regions in the outlook period.

Between 2011 and 2021, the products which experience the highest increase in consumption are expected to be poultry meat (37.2%), sugar (32.5%), vegetable oils (32.1%) and selected dairy products of butter (33.4%), cheese (27.8%) and skim milk powder (40.6%). Indirect consumption of products such as oilseed meals and coarse grains are also much higher in the developing countries as a result of expansion in domestic livestock industries and animal inventories associated with rising consumption of meats and dairy products.

Within the developed countries, food expenditures generally represent a small share of around 10-15% of disposable incomes and where basic dietary needs have long been satisfied. Consumption gains, in general, tend to be less striking. Diets in the developed countries are expected to reflect more variety in terms of a wider range and composition of foodstuffs (including highly processed and pre-cooked and convenience foods), changes in population structure (ageing), and a focus on healthier lifestyles. Consumption of products such as red meats, butter, milk powders, and sugar are expected to post little to moderate growth to 2021 as eating preferences continue to shift towards products such as poultry meat, fish, and cheese. Demand for fish products is expected to remain strong due to population and income growth, and as consumers continue to switch to healthier sources of animal protein.

Per capita consumption of agricultural products is generally higher in the mature markets of developed countries than in the developing world, although exceptions exist depending on traditional consumption habits *e.g.* rice. Apart from cheese, poultry meat, coarse grains and wheat, *per capita* consumption is expected to grow the most in the developing countries. Considerable variation in *per capita* consumption levels exists between countries in both the developed and developing world and these differences will persist over the outlook period.

Strong demand leads to increasing trade

Traditional exporters of a wide range of agricultural products, such as Australia, Argentina, Canada, the European Union, New Zealand and the United States, will remain important in global trade in the coming decade. But countries that have made significant investments in their agricultural sectors and which have sought to exploit their advantages in agricultural production, including Brazil, the Russian Federation, Ukraine and China, are expected to have an increasing presence on world markets as well.

Figure 1.12. **Higher consumption of crop products**

Percentage change in consumption: 2021 relative to average 2009-11

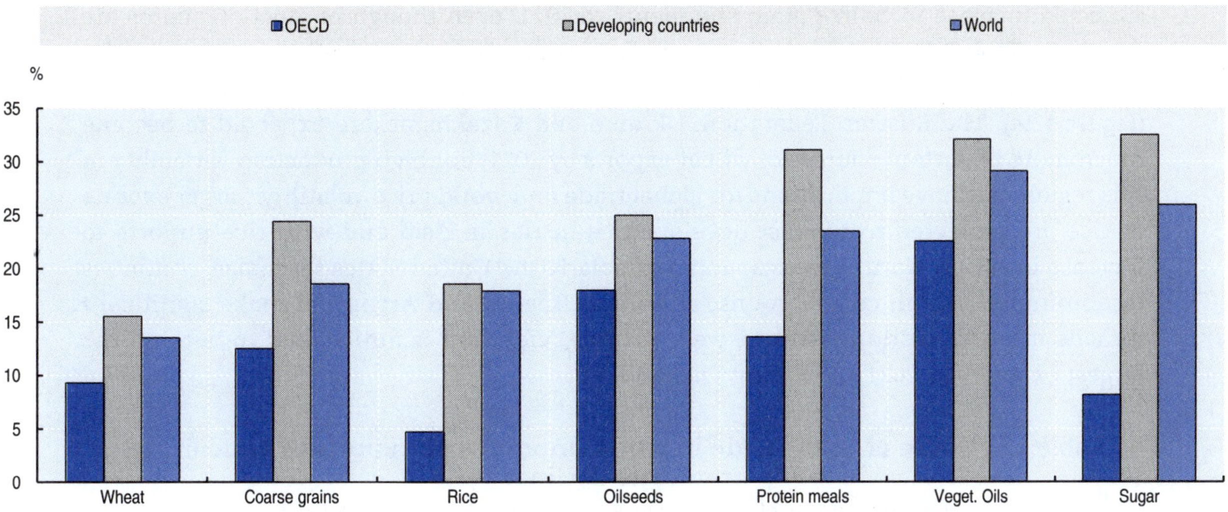

Source: OECD and FAO Secretariats.

StatLink http://dx.doi.org/10.1787/888932638982

Figure 1.13. **Higher consumption of livestock and fish products**

Percentage change in consumption: 2021 relative to average 2009-11

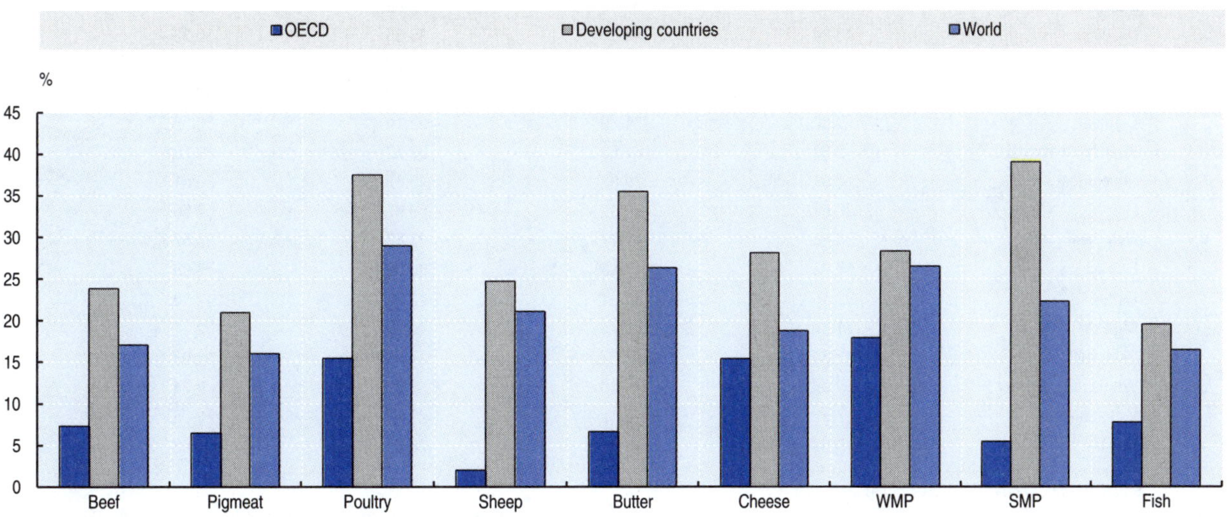

Source: OECD and FAO Secretariats.

StatLink http://dx.doi.org/10.1787/888932639001

Trade in biofuels is set to become more important to the future prospects of the industry. Global bio-ethanol trade is projected to account for an increasing share of world production, growing from around 10% of global production in the previous decade to about 18% by 2021. Most of this increase is accounted for by expected growth in two-way trade between Brazil and the United States, on the assumption that the US ethanol blend wall will not be a barrier to future low-blend ethanol demand growth until 2016.

Global export trade[6] volumes for wheat and coarse grains are expected to increase by 17% and 20%, respectively by 2021, while that of rice trade grows even more substantially by 30%, when compared to the base period. The developed countries continue to dominate world wheat and coarse grains trade in absolute volume terms, and account for most of the expected increase in coarse grain shipments to 2021, even though production shares are declining (Table 1.3). The developing countries dominate rice trade and are responsible for most of the projected expansion in rice and wheat trade over the outlook period (Figure 1.14). The Russian Federation, Ukraine and Kazakhstan are expected to become much more important sources of wheat exports by 2021, but high production variability in this region may have implications for global trade and world price volatility. Larger exports of rice are projected from least developed countries in Asia and with rice imports to increase in Africa. North America, and particularly the United States for maize, continues to dominate trade in coarse grains, although Ukraine and Argentina make significant inroads into this world market as well. For rice, Viet Nam is anticipated to become the largest exporter by 2021, displacing Thailand.

Table 1.3. **Share of OECD trade in world imports and exports of agricultural products declines**

Percentage share of world imports and exports: 2002-11 and 2012-21

Commodity	Export		Import	
	Average 2002-2011	Average 2012-2021	Average 2002-2011	Average 2012-2021
Wheat	66.37	56.50	23.55	20.20
Rice	13.17	10.78	14.96	14.47
Coarse grains	61.77	57.13	47.60	39.51
Oilseeds	48.24	44.00	39.86	30.33
Protein meals	15.40	18.46	62.20	53.34
Beef	49.87	49.20	54.32	47.89
Pigmeat	77.80	81.70	53.10	52.38
Poultry meat	9.88	7.53	23.72	19.72
Sheepmeat	78.91	76.62	42.48	36.81
Fish	36.07	34.07	59.87	54.50
Fish meal	38.38	28.77	42.87	36.57
Fish oil	50.05	46.48	86.59	77.88
Butter	82.46	81.24	18.92	12.50
Cheese	71.37	64.75	40.63	32.71
Skim milk powder	82.35	87.67	17.33	12.03
Whole milk powder	69.96	74.80	5.93	3.24
Vegetable oils	7.62	8.17	29.10	28.85
Sugar	19.02	11.65	26.68	27.66

Source: OECD and FAO Secretariats.

StatLink http://dx.doi.org/10.1787/888932640502

World trade in oilseeds, which grew at over 6% p.a. in the last decade, is projected to slow in the coming ten years from a larger base as export growth slows in the United States, Argentina and Brazil. On the import side, purchases of oilseeds by China are expected to continue to grow substantially and to be propelled by growing demand for vegetable oil and livestock feed.

In the case of sugar, exports are heavily skewed in favour of a single country, Brazil. Brazil is projected to remain the largest supplier of high quality raw sugar to the world market and to become a larger exporter of white sugar. In addition to Brazil, Thailand and Argentina also gain market shares while that of the EU contracts. China becomes the largest sugar importer, surpassing the European Union, Indonesia and the United States,

Figure 1.14. **Increasing exports of crop products**

Exports of agricultural crop products in million tonnes: 2009-11 to 2021

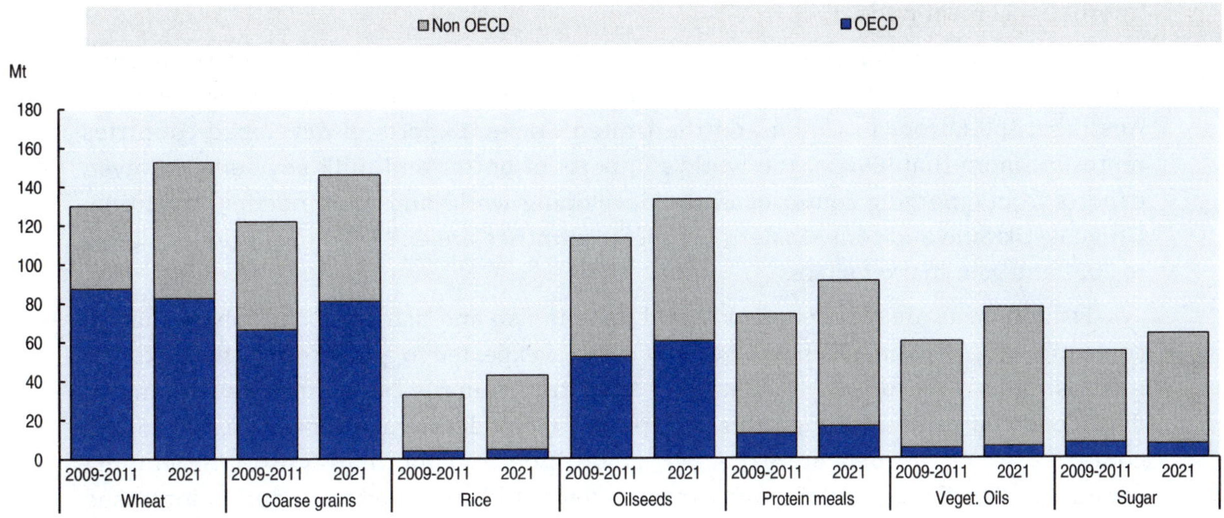

Source: OECD and FAO Secretariats.

StatLink http://dx.doi.org/10.1787/888932639020

with growing imports from Mexico. While rising domestic production and self-sufficiency leads to a substantial reduction in imports by the Russian Federation.

World meat trade, comprising beef, pigmeat, poultry meat and sheepmeat shipments, is expected to rise by 1.5% p.a. over the next decade, stimulated by strong demand from rapidly rising incomes and affluence in the developing countries and firmer demand in developed countries(Figure 1.15). World meat exports are projected to increase by 19%

Figure 1.15. **Increasing exports of livestock and fish products**

Exports of livestock and fish products in million tonnes: 2009-11 to 2021

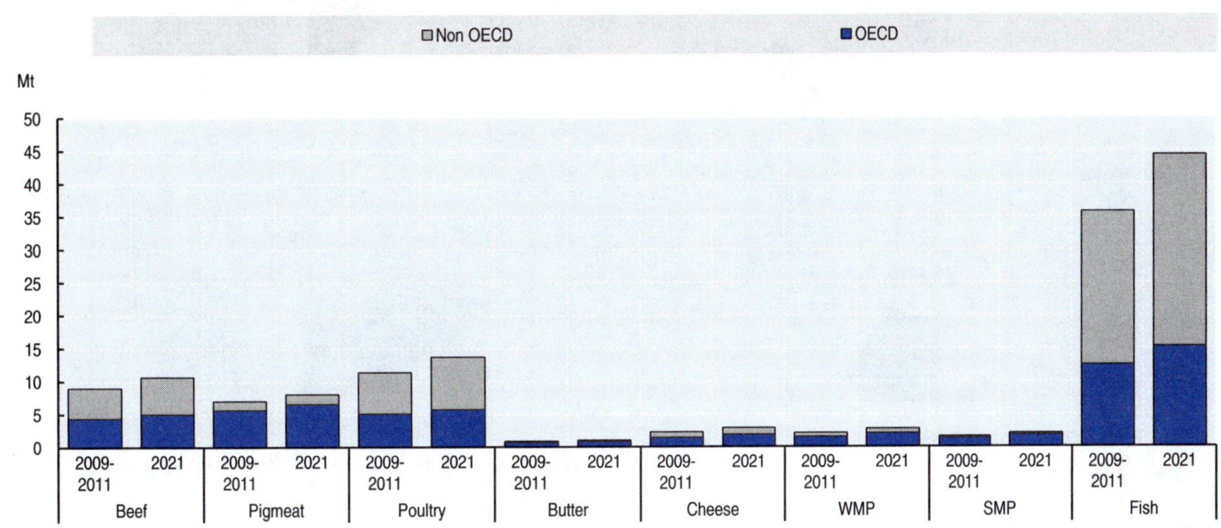

Source: OECD and FAO Secretariats.

StatLink http://dx.doi.org/10.1787/888932639039

to 2021, led mainly by higher poultry and beef shipments. Developing countries are expected to respond to higher meat prices and increase their exports and share of international meat trade.

World trade in dairy products continues to represent a small share of world milk production, and remains dominated by developed countries including New Zealand, Australia, the European Union and the United States. Exports of developed countries represent more than 85% of the world's imports of butter and milk powders. However, exports from emerging countries in the developing world and most notably Argentina, Uruguay, Ukraine and some Eastern European countries are expected to continue to grow to 2021 and gain market share.

Fish products are highly traded. World trade in fish and fish products (fish for human consumption and fishmeal on a live weight basis) is expected to grow strongly with exports increasing by 34% to 2021. Trade flows have traditionally been from developing to developed countries, although this is expected to moderate as demand in developing countries grows. Exports are expected to continue to come mainly from developing countries, with the centre of gravity moving towards Asian producers due to increased availability of aquaculture products.

Figure 1.16 illustrates the changes projected in regional shares for the value of agricultural imports and exports by 2021, relative to the base period. The Asia and Pacific region followed by North Africa and the Middle East account for the majority of the increase in the value of agricultural imports to 2021. In the case of exports, Latin America, East Europe and Central Asia and North America account for most of the increase to 2021.

Figure 1.16. **Regional shares of the change in value of agricultural imports and exports**

Percentage change 2021 compared to 2009-11

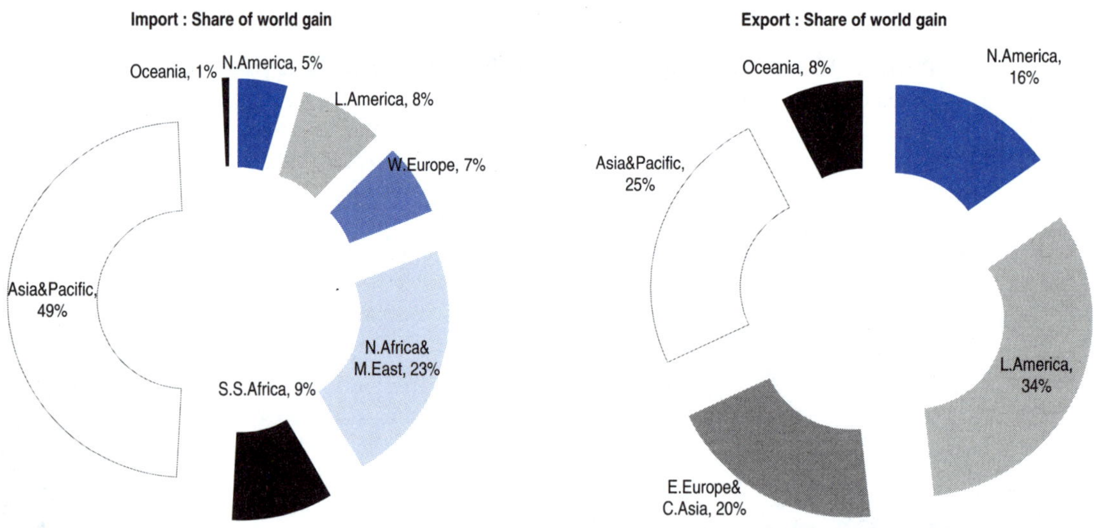

Source: OECD and FAO Secretariats.

StatLink http://dx.doi.org/10.1787/888932639058

Risks and uncertainties

The outcomes described in the baseline projections are conditioned by a specific set of assumptions on the environment affecting the sector, which are subject themselves to large uncertainties. The key assumptions in this *Outlook* pertain to crop yields, crude oil prices, developments in biofuel markets, policies settings for the main commodities in leading producing countries and macroeconomic assumptions such as income growth and exchange rates. Differences in the outcomes of these factors would change the baseline projection values, especially for commodity prices, but also for the supply, demand and trade in agricultural and food commodities. For example, a slowdown in economic growth would moderate international prices; a sudden exchange rate appreciation could rapidly erode a cost advantage of an emerging or established exporter. To illustrate some of these sensitivities, the AGLINK-COSIMO model is used to generate alternative scenarios around the baseline projections presented in this chapter. These additional scenarios complement the results of the scenario presented in Chapter 2 on higher yields, and those in the biofuels and dairy products chapters, to show how sensitive commodity prices are to various assumptions.

Impact of different crude oil price assumptions

The crude oil price is a key driver of the projections. In order to assess the magnitude of the effect of changes in crude oil prices, simulations were conducted with crude oil prices set at 25% above, and below, the level assumed in the outlook period. The impacts on commodity prices are presented in Figure 1.17. They underline the strong relationship between crude oil and commodity prices, arising from the supply and demand side.

Figure 1.17. **Crude oil prices affect agricultural commodity and biofuels markets**

Absolute impact of a 25% change of crude oil price on commodity prices
(average change relative to baseline, on the projection period, in absolute terms)

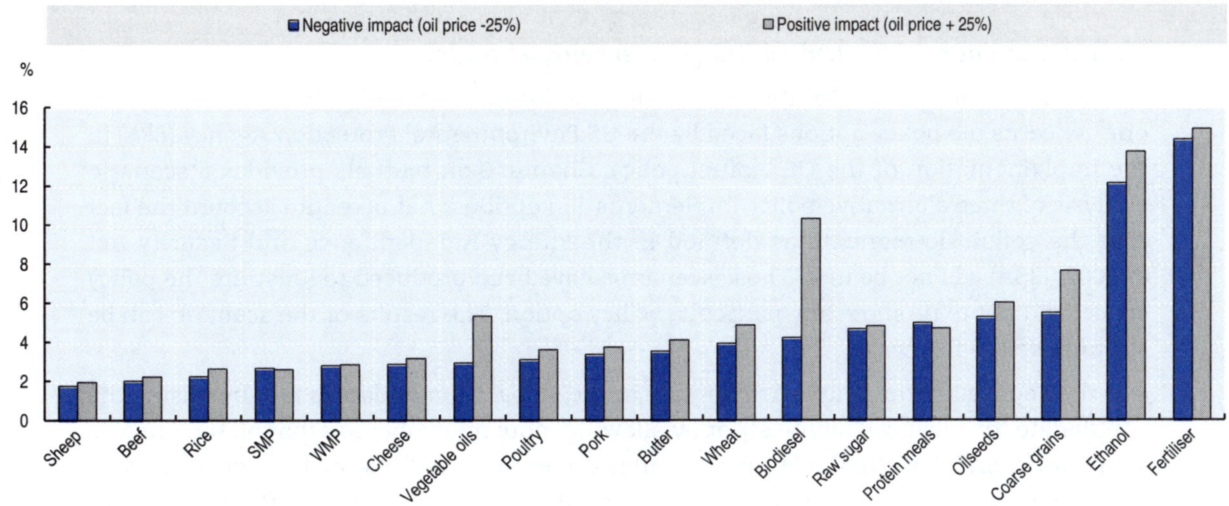

Source: OECD and FAO Secretariats.

StatLink http://dx.doi.org/10.1787/888932639077

The supply of agricultural products has been influenced by energy costs already for a long time. Higher energy prices reduce agricultural production given the rising costs of inputs such as fertilisers and fuels. For example, in the scenario, a 25% higher crude oil price would increase fertiliser prices by 14%. As a result, agricultural production would be lower, leading to higher agricultural commodity prices. The impact is more pronounced in the crop sector than in the meat sector where, on average, energy requirements are less important.

The demand side, abstaining from energy costs impacts on consumers' purchasing power, would be affected through the increasing role of biofuel markets and related policies. Higher crude oil prices, all other factors unchanged, would generate higher demand[7] and prices for biofuels and, in turn, higher demand and prices for agricultural feedstocks. In the scenario, a 25% higher crude oil price causes global ethanol production to rise by 6%, on average, and biodiesel production to rise by 7%.[8] Feedstock demand for wheat and coarse grains rises by 5%, for vegetable oil by 8%, and for sugar crops by 13%.

The combined effect of reduced supply and increased demand lifts agricultural commodity and biofuel prices. The effects of lower energy prices, according to the scenario analysis, are largely similar but of opposite sign, when compared to the higher oil price scenario. An important exception concerns biodiesel; biofuel targets in the European Union and in the United States are assumed to remain in place independently of the levels of energy prices. This would keep global demand for biodiesel high over the next ten years and would sustain strong world biodiesel prices, even if crude oil prices fall.

The different results for ethanol[9] prices compared to biodiesel ones are due to the combination of the following factors:

- the integration of the US and Brazilian ethanol markets since the 1 January 2012 with the elimination of the specific tariff;
- the large and rising consumption of ethanol by flex-fuel cars in Brazil;
- the prospective large two-way trade between these two countries created by the United States advanced biofuel mandate.

Biofuel mandates in the United States to encourage trade?

There are many uncertainties concerning the future of biofuel policies. An important one concerns the policy options faced by the US Environmental Protection Agency (EPA) in the implementation of the US biofuel policy. Chapter 3 on biofuels provides a scenario analysis of three alternative policy implementation options that take into account the fact that the cellulosic mandate as defined in the Energy Independence and Security Act of 2007 (EISA) will not be met. Those scenarios have been produced to illustrate the policy space, and not to promote any particular policy option. The results of the scenario can be summarised as follows:

- If the shortfall in the cellulosic mandate in the global US mandate is met by raising the mandate for advanced biofuels[10] or by allowing more maize-based ethanol, the impacts on world prices for biofuels (in particular ethanol) as well as for biofuel feedstocks (coarse grains, sugarcane) are likely to be important. Spill-over effects on other agricultural commodity prices (including meat and fish) would occur.
- Meeting the adjusted US biofuel mandate will require some adjustment in terms of ethanol production and consumption patterns, as well as in terms of ethanol feedstock use around the world. Food is likely to cost more as a result of such adjustments.
- An important policy driven two-way ethanol trade is likely to emerge between Brazil and the United States under certain conditions. Brazil is likely to be the sole country able to adapt and respond to US demand.[11] This is due to the nature of its ethanol production

based on sugarcane, its flexibility to switch between ethanol and sugar production, and to the rising demand for ethanol for flex-fuel vehicles in Brazil.

Price impact of a lower GDP growth in the BRIC countries

Along with the assumption of high oil prices, a higher plateau for commodity prices is maintained over the projection period in part due to strong demand arising from robust economic growth in emerging and developing countries. In this *Outlook* an annual GDP growth of 8% on average for China and India, and nearly 5% for Brazil and the Russian Federation have important implications for global markets. A scenario of a 10% change of their GDP growth was made to examine what impact higher and lower growth in these countries would imply for the baseline projection. Since the model results are largely symmetric to income changes up or down, only the scenario of a 10% lower GDP growth are discussed (Figure 1.18).

With lower growth the shift in diets from staple foods to more value-added and higher protein products will take place at a slower pace than shown in the baseline projection, and with lower demand transmitted to global markets causing prices to fall. The impacts are shown to be higher on meat, particularly pigmeat, and dairy consumption than on the grains.

Among the four countries, China represents about 47% of global pigmeat consumption. A lower demand for pigmeat would decrease feed requirements. This is important as China accounts for 7% of global coarse grains imports, but a massive 52% of global oilseed imports, mainly for crushing into vegetable oils for food consumption and protein meal for livestock feed. Lower pigmeat consumption in China leads to somewhat larger downward adjustment in the world protein meal price than for the world coarse grains price. Overall, the main effect of lower GDP growth of 10% is shown to be lower world agricultural prices of about 3.6% for oilseeds and oilseed products, 3.2% on average for meats, 3.1% for grains and 1.6% for dairy products.

Figure 1.18. **Impact on consumption and world prices of a 10% lower GDP growth in BRIC countries**

Average of percentage change over the forecast period, relative to baseline

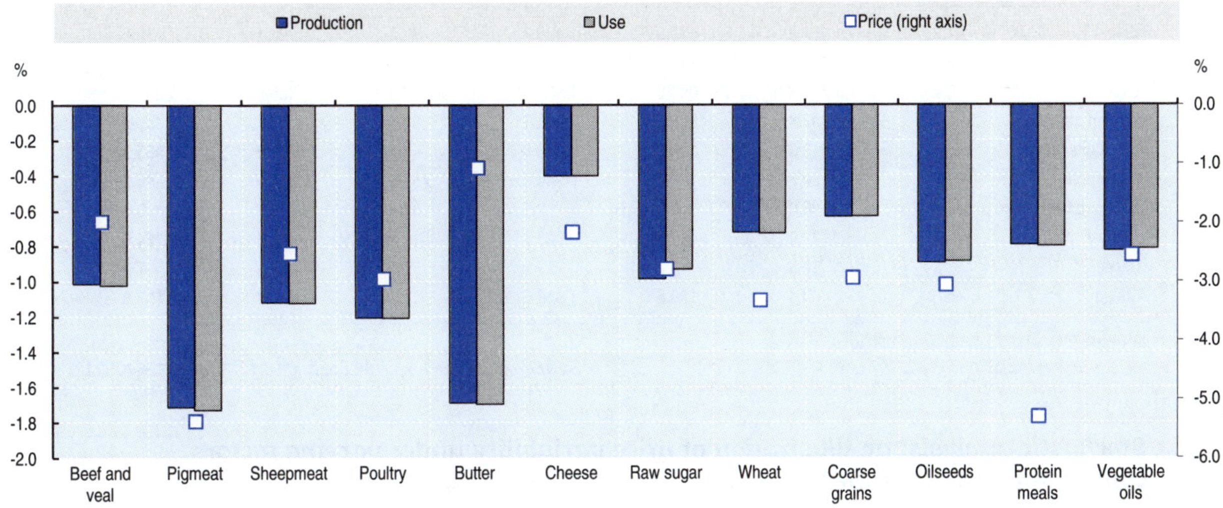

Source: OECD and FAO Secretariats.

StatLink http://dx.doi.org/10.1787/888932639096

Price impact of an appreciation of the US dollar

Exchange rates play an important role, particularly in trade. Assumptions for the movement of exchange rates are therefore important in determining the competitiveness of countries in both supplying exports, but also in demanding imports. The scenario presented in this section assesses the effect of an appreciation of the US dollar compared to the *Outlook* assumption. For the scenario, all exchange rates against the US dollar have been raised by 10% relative to the baseline for all years over the outlook period, with all other assumptions held constant. The results are presented in Table 1.4. The first column presents the effects of the appreciation of the US dollar on world prices compared to the baseline in percentage terms. The other columns show the differences in net exports (kt) for the various products.

A stronger US dollar compared to the baseline projection leads to lower world prices in US dollar terms. In general, as all other currencies depreciate against the US dollar, exporting countries (other than the United States) become more competitive, increasing their exports and reducing their imports. However, these results may be more complicated if, for example, meat exports increase for a country, but where grain exports may also decrease as national feed demand increases. This is the case in the scenario for the results presented for Australia and Brazil. In a similar vein, if the increase in exports occurs in a by-product of the oilseed complex, the appreciation of the US dollar may have a negative effect on exports of oilseeds, grains and meats. This is the case for Canada, although it is a net importer of poultry meat.

In contrast, for the United States, the appreciation of its currency reduces the price of its foreign country imports. The main effect occurs for ethanol imports from Brazil as less domestically produced maize-based ethanol is necessary to meet the total US biofuel mandate and, as a consequence, US grain exports increase. These results emphasise the important role that macroeconomic factors, which are not influenced significantly by the agricultural sector itself, can have on agricultural production and agricultural trade.

Table 1.4. **Impact in 2021 of a 10% appreciation of the USD on world prices and net trade**

Price in %, changes in net trade in kt

	World price	USA	Canada	E27	Australia	Brazil	China
Wheat	–7.2	–31.2	–972.0	–1 426.7	–96.3	292.1	–39.6
Coarse grains	–7.6	2 446.5	1 145.6	–102.7	–101.0	–1 939.5	–46.3
Oilseeds	–2.7	–4 829.7	–1 094.0	1 436.6	–136.4	888.0	5 539.7
Protein meals	–0.6	–1 942.4	688.4	–505.1	97.9	52.1	2 356.2
Vegetable oils	–8.9	–412.9	439.1	778.0	–8.7	–232.2	–883.3
Raw sugar	–7.8	129.7	52.4	–34.3	765.3	1 347.5	–46.0
Beef and veal	–5.7	92.0	–11.3	–20.1	–104.9	173.7	–8.5
Poultry	–5.2	122.9	49.3	23.7	11.3	–247.4	64.6
Cheese	–7.3	16.8	0.0	–11.2	1.0	–1.3	7.0
Ethanol	–5.5	–2 460.4	199.5	613.7	. .	859.4	–157.8

Source: OECD and FAO Secretariats.

StatLink http://dx.doi.org/10.1787/888932640521

Stochastic analysis: an illustration of price variability under varying factors conditioning the baseline projection

A partial stochastic analysis[12] has been conducted with the AGLINK-COSIMO model based on 500 draws of different cereal yields, crude oil and fertiliser prices, and macroeconomic variables. Such an analysis provides an indication of just how much prices might deviate from the baseline projection with the variation in assumptions that accord with historical experience. The results of the analysis for world prices are shown to remain within an average range of –15% and +19% for wheat, and –17% and +20% for coarse grains, around the median projection prices (10th and 90th percentiles[13]), while the world rice price is shown to stay on average between –8% and +10% (Figure 1.19). It is noteworthy that higher price outcomes predominate over lower outcomes.

Figure 1.19. **Simulated price results for cereals in 2021**

In USD/t

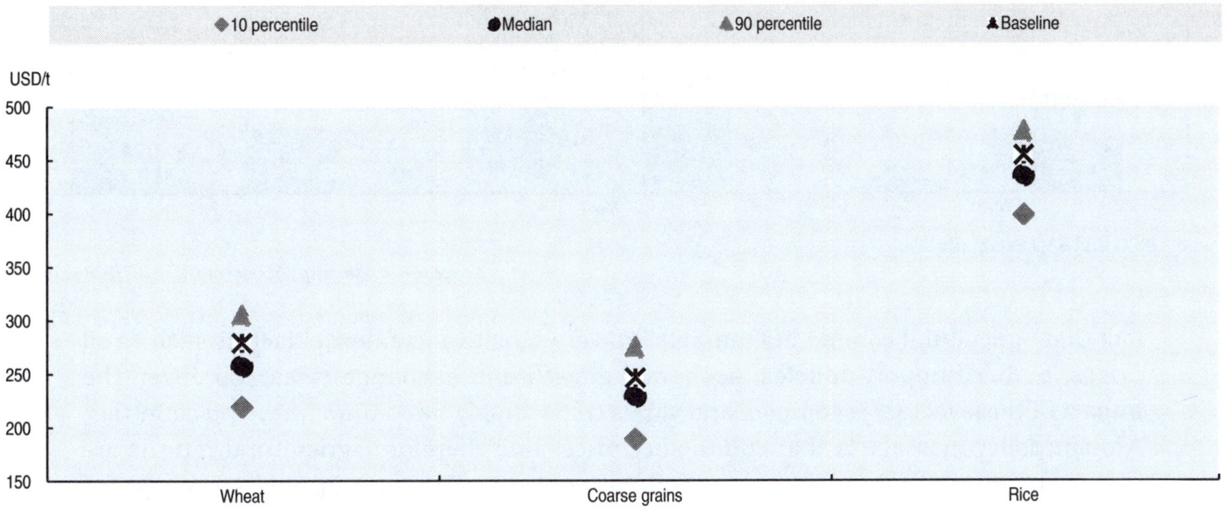

Source: OECD and FAO Secretariats.

StatLink http://dx.doi.org/10.1787/888932639115

Volatility is a characteristic feature of world agricultural markets and is caused by unexpected and unpredictable shocks in the context of agricultural market supply and demand characteristics. Volatility is of particular concern during periods of price surges. Historically high prices in the past few years have lead to *ad hoc* policy reactions that impeded the free flow of trade and exacerbated international food insecurity. It is possible to estimate measures of annual price volatility within this stochastic analysis as was done in OECD (2011). Price volatility here is measured as the median of the standard deviation of year-to-year changes of prices over 2017 to 2021. Figure 1.20 illustrates that volatility for maize may be higher than for wheat and rice. This may be due primarily to the greater sensitivity of maize prices to shocks in crude oil prices. Biofuel mandates may mitigate downward price variation for maize under a crude oil shock, but they may exacerbate price variation stemming from crop yield shocks. It is worth noting that the 90th percentile measures of price volatility for maize and wheat are relatively close to the measure of historical volatility. The estimated volatility for rice is much lower than for the other cereals as the rice market is dominated by policies and is less responsive to crude oil and income shocks. For the three cereals, higher volatility is observed for the 90th percentile than for the 10th percentile, which means that risk on the markets (as a result of price volatility) seems to be increased when prices are high.

This last section has provided a quantitative assessment of the sensitivity of the baseline projection to several factors conditioning the *Outlook*. The analysis underlines the

Figure 1.20. **Simulated price volatility for cereals**

In percentage

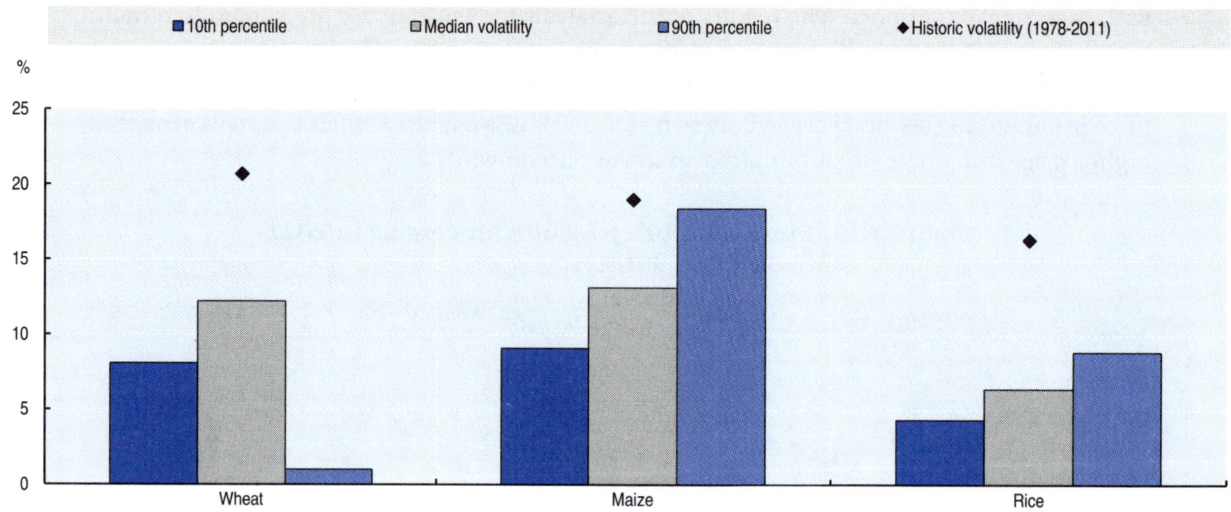

Source: OECD and FAO Secretariats.

StatLink http://dx.doi.org/10.1787/888932639134

fact that agricultural commodity markets are very sensitive to external factors such as oil prices, biofuel support policies, economic growth and exchange rates. Moreover, the impact of these factors is complex and varies considerably by commodities and countries. A main policy message is that commodity prices and therefore agricultural returns are highly uncertain and subject to periods of volatility. The appropriate response for producers is to manage the risk wherever possible (*e.g.* diversified production, alternative income streams, optimal agronomic practices, accessing available crop insurance schemes). Governments can help by reducing policy-based market distortions, promoting market information (*e.g.* AMIS initiative) and facilitating market-based risk management schemes (*e.g.* futures contracts, option contracts on food imports, transparent emergency assistance programmes).

Over the longer term, developing a more sustainable global agricultural system which can meet the rising demand for food, feed, fuel and fibre while taking into account the growing resource constraints (land, water, nutrients) and environmental pressures (soil degradation, preservation of ecosystems, climate change) facing the planet will be essential for ensuring the resilience of the sector and global food security. The following chapter focuses on the challenge of increasing agricultural productivity growth in a sustainable manner.

Notes

1. The aggregate includes nearly 95% of the world's population and account for approximately 90% of world GDP. Table B.1 in the Statical Annex provides detailed information for countries included in these aggregates.
2. Year-on-year (YoY) percentage change in CPI – food for the month of January is defined as the food inflation rate. Calculations of inflation using year-on-year changes for a given month provide one representative measure of inflation in food prices and total inflation.
3. There are slight differences in YoY figures compared to the average annual rate; the former underline the differences across countries and over time.

4. The food expenditure data in many low income countries refers to urban areas only.
5. See *www.fao.org/worldfoodsituation/wfs-home/foodpricesindex/*.
6. The trade numbers here represent exports, imports or net trade (exports minus imports) for the countries and regions explicitly considered in the *Outlook*.
7. Low blend ethanol demand cannot exceed the constraints created by the blend wall.
8. Ethanol can be used as a low blend in gasoline or in flex fuel vehicles; ethanol demand depends on the ethanol to gasoline price ratio as well as on the related support measures. Biodiesel demand depends on the biodiesel to diesel price ratio as well as on the related support measures. A differentiation is made between the price responsive demand for biofuels and the minimum set by public mandates in several countries.
9. The results presented in Figure 1.17 are for the Brazilian ethanol price but the impact on the US ethanol price would be of the same magnitude.
10. Sugar cane based ethanol is considered an advanced biofuel.
11. US low-blend ethanol demand growth is expected to be constrained by the blend-wall, *i.e.* the maximum share of ethanol that can be blended into gasoline, currently fixed at 15% by EISA.
12. The Methodology Annex presents the methodology used to undertake partial stochastic analysis with the AGLINK-COSIMO model.
13. The 10th and 90th percentiles do not represent low and high extremes, but rather plausible alternatives based on past variations.

OECD-FAO Agricultural Outlook 2012

Chapter 2

Achieving Sustainable Agricultural Productivity Growth

In the previous three issues of the *Outlook*, the rapid rise in agricultural commodity prices since 2006 has drawn significant attention. It has been projected that commodity prices will remain not only high but also highly volatile during the next decade, if not beyond. The policy landscape has also changed considerably. A decade ago, cereal prices, adjusted for inflation, stood at historic lows (for at least 50 years), OECD governments were providing producers with large subsidies to support incomes, and developing countries were protesting that subsidies were depressing commodity prices. Within the last ten years, however, real food commodity prices have doubled, underpinned by high economic growth in emerging developing countries, and higher global prices for energy and associated inputs. Initially, relatively low prices for agricultural commodities, in combination with stimulative bioenergy policies, kick-started the rapidly growing demand for agricultural feedstocks. But the resulting high and volatile food prices generated not only concern for food security but raised the spectre of future shortages and focus about the inability to feed the world, in a context of climate change, resource scarcity (*i.e.* land, water and nutrients) and disparate economic opportunities. There have been rising demands by civil society for "green growth" in agriculture, given its important impact on the environment. In 2011, agricultural heads from the G20 countries requested advice on improving productivity growth in agriculture and how to address rising concerns for food security and sustainability, and the resilience of the agri-food system, particularly for small family farmers and low income countries. As a follow-up a 2012 Interagency Report to the Mexican G20 Presidency entitled *Sustainable Agricultural Productivity Growth and Bridging the Gap for Small Family Farms* has been prepared.

Can supply keep up with rising demand?

The growing demand for food, feed, fuel and fibre is well documented. The United Nations estimates that world population will reach 9.1 billion by 2050, an increase of 2.3 billion or 34% from today. This increase will take place almost entirely in the developing world, with the greatest relative increase in the least developed countries (120%) (FAO, 2012 forthcoming). Moreover, accelerating rural migration will lead to a significant majority of the population living in urban areas (compared with about 50% today), and which will be dependent on purchased rather than home-produced food.

Moreover, incomes in the developing world are increasing rapidly and a high proportion of any additional income is being spent on food. Income growth has been particularly high in Asia and sub-Saharan Africa, although the 2008-09 financial crisis dampened this trend. With higher incomes, the shift to western diets to include more protein-rich food (*e.g.* meat and fish consumption) has major implications for the environment as well for the composition of the food basket. This rising food demand will be dampened by a slowdown in population growth and because an increasing share of the world population is reaching medium to high levels of food consumption. Still, although estimates vary, over 900 million people in the world today do not have access to sufficient and nutritious food.

The FAO estimates that agricultural production will need to increase by 60% globally (and nearly 77% in developing countries) by 2050 to cope with a larger, more urban and wealthier population, and to raise average food consumption to 3 070 kcal per person per day (FAO, 2012 forthcoming). This translates into additional consumption of 940 million tonnes of cereals and 200 million tonnes of meat a year by 2050 (compared with 2005/2007 production levels), raising concerns about the additional land and other resources that will need to be brought into production. These estimates do not include additional demands coming from the growing biofuel sector.

Increasing amounts of agricultural feedstocks are being used in the production of liquid biofuels. World ethanol production has doubled since 2005, while biodiesel production has increased fivefold. The production of both types of biofuels is expected to expand rapidly. This *Outlook* projects world production of ethanol and biodiesel will both increase by 5% p.a. over the next ten years. Despite the developments in second generation feedstock (*i.e.* woody crops, agricultural residues or waste), the share of total production of key food crops used for biofuel production will continue to increase through 2021: sugarcane (8% p.a.), corn (1.5% p.a.) and vegetable oil (4.4% p.a.). Although biofuels are predominately produced in food exporting countries, several low income, food and energy importing countries are also investing in biofuel production facilities.

Overall, production gains in agriculture have kept pace with demand growth over time. Real agricultural commodity prices have declined on trend over history; even if there have been significant periods of real price increases, such as in the 1970s and recently since 2000. Relative to global *per capita* incomes, food prices remain low in a historical context (FAO, SOFA 2008). Increased production has been attained by increases in resource use and factors of production – land, capital and variable inputs (such as fertiliser) – as well as by increases in the productivity of those inputs. In both instances, however, there is evidence suggesting that increasing the use of resources and their productivity will face challenges in the future. In many countries, it may be difficult to further exploit the resource base on a sustainable basis. While productivity growth[1] may be slowing in some regions of the world, the potential to increase productivity appears substantial, if not by pushing out productivity frontiers, by helping close productivity gaps[2] where they are large.

According to World Bank and OECD-FAO estimates, yield improvements of the three most important cereals (rice, wheat and maize) rather than area expansion has been the basis for production increases over the last 50 years (Table 2.1). In the previous decade, however, area expansion has been more important primarily for maize, through extension into new lands and by shifting land from other crops (see Chapter 4 on cereals).

Table 2.1. **Average annual growth rates in world rice, wheat and maize production**

	1960-2011	1992-2001	2002-2011
Production	2.4	0.9	2.5
of which yield	1.9	1.4	1.9
of which area	0.5	–0.5	0.7

Source: Robert Townsend, World Bank, using USDA data (for years 1960-2011); OECD and FAO Secretariats. (for years 1992-2011).

StatLink http://dx.doi.org/10.1787/888932642421

According to this *Outlook*, which is based largely on the views of national experts and commodity experts at OECD and FAO, and assumptions of "normal" growing conditions, the average annual growth in global agricultural production through 2021 (1.7%) will be lower than that of the previous ten years (2.6%) (Figure 2.1).[3] This is despite the relatively strong medium-term commodity price outlook and the fact that several natural disasters in major producing countries served to lower global output in recent years. However, even at this lower rate of growth, production gains will still outpace expected population growth, such that output *per capita* would continue to increase by 0.7% annually. In Latin America and Eastern Europe,[4] production growth is expected to increase more rapidly through increased area, higher yields and greater investment.

Figure 2.1. **Annual growth in global agricultural production**

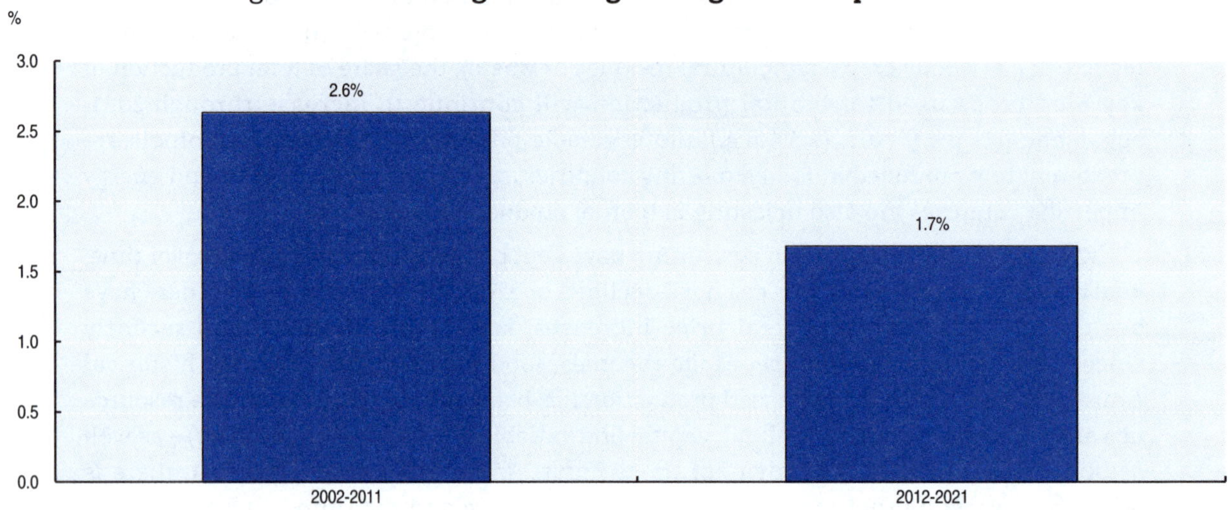

Source: OECD and FAO Secretariats.

StatLink http://dx.doi.org/10.1787/888932639153

This projected lower rate of production growth could be threatened by greater market volatility. Future agricultural investment and expansion depends to a large extent on the level and stability of returns to farming. The projected higher average real commodity prices should provide the right market signals. However, several reports (OECD 2010b, 2011c) have suggested that market volatility can have extensive negative impacts on the agriculture sector, food security, and the wider economy in both developed and developing countries. Many of the driving forces behind the increased market volatility, such as weather, energy prices and macroeconomic conditions, are themselves likely to remain volatile in the foreseeable future. Amongst different measures to address and mitigate market volatility in agriculture (FAO *et al.* 2011), there is a need for the agricultural sector to become more resilient. Improving resilience is particularly important to stabilise small holders' income and, therefore, mitigate the negative effects of price volatility (FAO 2011b).

What are the land and water constraints for agriculture?

Analysis of Global Agro-Ecological Zones (GAEZ) data shows that ample land resources with potential for crop production remain, but this assertion needs to be heavily qualified. Much of the potentially arable land is located in Latin America and sub-Saharan Africa, often in remote locations, far from population centres and agricultural infrastructure.

Agro-climatic suitability can be limited to certain crops and development for production can incur considerable costs because of soil, terrain or infrastructure constraints (Figure 2.2). Although a number of countries, particularly in the Near East, North Africa, and in South Asia, have reached or are about to reach the limits of their available land for agriculture, at the global level there is still sufficient land resources to feed the world population for the foreseeable future under current yield growth assumptions (Figure 2.1).

Figure 2.2. **World vulnerable areas regarding land availability and quality**

Note: The small circles highlight vulnerable areas, including land erosion, land scarcity and problems with soil fertility. *This map is for illustrative purposes and is without prejudice to the status of or sovereignty over any territory covered.*
Source: OECD and FAO Secretariats, based on EU (2011) and FAO (2011c).

The potential to expand agricultural land use exists, but there is competition for much of this land for urban growth, industrial development, environmental reserves, and recreational uses, while other areas are not readily accessible or are of poorer quality. There are other important resource constraints as well, especially water. The effects of climate change, while uncertain, could also imply significant changes in water availability and possible shifts in production zones. For instance, the incidence and severity of flood droughts has been increasing for the majority of OECD countries, which has put increasing pressure on irrigated farming in drier and semi-arid areas (OECD 2010a)

Given commodity prices, technology and competing demands, the feasible scope for area expansion is limited. FAO predicts that from the 2005-07 base period to 2050 only 10% of the global growth in crop production (21 % in developing countries) is expected to come from land expansion, with the remainder coming from higher yields and increased cropping intensity. Arable land is projected to expand by 69 Mha (less than 5%), with an expansion of about 107 Mha in developing countries being offset by a decline of 38 Mha in developed countries (Figure 2.3). Almost all of the land expansion in developing countries is projected to occur in sub-Saharan Africa and Latin America.

The capacity to increase land in agriculture is closely linked to available water supplies. At present, agriculture accounts for over 70% of water use globally, but both the absolute amount of water available for agriculture and its share are expected to decline (to 40% by 2050) (OECD 2012a). The availability of freshwater resources shows a similar picture to that of land availability, with sufficient resources at the global level being unevenly

Figure 2.3. **Arable land and land under permanent crops, past and future**

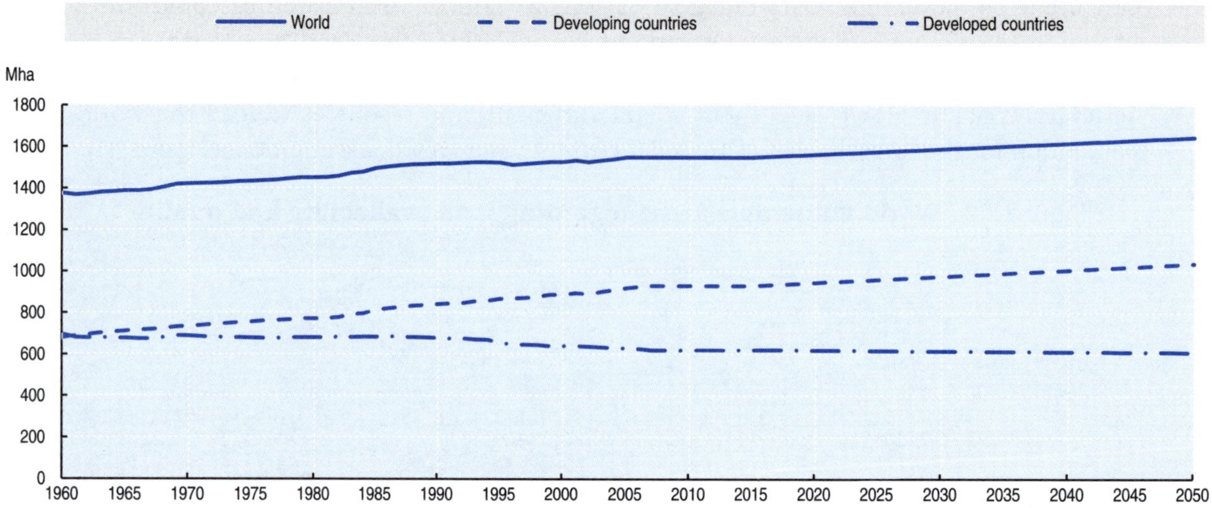

Source: FAO (2012).

StatLink http://dx.doi.org/10.1787/888932639172

distributed and an increasing number of countries or parts of countries reaching critical levels of water scarcity. Many of the water-scarce countries in the Near East and North Africa and South Asia also lack land resources. A mitigating factor could be that there are still ample opportunities to increase water-use efficiency, *e.g.* by providing the right incentives to use less water. Figure 2.4 shows the geographical distribution of priority areas in terms of freshwater scarcity in agriculture. Due to their vulnerability, coastal areas, the Mediterranean basin, Middle East and North African countries, and dry central Asia appear as regions where investment in water management techniques should be considered a priority when promoting agricultural productivity growth.

Figure 2.4. **World vulnerable areas regarding freshwater availability**

Note: The small circles highlight vulnerable areas, including water quantity and quality problems. *This map is for illustrative purposes and is without prejudice to the status of or sovereignty over any territory covered.*

Source: OECD and FAO Secretariats, based on EU (2011) and FAO (2011c).

What has been the performance in agricultural productivity growth?

Increasing agricultural productivity is critical to meet expected rising demand and, as such, it is instructive to examine recent performance. It is relatively complex to track trends in agricultural productivity; there are various indicators (partial and total measures), data sources and estimation methods, and agricultural production is highly variable. Major efforts have been made to provide more consistent estimates (OECD 1995). While numerical estimates should be interpreted with caution, historical trends can provide useful insight into relative changes over time and across regions. Available measures reveal a complex picture over time and across countries.

In many OECD countries, labour productivity has increased faster than land productivity as labour continues to move out of the sector. Latin America and China recorded strong growth in both land and labour productivity over the 1990-2005 period. This contrasts with the rest of Asia and with Africa, where gains in land productivity have outpaced labour productivity ones ones. At the global level, there is evidence that the annual growth rate in land productivity is slowing. If China's performance is excluded, annual land productivity growth fell from 1.9% in 1961-90 to 1.2% in 1990-2005 (Alston, 2010; OECD, 2011a).

Total factor productivity (TFP) is broadly defined as total outputs over total inputs.[5] Measured over time for a particular sector such as agriculture, TFP can be seen as a proxy for changing efficiency or industry competitiveness and technological change. According to Fuglie (2012), TFP grew strongly in developed countries from the 1960s to the mid-1990s at an average rate well above 1% per year in the 1970s and 1980s, and at rates above 2% per year in the two following decades (Table 2.2).

Table 2.2. **Total Factor Productivity growth of agriculture in world regions, 1961-2009**

Average annual growth rate by period

	1961-70	1971-80	1981-90	1991-2000	2001-09
All developed countries	**0.99**	**1.64**	**1.36**	**2.23**	**2.44**
All developing countries	**0.69**	**0.93**	**1.12**	**2.22**	**2.21**
North Africa	1.32	0.48	3.09	2.03	3.04
Sub-Saharan Africa	0.17	–0.05	0.76	0.99	0.51
Latin America – Caribbean	0.84	1.21	0.99	2.30	2.74
Caribbean	–1.00	0.57	–0.26	–0.55	–0.16
Central America	2.83	1.95	–1.69	3.05	2.33
Andean Countries	1.49	1.18	0.55	2.12	2.60
Northeast	0.25	0.60	3.02	2.62	4.03
Brazil	0.19	0.53	3.02	2.61	4.04
Southern cone	0.58	2.56	–0.82	1.61	1.29
Asia (except West)	0.91	1.17	1.42	2.73	2.78
Northeast Asia	0.94	0.67	1.71	4.10	3.05
China	0.93	0.60	1.69	4.16	2.83
Southeast Asia	0.57	2.10	0.54	1.69	3.29
South Asia	0.63	0.86	1.31	1.22	1.96
West Asia	1.21	2.21	0.95	1.70	1.34
Oceania	–0.14	0.47	–0.73	0.54	1.33
Transition countries	**0.57**	**–0.11**	**0.58**	**0.78**	**2.28**
Russian Federation	0.88	–1.35	0.85	1.42	4.29

Note: Estimated using FAOSTAT data The average annual growth rate in series Y is found by regressing the natural log of Y against time, *i.e.* the parameter B in ln(Y) = A + Bt.

Source: Fuglie (2012).

StatLink http://dx.doi.org/10.1787/888932642440

The TFP annual growth rate in some countries, such as Australia, has slowed in the 2000s, even when correcting the trend for bad climatic conditions (Sheng, 2011). Other studies using different time periods, data sources and estimation methods indicate that TFP growth has slowed since the mid-1990s in countries that have already attained very high levels of productivity (*e.g.* Alston *et al.*, 2010).

The situation is diverse in other countries over the same time period (Table 2.2). Growth in agricultural TFP resumed in some transition economies after a temporary slowdown in the 1990s and is high in some large producing countries of Central and Eastern Europe. Productivity growth has been particularly strong in some emerging economies like Brazil and China. However, agricultural productivity growth is still low in most least-developed countries, in particular Sub-Saharan Africa, although some countries achieved annual TFP growth rates of about 2-3% in the 2000s. Overall, there is no clear evidence that TFP growth is decreasing at the global level, (Fuglie, 2010b, 2012; OECD, 2011c; Piesse and Thirtle, 2010) although sustainability concerns should be explicitly addressed and large productivity growth differences between countries could be reduced if appropriate investments were made and appropriate technology adopted. This is further analysed in the forthcoming report *Agricultural Policies: Monitoring and Evaluation 2012* (OECD, 2012c). Increasing productivity also has social dimensions. *The State of Food and Agriculture 2010-11* (FAO 2011d) makes the business case for addressing gender issues, noting that the agricultural sector in developing countries is underperforming, in part, because women do not have the access and opportunities they need to be productive. The report critically evaluates experiences in many countries and shows how agricultural policies and programmes aimed at closing the gender gap can generate significant gains (Box 2.1).

Future productivity gains will depend on protecting the resource base, investments in research and development, and on the industry's ability to adopt the latest technologies. Agricultural R&D is the main source of innovation, which is needed to sustain agricultural productivity growth in the long-term. According to OECD data (OECD, 2011c), public R&D expenditure on agriculture across OECD countries has increased over the last decade both in real terms and as a share of agricultural GDP, ranging from less than 1% to 4% (United States). Public expenditures still dominate although private sector expenditures have also been increasing in some countries. Most studies argue that the estimated benefits of agricultural R&D generally far exceed costs, and Alston (2010) suggests the benefits are often underestimated, suggesting there may have been underinvestment.

Public R&D expenditures on agriculture in developing countries are increasing in real terms at a lower rate of growth than in the past (Beintema *et al.* 2008; Fuglie 2011) and declining as a percentage of agricultural GDP (OECD, 2011a). Moreover, funding is often dependent on foreign aid, granted for time-limited projects, and this may hamper the development of national R&D institutions and capacity building. However, research in some developed and emerging economies have spill-over effects and technology is being transferred to developing countries. An important challenge is to better adapt research results to local conditions and to foster the adoption of technologies able to improve sustainable productivity growth in diverse conditions.

Box 2.1. **Women in agriculture: Closing the gender gap for development**

On average, women comprise 43% of the agricultural labour force in developing countries, ranging from 20% in Latin America to 50% in Eastern Asia and sub-Saharan Africa. But agricultural productivity is reduced because women have less access to productive resources, namely inputs (such as fertiliser and improved seed), land, water, equipment, extension, and credit.

Figure 2.5. **Gender gaps regarding fertiliser use**

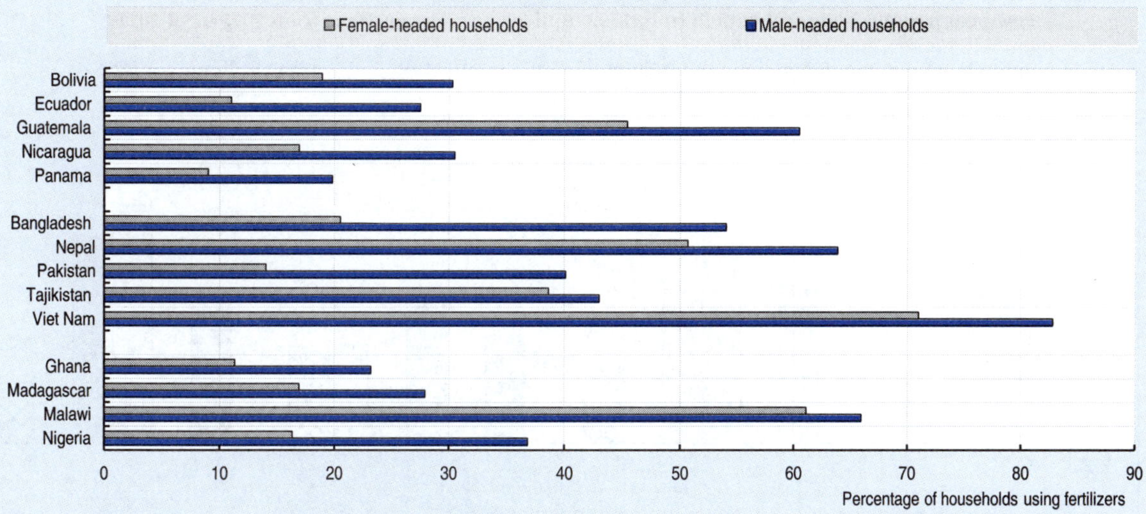

Source: FAO, data from Rural Income Generating Activities project.

StatLink http://dx.doi.org/10.1787/888932639191

Studies show that if women had the same access as men to productive resources, they would increase yields on their farms by 20-30%, raise total agricultural output by 2.5-4% in developing countries, improving food access. For this to happen, effective policy interventions to close the gender gap in agriculture are needed. These include: eliminating discrimination against women in access to resources, education, extension, financial services and labour markets; investing in productivity enhancing technologies and infrastructure; and, facilitating the participation of women in flexible, efficient and fair labour markets.

For example, Davis *et al.* (2010) examined the impact of farmer field schools on agricultural productivity and income in East Africa (1 126 farm households in Kenya, Tanzania and Uganda). They found that the schools attracted a high number of females with female farmers in Tanzania and Uganda benefiting more than their male colleagues. The value of crop production per unit area increased significantly. These results suggest that farmer-centered extension approaches can facilitate access to information on modern technologies and raise crop productivity while offering a potential element in strategies that seek to increase the access of women to innovations.

Source: FAO (2011d); Davis *et al.* (2010).

What is the potential for future partial and total factor productivity growth?

Even if there is no clear evidence that TFP growth is decreasing at the global level, there are concerns about a future slowdown of productivity growth rates when looking at the evolution of trends in partial indicators such as yields. When looking at individual crops, there has been a distinct slowdown in yield growth rates in recent decades. The *World Bank Development Report 2008* (World Bank, 2007) highlights the decrease of annual average yield growth rates for maize, wheat, rice and soybeans globally and for most country groupings, except in Eastern Europe for wheat and soybeans. Alston *et al.* (2008) report similar results for developing and developed countries, in particular for cereal yields in the majority of large producing countries. This decrease in yield growth is partly linked to the increasing pressure on resources. Based on the OECD-FAO agricultural projections, the decrease of yield growth in the coming decade is most noticeable for wheat and oilseeds (Figure 2.6).[6]

Figure 2.6. **Annual growth rate in yields for selected crops at world level**

Five-year periods, historical data in from 1992 until 2011 and projections from 2012 until 2021

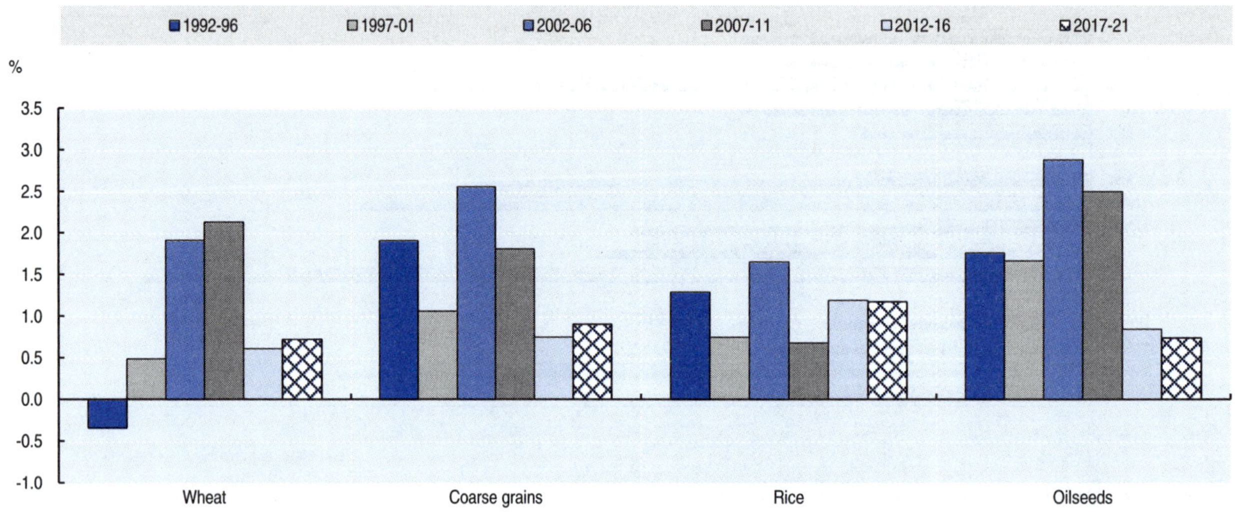

Source: OECD and FAO Secretariats.

StatLink http://dx.doi.org/10.1787/888932639210

As noted above, even if not capturing sustainability constraints, TFP is a more complete measure of resource efficiency, since it includes multiple measured inputs. Ludena *et al.* (2007) provide projections of agricultural TFP by region and sector to 2041. The authors point out that increased productivity growth in the last two decades of the twentieth century is due to accelerating productivity growth in those developing regions where substantial economic reforms have taken place since 1980: China, Eastern Europe and the former Soviet Union, Sub-Saharan Africa, and Latin America.

The analysis decomposed productivity growth into two components: technical change (moving the technology frontier) and technical efficiency (reducing the productivity gap). At the world level, agricultural TFP growth for the 2001-40 period is projected to exceed that of the previous decades due to an outward shift in the production possibilities frontier and not driven by a reduction of the productivity gap (Ludena *et al.* 2007). Taking into account the production-weighted averages of different regions and subsectors in industrialised countries, the frontier in agriculture advanced more rapidly in this period (0.99%) than TFP, thereby leading to negative technical efficiency growth (–0.21%).

TFP growth was also calculated by sector (Table 2.3). For developing countries, most regions were expected to achieve larger productivity gains in livestock than in crops, with the non-ruminant sectors (pigs, poultry) more dynamic than the ruminant sectors (cattle, sheep, goats), fuelled by higher technical efficiency. China, for example, was expected to maintain a rapid TFP growth rate due to an anticipated transition to modern pig and poultry production systems. The authors suggest the projected faster livestock TFP growth in developing countries is a positive development for consumers given the relatively high income elasticities of demand for livestock products in the developing world. Conversely, lower TFP growth was expected for the livestock sectors in industrialised countries.

In Sub-Saharan Africa, projected modest gains in technical efficiency and technical change result in relatively slow rates of TFP growth over the 2001-40 period. Given the current low rates of productivity and the expected high rates of population growth, this scenario is not encouraging. The authors suggest significant investments in research and extension infrastructure will be required to improve this trend.

Table 2.3. **Average total factor productivity growth rates by region and sector, 2001-40 (%)**

	TFP average*	EFF average	TCH average	TFP crops	TFP ruminants	TFP non-ruminants
World	1.38	0.34	1.04	0.94	0.82	3.60
Industrialised countries	0.77	–0.21	0.99	1.14	0.27	0.63
Economies in Transition	1.24	0.38	0.85	1.39	0.53	2.09
China	3.11	1.33	1.75	1.45	3.01	6.6
East and South East Asia	–0.08	–0.83	0.75	–0.66	–1.24	3.67
South Asia	1.16	0.68	0.48	0.96	1.48	3.48
Middle East and North Africa	0.22	–0.12	0.34	0.45	–0.31	–0.28
Sub-Saharan Africa	0.78	0.49	0.29	0.91	0.57	–0.05
Latin America and Caribbean	1.14	0.13	1.28	0.62	1.50	4.55

Note: Weighted average productivity growth rates are estimated using output shares of each subsector in agriculture in 2001. Projections based on a business as usual scenario. The change in efficiency (EFF) and technical change (TCH) are derived from a directional Malmquist index.
Source: Ludena et al. (2007).

StatLink http://dx.doi.org/10.1787/888932642459

What would be the impact of reducing the yield gap in developing countries?

Discussions of productivity growth in agriculture inevitably make reference to the evolution of crop yields. Other things equal, an increase in crop yields has immediate implication for everything from farm household income to commodity prices and to resource use in agriculture. Yields have grown remarkably over the past 50 years. Indeed, over the period from 1961 to 2011, the 2% annual increase in wheat production has been met almost entirely from yield increases (FAOSTAT database). The figures for maize are 2.7% annual increase in production (1.9% attributed to yield and 0.8% to area expansion) and 2.2% for rice (1.7% yield and 0.5% area).

Despite rapid yield growth, realised yields are still well below both their genetic potential and what may be considered their potential in an economic sense, in the context of prices, costs and other constraints that must also be considered to attain a potential output. Deviations from potential yields appear to vary remarkably among countries and regions even after adjusting for different soil, moisture and temperature environments. Other conditioning factors, such as different farm sizes and management capacities, access to markets, and legislative/institutional factors, play heavily in determining yield performance. Access to productive resources by gender has also shown to be an important factor (Box 2.1). The state of economic development, including human capital, fertilisers and other inputs, make a huge difference to attaining potential yields (Figure 2.7).

Figure 2.7. **Determinants of crop yields**

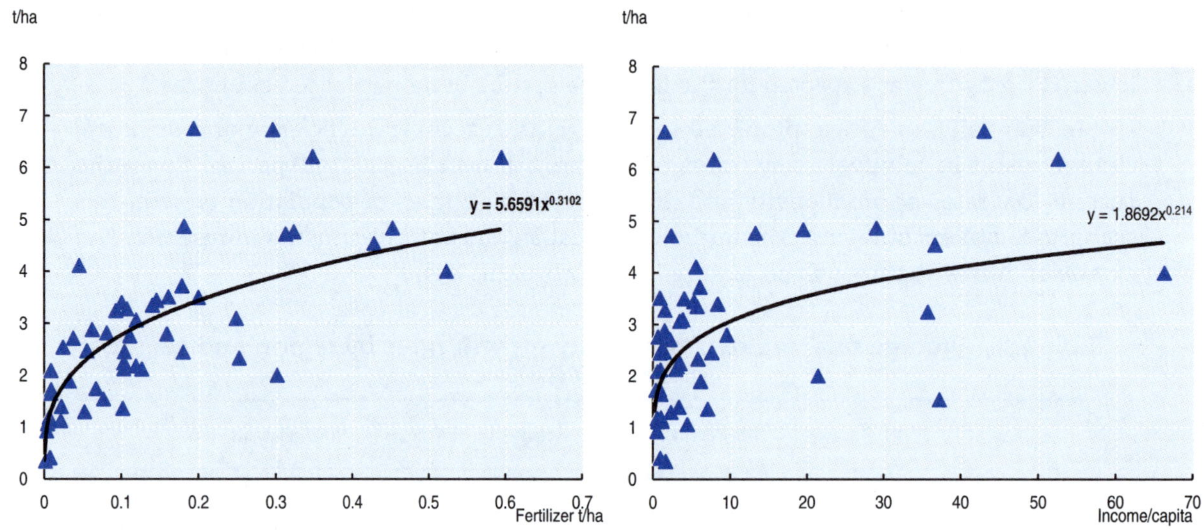

Note: Scatter plots are average cereal yields 2005-09 for Aglink-Cosimo model countries and regions against GDP/person and fertiliser per hectare.

Source: OECD and FAO Secretariats.

StatLink http://dx.doi.org/10.1787/888932639229

The important question is if actual gaps from crop yield potentials can be closed at low or minimal cost in a geographical zone, taking into account regional specific conditioning aspects such as climate and farming systems. The implications for farm and rural incomes, as well as national and global food security, and sustainability, may be significant. From a policy perspective, the key questions concern the micro- and macroeconomic effects of bridging these gaps, and it is important to identify what these are. In this context, this section undertakes analysis using the Aglink-Cosimo model to look at the possible impacts on commodity prices, production and trade to reduce measured yield gaps for the crops[7] considered in the *Outlook*. The analysis addresses some important questions such as: "how significant are the impacts of reducing yield gaps on production and land use?", "where are those located?" and, "how are commodity prices affected?". The analysis is not meant to be definitive, but indicative of some of the possible impacts of greater investments to reduce observed gaps. The effects of productivity on food prices (and farmers' incomes) have major food security implications for poor people. In addition, sustainability is essential, to ensure that productivity can be maintained in the face of depleting non-renewable resources and that production systems do not degrade the environment (Fischer *et al.*, 2011).

Table 2.4 provides some basic definitions of various measures of crop yields, which in basic terms is "the weight of a crop harvested with a standard moisture content per unit of field area" (Fischer *et al.*, 2009). The starting point is an average farm yield (FY), which forms the basis for calculating gaps relative to the economically attainable yield (AY) or relative to a potential yield (PY). Water-limited potential yield (PY_W) is included as a sensible yardstick where crops receive on average only low to moderate water supplies. For increasing FY, both increasing PY (or PY_W) and reducing the yield gap are important, and somewhat different interventions operate on each.

Progress in potential yield (or PY_W) through genetic and agronomic research has been an important source of yield growth because raising the yield frontier provides scope to

Table 2.4. **Definition of yield measures**

Measure	Definition	Estimation
Average farm yield (FY)	Average yield achieved by farmers in a defined region over several seasons.	Regional or national statistics, ground or satellite surveys of fields.
Economically attainable yield (AY)	Optimum (profit-maximising) yield given prices paid/received by farmers, taking account of risk and existing institutions.	On-farm experiments, varying inputs, sometimes crop models, disaggregated farm surveys.
Potential yield (PY)	Maximum yield with latest varieties, removing all constraints including moisture, at generally prevailing solar radiation, temperature and day light.	Highly controlled on-station experiments or crop models calibrated with latest varieties, well-monitored crop contests.
Water-limited potential yield (PY_W)	Maximum yield under normal rainfed conditions, removing all constraints except moisture.	Highly controlled on-stations experiments, crop models or crop contests.

Source: Evans and Fischer (1999).

increase the other yield measures. Progress in potential yield enhancement has also exploited positive interactions between genetic and agronomic routes to improve yields. In advanced systems, however, yield increases from agronomy alone appear to be slowing (Evans and Fischer, 1999; Sinclair, 1998).

The literature distinguishes two components of yield gaps: agro-environmental and other non-transferable factors, which create gaps that cannot be narrowed, and crop management practices, such as suboptimal[8] use of inputs which may occur for a variety of reasons. This second component can be narrowed – provided it makes economic sense to do so – and is therefore termed the "exploitable" or "bridgeable" yield gap (Bruinsma, 2011). In general, yield gaps at the lower end, such as AY-FY, are explained mainly by farmers' access to information and technical skills, while high-order yield gaps reflect opportunities for research and broader policy and institutional/social constraints.[9] Both farmer characteristics and system-wide constraints explain these various yield gaps and suggest how they might be reduced.

The FAO report, *The State of the World's Land and Water Resources for Food and Agriculture* (FAO, 2011c), provided indications of the size of yield gaps by region in 2005. The results in Table 2.5 show that yield gaps vary markedly by region and in most cases are sizeable. They are greatest in Sub-Saharan Africa, where yields are estimated to be only about 24% of what could be produced. The gap is the lowest in East Asia at only 11%. The potential to increase output is particularly high in the regions where the yield gap is larger than 50%: Northern Africa, Sub-Saharan Africa, Central America, and the Caribbean, Southern America, Western Asia, Central Asia, South Asia, Eastern Europe and Russian Federation, and the Pacific Islands. While farming in East Asia is running much closer to potential and reducing gaps would not affect production by as much, the sheer size of this region means that an increase in output could have significant implications.

A survey of the literature on yield gaps for wheat, rice and maize cropping systems done by Lobell *et al.* (2009) confirms the wide range of estimated yield gaps throughout the world. For tropical maize in Africa, where biophysical and management conditions result in frequent nutrient, water, pest, and disease stresses, average yields are commonly less than 20% of yield potential. In contrast, average yields in irrigated wheat systems in northwest India can reach 80% of potential. Although their study reports the full range of values extends from 16% to 95%, they consider a range of 20% to 80% to include nearly all of the major cropping systems of the world owing to measurement errors. Similar findings can be found in the analysis of three major cereal yields and their productivity in key

Table 2.5. **Estimates of yield gaps by region**

Region	Actual yield in 2005 (% of economically attainable yield)	Yield gap(%)
Northern Africa	40	60
Sub-Saharan Africa	24	76
Northern America	67	33
Central America and Caribbean	35	65
Southern America	48	52
Western Asia	51	49
Central Asia	36	64
South Asia	45	55
East Asia	89	11
Southeast Asia	68	32
Western and Central Europe	64	36
Eastern Europe and Russian Federation	37	63
Australia and New Zealand	60	40
Pacific Islands	43	57

Note: Percentage of potential for cereals, roots and tubers, pulses, sugar crops, oil crops and vegetables combined for rainfed cultivated land across regions in 2005.
Source: Fischer *et al.* 2010.

StatLink http://dx.doi.org/10.1787/888932642478

regions done by Fischer *et al.* (2010). They found that in most wheat and rice farming systems, there is a gap that exceeds 30% between average and potential yields. The gap is usually larger in the case of maize, notably in Sub-Saharan Africa.

Impacts of partial and proportionate reducing of yield gaps

Future crop yields and global food security may well hinge on the ability of farmers around the world to narrow the gap between current and potential yields, especially as progress in the latter may slow down because of climate change and diminishing returns in breeding. Because average crop yields are critical drivers of food prices, farmers' incomes, food security, and cropland expansion, there is value in better quantification and understanding of yield gaps. With the purpose of examining the sensitivity of market outcomes with respect to yields, a stylised scenario was conducted with the Aglink-Cosimo model. In this scenario, estimated crop yields were increased over the baseline period in a manner that reduces the gaps proportionately in all developing countries by one-fifth of the estimated gap (Table 2.5)[10] by the end of the projection period. For example, this implies reducing the gap for North Africa to 48% and for Sub-Saharan Africa to 61% by 2021. For these two cases, this involves a scenario of increasing yields for crops covered in the model by 28% and 63% for North Africa and for Sub-Saharan Africa, respectively.

Table 2.6 presents the regional impacts of the scenario against the baseline for area harvested, supply and yield of cereals. The overall effect on yields at world level varies from 12% for rice (from 3.3 to 3.7 t/ha) to 7% for wheat (from 3.2 to 3.4 t/ha). It should be noted that the proposed shock on yields, even if limited, boosts world cereal production by 5.1%. As a result of this intensification of crop production, world area harvested contracts by 2.7% since marginal production areas are taken out of production. From a regional perspective, production in North Africa increases the most (26%), followed by developing countries in Asia (8.7%). Due to substitution effects, supply and harvested area contracts in developed countries, especially in coarse grains and rice, crops for which developing countries hold a larger production share than in the case of wheat.

Table 2.6. **Supply details in the baseline and scenario results on reducing the yield gap in developing countries**

Area in million hectares, yield in tonnes per hectare, and supply in million tonnes

		Baseline (year 2021)			20 reduction of the yield gap in developing countries (year 2021, % change *vs.* the baseline)		
		Coarse grains	Rice	Wheat	Coarse grains	Rice	Wheat
						%	
World	*Area*	332	166	227	–3.6	–5.9	–2.6
	Yield	4.1	3.3	3.4	7.1	12.2	6.9
	Supply	1 359	542	761	3.4	5.7	4.1
Developed countries	*Area*	119	4	123	–1.8	–3.1	–2.9
	Yield	6.0	5.1	3.3	–0.8	0.4	3.0
	Supply	718	19	406	–2.7	–2.7	–0.1
North America	*Area*	44	1	27	–3.1	–3.3	–2.7
	Yield	10.0	6.0	3.1	–0.7	–0.5	
	Supply	439	7	84	–3.8	–3.8	–2.8
Western Europe	*Area*	30	0	26	–0.6	–2.1	–1.9
	Yield	5.2	3.8	5.7	–0.8	–1.6	–0.9
	Supply	155	2	150	–1.3	–3.8	–2.7
Oceania Developed	*Area*	6	0	13	–3.7	–18.7	–3.1
	Yield	2.5	7.1	2.0	0.4	–0.4	
	Supply	15	1	25	–3.5	–19.0	–3.1
Other Developed	*Area*	4	2	6	–4.8	–0.1	–2.5
	Yield	3.9	4.8	3.4	4.1	1.5	28.6
	Supply	16	8	19	–0.8	1.3	25.1
Eastern Europe	*Area*	35	0	51	–0.7	–12.8	–3.5
	Yield	2.6	4.1	2.5	0.4	3.9	5.2
	Supply	93	1	128	–0.4	–9.4	1.7
Developing Countries	*Area*	212	162	104	–4.6	–5.9	–2.2
	Yield	3.0	3.2	3.4	15.6	12.7	11.1
	Supply	641	523	354	10.3	6.0	8.9
Africa	*Area*	89	16	29	–9.8	–9.6	–4.9
	Yield	1.6	1.9	2.7	47.2	47.1	13.4
	Supply	143	31	77	32.6	32.6	8.0
North Africa	*Area*	5	0	6	–2.7	–14.0	1.7
	Yield	1.6	4.8	2.3	29.0	27.1	28.8
	Supply	7	0	13	24.9	9.4	30.7
Sub-Saharan Africa	*Area*	74	13	3	–10.9	–11.0	–19.5
	Yield	1.4	1.9	2.2	58.7	58.5	54.8
	Supply	107	24	7	41.2	40.8	24.7
Near East	*Area*	10	3	20	–4.8	–4.5	–4.4
	Yield	3.0	2.0	2.9	8.8	8.9	5.6
	Supply	30	7	57	3.6	3.9	0.8
Latin America and Caribbean	*Area*	39	6	10	–3.9	–17.4	–0.3
	Yield	4.2	3.8	3.3	6.9	7.6	2.1
	Supply	163	21	33	2.7	–11.2	1.8
Asia	*Area*	84	140	65	0.7	–5.0	–1.2
	Yield	4.0	3.4	3.8	3.8	10.7	11.7
	Supply	335	471	245	4.4	5.0	10.1

Source: OECD and FAO Secretariats.

StatLink http://dx.doi.org/10.1787/888932642497

The fairly inelastic demand for food and feed exerts downward pressure on world prices (Figure 2.8), with rice and coarse grains experiencing the largest projected changes. This can be explained by a large proportion of these products being produced in developing countries (especially rice) and, therefore, much more affected by the reduction in the yield gap. The net effect of increased productivity on farmers' incomes is indeterminate (as yields increase but prices fall), and will likely vary across farms of different types and sizes.

Figure 2.8. **World price effects of reducing the yield gap by 20% in developing countries**

% changes compared to the baseline

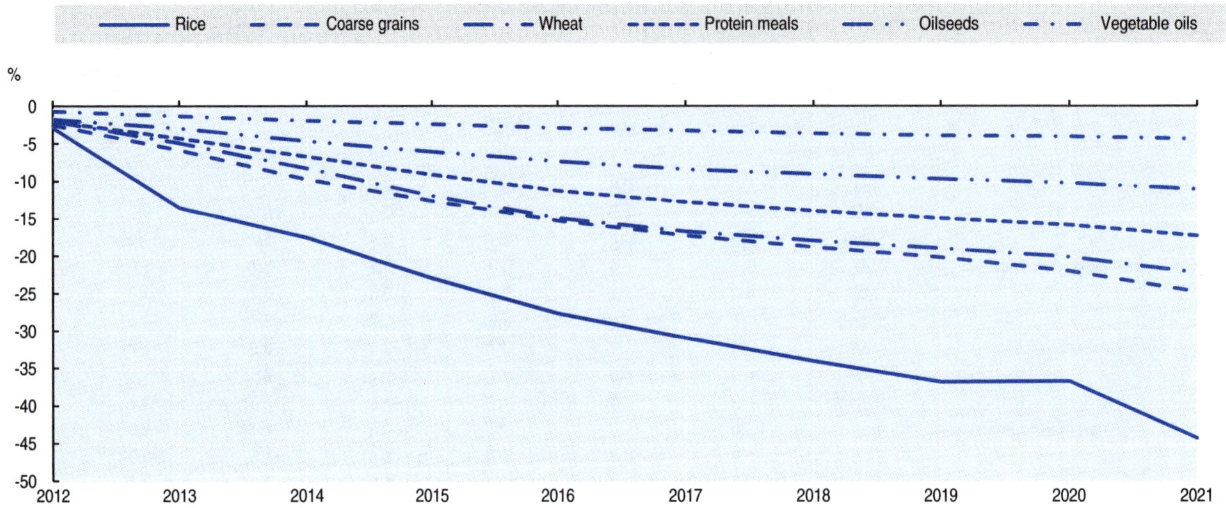

Source: OECD and FAO Secretariats.

StatLink http://dx.doi.org/10.1787/888932639248

This scenario is for illustrative purposes and the results are to be interpreted with caution since the proposed reduction in yield gaps is achieved "at zero cost", assuming better management practices and use of improved seed varieties.[11] Therefore, no assumption of increased fertiliser input to achieve the reduction in the yield gaps is taken into account. This stylistic design of the scenario allows for a straightforward interpretation of results.[12]

Focusing on the food/feed *versus* fuel discussion, and in light of the lower commodity prices, it is important to analyse where the increase in coarse grain and sugarcane production goes. Figure 2.9 shows that 33% of each additional tonne of world grains is

Figure 2.9. **Global coarse grains and sugarcane use**

Additional million tonnes in selected years of the projection period compared to the baseline

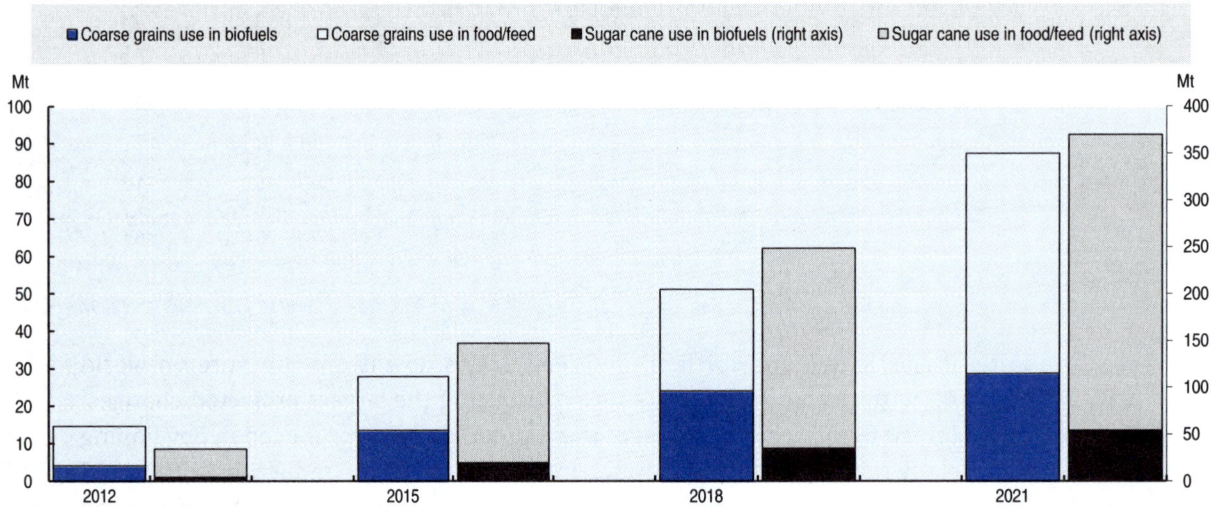

Source: OECD and FAO Secretariats.

StatLink http://dx.doi.org/10.1787/888932639267

projected to go into biofuel production in year 2021. Similarly, 15% of additional sugarcane production is projected to be converted into ethanol. This has to do with the fact that coarse grains and sugarcane demand for food and feed in developing countries is fairly inelastic compared to biofuel demand by the US ethanol industry. For instance, whereas cereal food and feed consumption increases by 1.5% at world level (2% for developing countries), the industrial use of cereals increases by 7%.

Why the need for *sustainable* agricultural productivity growth?

As noted in the previous sections, the food and agricultural sector has been successful in providing for an increasing and wealthier population at the global level, despite yield gaps in many regions. TFP growth in agriculture has been strong and has exceeded the population growth rate. Innovation and good management practices have boosted crop yields and livestock productivity[13], aquaculture supplies an increasing share of total fish and seafood consumption, and the real price of food has declined over the long term. Agricultural growth has been an important engine of economic growth in many developing countries and as such, a major driver of poverty reduction and food security (World Bank 2008).

Nevertheless, progress has been uneven and food security problems remain an issue of concern. While in some countries and regions productivity growth has been low, in others it may conceal depletion of resources that cannot be sustained over time. There is growing pressure on natural resources, including land, water, marine ecosystems, fish stocks, forests, and biodiversity, which are fundamental to sustainable production (Tillman 1999). A recent FAO report warns of "the creeping degradation of the land and water systems that provide for global food security and rural livelihoods" (FAO, 2011c). Approximately 25% of the world's agricultural land area is highly degraded. Degraded soils, lacking in soil organic matter and nutrients, are more vulnerable to temperature extremes, droughts and floods. These pressures have reached critical levels in some areas and climate change is expected to exacerbate vulnerabilities (SRES 2011; Easterling *et al.* 2007). Nearly all researchers conclude that increased climate variability and extreme weather events are projected to increase in the near term, affecting all regions (Cline, 2007; SREX 2012). Changes in the frequency and severity of individual extreme weather events will probably have a much bigger impact on food production and food security than mean changes in climate, especially in the period to 2050 (The Government Office for Science, 2011; FAO 2012).

Environmental pressures are also caused by agriculture, which is a net contributor of greenhouse gases. The agricultural sector emits around 14% of total anthropogenic greenhouse gas emissions (IPCC, 2007), being the 4th largest sectoral contributor after energy, industry and forestry (including deforestation). Moreover, methane and nitrous oxide emissions are projected to increase to 2050 (Figure 2.10). Although agricultural land is expected to expand only slowly, the intensification of agricultural practices (especially the use of fertilisers) in developing countries and the change of dietary patterns (increasing consumption of meat) are projected to drive up these emissions.

Given increasing resource constraints and growing environmental pressures, the challenge is not just to increase agricultural productivity, but to do so in a more sustainable manner. Increasing costs of environmental damage associated with agricultural production coupled with rising scarcities of natural resources essential for agricultural production are two key motivations for prioritising sustainability in agricultural productivity growth scenarios. A third motivation is the increasing importance of

Figure 2.10. **Share of different sectors in total anthropogenic greenhouse gas emissions (2004, CO_2-equivalents) and projections for different greenhouse gases (2010-50)**

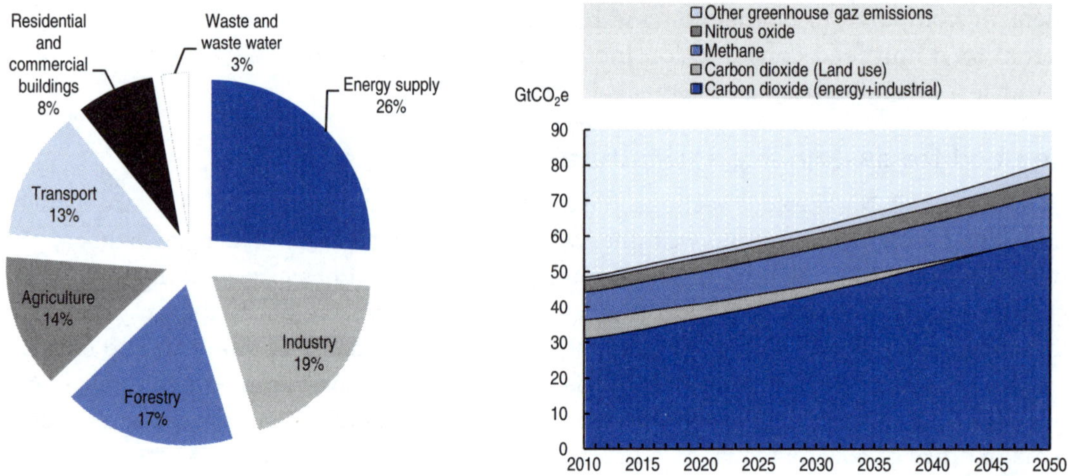

Source: IPCC, 2007 (left figure) and OECD, 2012a (*right figure*).

StatLink http://dx.doi.org/10.1787/888932639286

maintaining and augmenting resilience in productivity growth, and the key role of ecosystem services in achieving it. The need to increase productivity sustainably has been reinforced and extensively documented in a number of recent reports and papers (EC, 2011; UK, 2011; Foley *et al.* 2011; FAO 2012). The pressing nature of the issue was explicitly recognised in FAO *et al.* (2011), a joint report of international organisations to the G20 in which the first recommendation addresses measures "to strengthen the longer term productivity, sustainability and resilience of the food and agriculture system world-wide" as part of "an enduring solution to global food insecurity".

To avoid a future in which unsustainable patterns of growth put food security at risk, well-functioning markets must provide clear price signals that reflect the scarcity value of natural resources, and property rights must be defined so as to encourage optimal use of resources, both individually and collectively. The over-arching policy challenge is to create the enabling environment and incentives that would optimise resource use from an economic, environmental and social perspective.

Progress is being made. Governments have started to orient their policy priorities to take account of the environmental consequences of food and agriculture production and consumption, and to improve incentives for optimal resource use by farmers. As a result, in many countries there have been improvements in the environmental performance of agriculture over past decades, although progress is quite varied (OECD 2008). Some countries are already moving towards "greener" growth strategies in agriculture (Box 2.2).

Nutrient surpluses are a relevant indicator for environmental performance of agricultural systems, since excess of nutrients are likely to be leached into the groundwater, run off the fields into watercourses, or be lost to the atmosphere through conversion to ammonia (volatilisation). Amongst all nutrients applied mainly through fertilisation to crop production, nitrogen is the most-important one. Even if there is no optimal nutrient surplus level per hectare, since it depends on many agronomic and farm management factors, it is important to see what its evolution is over time. Nitrogen

Box 2.2. **A green growth strategy for food and agriculture**

The term "Green Growth" refers to a broad strategy that seeks to define an economic development path that is consistent with long-term environmental protection, using natural resources sustainably, while providing acceptable living standards and poverty reduction in all countries. The relationship between agriculture and green growth is complex. Resource endowments and environmental absorptive capacities vary widely across countries and regions, and impacts can differ in the short and long run and at different stages of production and consumption. Green growth for food and agriculture is not only desirable and achievable, it is also essential if the food and nutrition requirements of future generations are to be met.

The specific approach varies by agro-ecology, farming system and market conditions but consistently will involve increasing the conservation and sustainable use of natural resources in agricultural production systems, as well as the reduction of waste and pollution associated with inefficient input use and degraded ecosystems (FAO 2011b). Sustainable agricultural productivity growth implies lower costs through efficiency gains, however there can be significant investment costs required to achieve such gains. Green growth strategies require re-orientation of key institutions such as markets to provide the right signals (*e.g.* prices that reflect the scarcity value of natural resources); and establishing and enforcing secure property rights, in particular for marine resources, land and forests, greenhouse gas emissions, and air and water quality.

Source: OECD (2011a).

surpluses are projected to decrease in most OECD countries by 2050 (Figure 2.11) due to the efficiency of fertiliser use improving more rapidly than productivity increases.

In China, India and most developing countries, the trend goes in the opposite direction: nitrogen surplus per hectare is likely to increase as production grows more rapidly than efficiency. In China and India, crop production is expected to grow by more than 50% between 2000 and 2030 and 10% to 20% between 2030 and 2050. The growth in

Figure 2.11. **Nutrient surpluses per hectare from agriculture (1970-2050)**

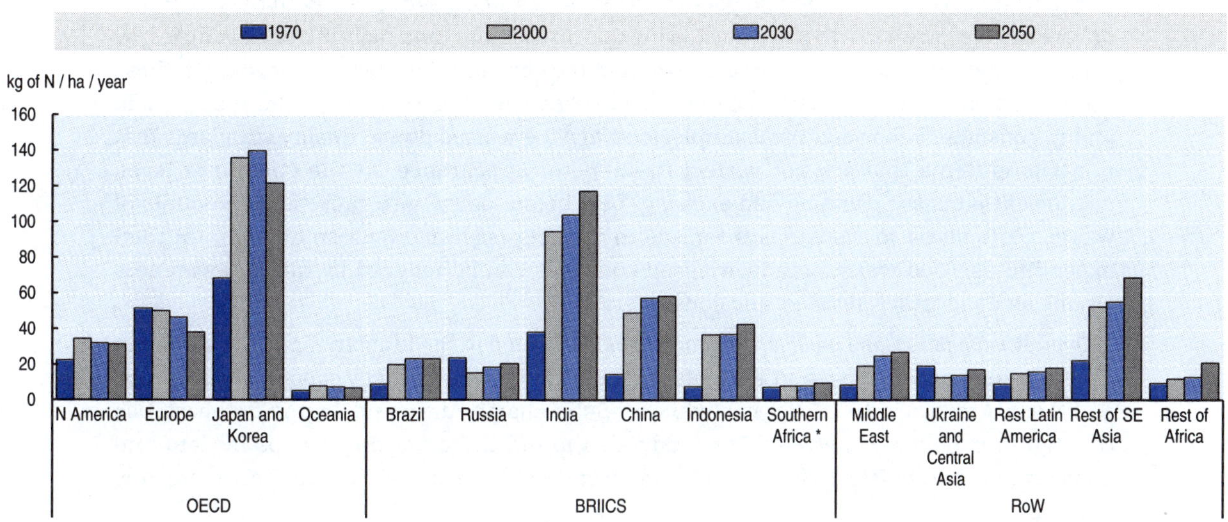

Source: OECD, 2012a.

StatLink http://dx.doi.org/10.1787/888932639305

fertiliser surpluses in Africa (excepting Southern Africa) is dominated by North Africa, which is projected to contribute 20% of Africa's total nitrogen surplus. Surpluses in Sub-Saharan Africa are smaller than in many other developing countries. As soils are often deficient in phosphorus, increased fertilisers are needed to restore and improve soil fertility and sustain crop production. Overall, total crop production in Africa is projected to increase between 2000 and 2050 (North Africa by 150%; West Africa, 375%; East Africa, 265%) (OECD, 2012a). This is assumed to be achieved through a considerable expansion in agricultural land and increased yields. If this production increase is to be sustained without expanding agricultural land any further, restoration and improvement of soil fertility, technological improvements and higher fertiliser application rates –especially phosphorus – are likely to be needed. More ecological farming techniques will be needed as well.

While not the focus of this chapter, it should be noted that reducing waste is another aspect of a more sustainable food and agriculture system. The need to increase production and productivity would be greatly reduced by reducing food losses and food waste – both post-harvest losses at the farm gate (mainly in developing countries) and food waste further along the food chain (mainly in developed countries). A recent FAO study (FAO 2011a) suggests that roughly one-third of food produced for human consumption is lost or wasted globally, which amounts to about 1.3 Bt per year. This inevitably means that significant amounts of the resources used in food production are also wasted (Box 2.3).

Box 2.3. **Reducing food losses and food waste**

Food is lost or wasted throughout the supply chain, from initial agricultural production down to final household consumption. In low-income countries, food is lost mostly during the production and processing stages of the food supply chain. The causes of food losses and waste in these countries are mainly connected to financial, managerial and technical limitations in harvesting techniques, storage and cooling facilities in difficult climatic conditions, infrastructure, and packaging and marketing systems. Food supply chains in developing countries need to be strengthened by, *inter alia*, encouraging small farmers to organise, diversify and scale up their production and marketing. Investments in infrastructure, transportation, food industries and packaging industries are also required. Both the public and private sectors have a role to play in achieving this. In medium- and high-income countries, by contrast, food waste occurs to a greater extent at the consumption stage. The causes in these countries mainly relate to a lack of co-ordination between different actors in the supply chain and to consumer behaviour. For example, food may be wasted due to quality standards that reject food items that are not perfect in shape or appearance. At the consumer level, insufficient purchase planning and expiring "best-before-dates" also cause large amounts of waste, particularly for consumers for whom food represents only a small share of total expenditures. Food waste in industrialised countries can be reduced by raising awareness among food industries, retailers and consumers.

Overall, on a per-capita basis, much more food is wasted in the industrialised world than in developing countries. The report estimates that per capita food waste by consumers in Europe and North-America is 95-115 kg/year, while for Sub-Saharan Africa and South/Southeast Asia this figure is only 6-11 kg/year. The study also found there are major gaps in data and knowledge regarding global food losses and waste. Further research in this area is urgently needed.

Source: FAO (2011a).

What farm practices can increase sustainable agricultural productivity growth?

There has been much discussion in many fora about the need to increase productivity and sustainability in agriculture in the medium to long term, but much less information is available on specific means to achieve these aims. The purpose of this section is to provide examples of practices that can contribute to increased sustainable agricultural productivity growth. The key message is that the two aims are compatible: it is not necessarily a trade-off between producing more and producing in a more sustainable manner. This section offers several examples of such "no-regrets" options, that is agricultural practices than can strengthen productivity while using specific scarce resources in a more efficient manner in both developed and developing countries.

More efficient use of land and water

Intensive production in some regions has been associated with degradation of land and water resources, and the deterioration of related ecosystem goods and services (*e.g.* biomass, carbon storage, soil health, water storage and supply, biodiversity, and social services). Sustainable intensification of production will require widespread adoption of sustainable land management practices and more efficient use of irrigation water through enhanced flexibility, reliability and timing of irrigation water delivery.

There is potential to expand production efficiently to address food security and poverty while limiting the impact on other ecosystem values. There is scope for governments and the private sector, including farmers, to be more proactive in advancing the general adoption of sustainable land and water management practices, which are inextricably linked in agriculture. Actions include not just technical options to promote sustainable intensification and reduce production risks, but also comprise a set of conditions to remove constraints and build flexibility. These include: the removal of distortions (*i.e.* getting the prices right); improvement of land tenure and access to resources; strengthened and more collaborative land and water institutions; efficient support services including knowledge exchange, adaptive research, and rural finance; and better and more secure access to markets (Boxes 2.4 and 2.5).

In dry areas, farmers dependent on rainfall for cereal production can increase yields using supplemental irrigation (SI), which entails harvesting rainwater run-off, storing it in ponds, tanks or small dams, and applying it during critical crop growth stages. One of the main benefits of SI is that it permits earlier planting; while the planting date in rainfed agriculture is determined by the onset of rains, supplemental irrigation allows the date to be chosen precisely, which can improve productivity significantly. In Mediterranean countries, for example, a wheat crop sown in November has consistently higher yield and shows better response to water and nitrogen fertiliser than a crop sown in January. The average water productivity of rain in dry areas of North Africa and West Asia ranges from about 0.35 to 1 kg of wheat grain for every cubic metre of water. ICARDA has found that, applied as supplemental irrigation and along with good management practices, the same amount of water can produce 2.5 kg of grain. The improvement is mainly attributed to the effectiveness of a small amount of water in alleviating severe moisture stress.

In the Syrian Arab Republic, SI helped boost the average grain yield from 1.2 t/ha to 3 t/ha. In Morocco, applying 50 mm of supplemental irrigation increased average yields of early planted wheat from 4.6 tonnes to 5.8 tonnes, with a 50% increase in water productivity. In Iran, a single SI application increased barley yields from 2.2 to 3.4 t/ha. When integrated

Box 2.4. Partnerships and land productivity growth by smallholders in Indonesia

In Indonesia, various forms of partnerships have been developed between large companies and smallholders. The best known is the nucleus-plasma model initiated by the Ministry of Agriculture in the late 1970s to expand perennial crop production, especially in Kalimantan, Sulawesi and Sumatra. It has been used mostly for palm oil, but also for rubber and sugarcane. The programme provided large companies (nucleus) with both subsidised capital and long-term leases to state land for perennial crop production, under the condition that these companies provide inputs, credit, and technical and marketing services to the smallholders (plasma) surrounding the company plantations. The current legislation requires large investors to build partnerships with smallholders on at least 20% of the plantation area. New schemes based on the nucleus-plasma model have been launched in 2006, including the plantation revitalisation programme which offers government-subsidised credit to plasma farmers.

These and other contractual arrangements might be one reason for the spectacular increase in land productivity of smallholders' plots compared to large estates. The yield gap has diminished as smallholder land productivity approached yields on large estates for palm oil, sugarcane and cocoa (Figure 2.12). The productivity gap remains large for tea and rubber. In the case of rubber, this may partly result from lower tree density per unit of land as smallholders practice mixed cropping, *i.e.* rubber trees are planted in such a way that allows accommodating other types of crops on the same area. This contrasts with the cropping system of large estates based on monoculture (Fuglie, 2010a).

Figure 2.12. **Yields of large and small perennial crop producers, 1971-2007**

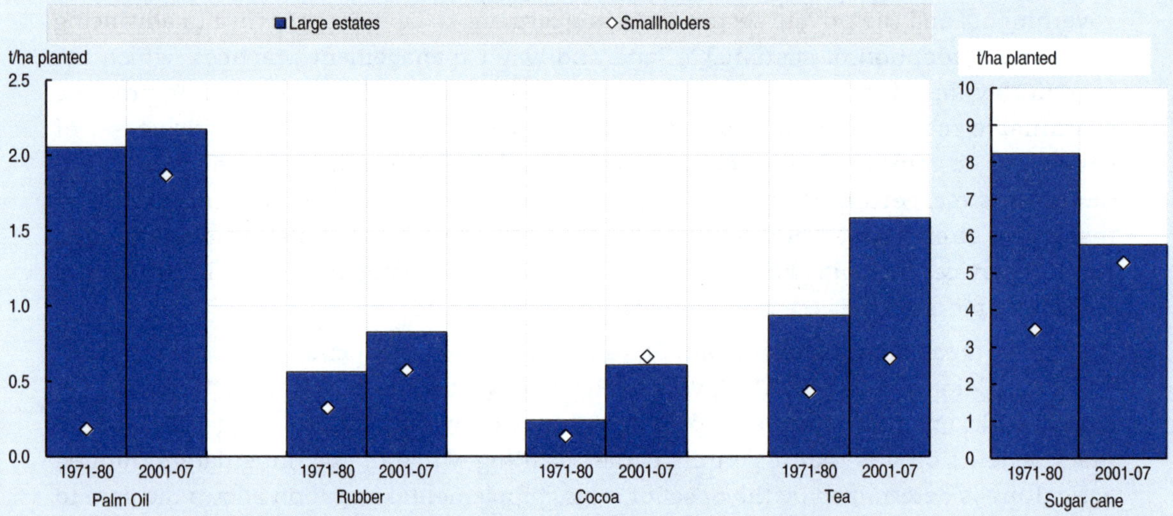

Source: OECD, 2012b based on Fuglie, 2010a.

StatLink http://dx.doi.org/10.1787/888932639324

with improved varieties and good soil and nutrition management, supplemental irrigation can be optimised by deliberately allowing crops to sustain a degree of water deficit (FAO, 2011b).

Better water management can generate benefits for health, agricultural and industrial production, and can preserve ecosystems and the watershed services they provide, thereby avoiding the enormous costs that can be imposed by flooding, drought or the degradation of watershed services. There is a need for policies to further reduce nutrient discharges in order to decrease eutrophication of lakes and oceans. In order to address the water

Box 2.5. **Water management in Israel**

Agricultural use of water in Israel decreased almost continuously from the mid-1990s to the early 2000s, but subsequently stabilised up to 2008 and is projected to rise by 2025, although at a lower rate than the increase in urban and industrial demand. As a consequence, the share of agriculture in total water use fell from over 70% in 1980 to 57% by 2005, and is projected to decline to 52% by 2025. A further development, induced by government water quota policies, has been the decline in use by agriculture of freshwater resources compared to an increase in the sector's use of recycled effluent and desalinated water.

A notable feature of Israeli agriculture has been its capacity to increase the efficiency of water use in agriculture. Efficiency has been improved in physical (technical) terms of water use per tonne of output (or hectare irrigated) (Figure 2.13). As a result of these improvements in agricultural water use efficiency, Israel is now a world leader in the management and technologies related to irrigation in arid environments (World Bank, 2006). The invention and development of drip irrigation in Israel in the 1960s has been the key innovation behind the rise in technical water use efficiency, as well as the shifting to other pressurised irrigation systems (*i.e.* sprinklers, micro-sprinklers, micro-jets). Flood irrigation is no longer used by farmers.

Figure 2.13. **Technical efficiency of agriculture water use**

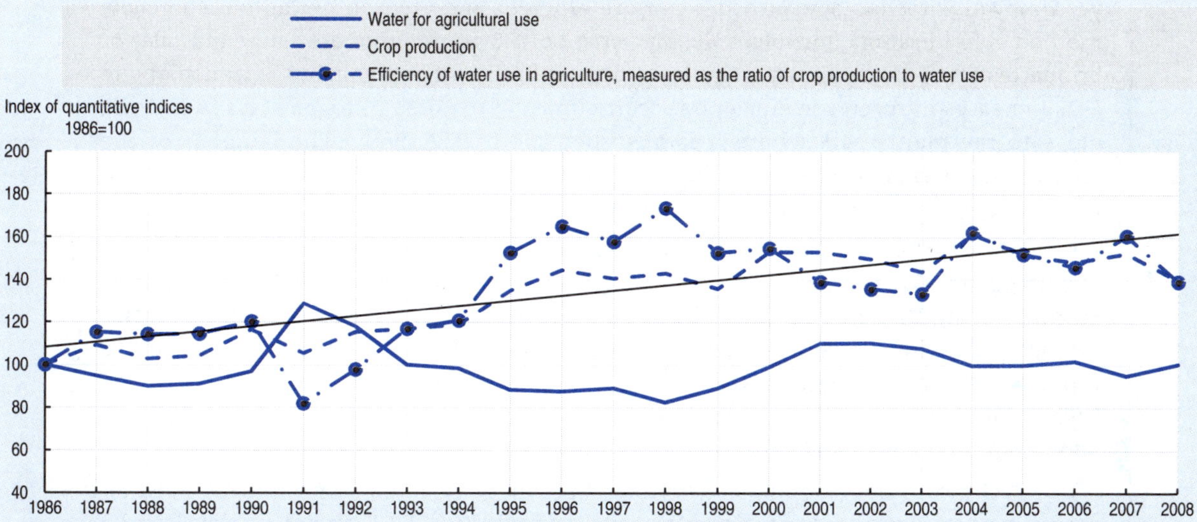

Note: Crop production and water use represented as quantitative indices. The linear regression is done on efficiency of water use in agriculture ($y = 2.4293 \times + 105.8$, $R^2 = 0.4548$).

Source: OECD (2010b).

StatLink http://dx.doi.org/10.1787/888932639343

challenges, countries have adopted a range of policy approaches, including regulatory measures (*e.g.* norms and standards for water quality), economic instruments (*e.g.* user tariffs, insurance schemes or tradable water rights and quotas), and information-based instruments (*e.g.* certification, integrated water management plans and cost-benefit analysis of water management policies) (OECD, 2012a). When pursuing productivity growth under water availability and water quality constraints, it is important to set up the right incentives for technological change to take place.

More efficient use of nutrients

Agricultural productivity can be sustainably increased through better plant nutrient management by improving the nutrient use efficiency (IFA, 2012). Nutrient use efficiency is particularly problematic for nitrogen fertilisation, since 50% to 70% of the applied nitrogen may not be taken up by the plant. Biotechnology research[14] may bring significant progress in the long term, but in the medium term increasing nitrogen use efficiency will be achieved through improvements in agronomic practices (Boxes 2.6 and 2.7).

Box 2.6. **More efficient fertiliser use in Bangladesh**

Like most smallholder farmers in Asia and Africa, farmers in Bangladesh are resource-poor and risk-adverse. Technology introduction in such an environment often has a slow return to invested capital, a deterrent to major private sector investment. In this environment, rice farmers depend on urea fertiliser to meet the nitrogen needs of their irrigated and rain-fed rice cultivation. Many farmers still spread urea into floodwaters to fertilise paddy plants, which is highly inefficient: about two-thirds of the fertiliser is lost as greenhouse gas or becomes a groundwater pollutant. Urea deep placement (UDP), a simple yet innovative technology, provides a more efficient and environmentally responsible method of fertilisation. It involves the placement of 1-3 grammes of urea supergranules or briquettes at a 7-10 centimetres (cm) soil depth shortly after the paddy is transplanted. UDP increases nitrogen use efficiency because most of the urea nitrogen stays in the soil, close to the plant roots where it is absorbed more effectively. The benefits of the technology are significant – a 25 % increase in crop yields and a 40% decrease in nitrogen losses.

By 2008/09, the Bangladesh Department of Agricultural Extension spread UDP technology to 500 000 ha of paddy fields, increasing production by 268 000 Mt annually. Farmers using UDP had additional annual net returns of USD 188/ha. (Bangladesh's average per capita annual income is about USD 500). UDP use reduced Bangladesh's urea import costs in 2008 by 50 000 Mt, saving USD 22 million in fertiliser imports and USD 14 million in government subsidies. The additional rice provided food security to an additional 1.5 million Bangladeshis.

UDP technology not only improves farmers' productivity and income, but the need for urea supergranules also creates employment opportunities. IFDC engineers developed a simple machine to mold urea into briquettes and helped establish village-level businesses to manufacture and distribute the machines. Nearly 2 500 urea briquette machines are in use across Bangladesh. UDP technology is also successfully being used in Nepal and Viet Nam.

Other fertiliser deep placement trials have also been conducted in Afghanistan and India. In India, scientists examined the effect of deep placement of briquettes made of urea, diammonium phosphate and potassium chloride compared with broadcasting NPK (nitrogen, phosphate, potassium). Deep placement resulted in similar or higher grain yields obtained with 40 kg/ha less nitrogen used. All farmers seek gains in efficiency and productivity, but nowhere is the need greater than in Africa. FDP technology has been introduced and is being tested in Burkina Faso, Madagascar, Malawi, Mali, Niger, Nigeria, Rwanda, Senegal and Togo.

Sources: IFDC (2011).

Box 2.7. **Zinc fertiliser boosts yields in Central Anatolia, Turkey**

Across the globe, soils and plants suffer from micronutrient deficiencies, including zinc deficiency. By adding Zn to fertilisers, this deficiency can be addressed, and yields increased, in regions that have low levels of available soil Zn, which is the case for 50% of all global cereal growing regions.

Zn deficiencies were identified in the early 1990s as reducing wheat yields in Central Anatolia. A research project was started in 1993 which showed that Zn applications led to significant increases in grain yield. In certain areas (DTPA-Zn ≤ 0.1 mg/kg soil) where cereal production was not previously economical and yield was extremely low (0.25 t/ha), grain yield increased 6- to 8-fold.

The effects of Zn fertilisers on plant growth and yield became more pronounced under rainfed conditions. Plants grown on Zn-treated soils or derived from seeds with very high Zn concentrations showed much better seedling establishment and higher tolerance to environmental stress factors ("winter killing"). Soil or foliar Zn application also improved grain quality by increasing grain Zn concentration and reducing phytic acid, a compound involved in the impairment of Zn bioavailability in humans.

The results were so positive that fertiliser industries in Turkey, in particular TOROS Fertiliser and Chemical Industry, produced increasing amounts of NP and NPK fertilisers containing 1% Zn at the same price as those containing just the three main plant nutrients.

Convinced by the results, Turkish farmers significantly increased the use of the Zn-fortified fertiliser within a few years, despite the re-pricing of the products to reflect the added-value of the content.

Nearly ten years after the Zn deficiency problem was first identified, the total amount of Zn-containing compound fertilisers produced and applied in Turkey has reached a record level of 300 000 tonnes per annum. It is estimated that the economic benefits associated with the application of Zn-fertilisers on Zn deficient soils in Turkey is around USD 100 million per year.

Sources: Alloway (2008); Cakmak (2005); Cakmak (2006); CIMMYT (2005).

The improvement of fertilisers combined with the use of precision farming techniques is a way to enhance agricultural productivity growth while at the same time reducing its environmental footprint at field level. The key to increased efficiency on a global scale is the establishment of a commercial, technical and regulatory environment that encourages site-specific agronomic improvements on farms, particularly in developing countries. In the United States, where maize growers have access to appropriate knowledge and technology, significant reductions in nitrogen losses have been achieved with important knock-on effects for the mitigation of GHG emissions (Figure 2.14). By contrast, in most developing countries, particularly China and India, (the two largest N fertiliser markets), nitrogen use efficiency has been on a downtrend.

Phosphorus (P) is an essential nutrient element for plants and animals and its deficiency restricts crop yields severely. Tropical and subtropical soils are predominantly acidic, and often extremely P deficient with high fixation capacities. Therefore, substantial P inputs are required for optimum plant growth and adequate food and fibre production. When phosphate fertilisers are applied to the soil, the majority (more than 70-75%) is bound in the soil in forms not immediately accessible to plants but that become available

Figure 2.14. **Evolution of nitrogen use efficiency in China and United States**

China (left figure) and United States (right figure)

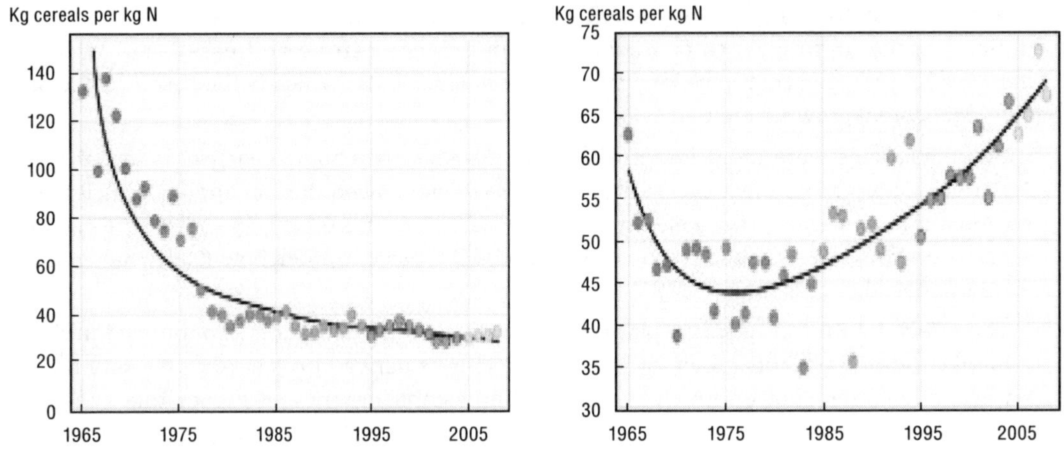

Source: International Fertiliser Association.

over time. New technologies are being developed to increase the immediate use-efficiency of applied P-fertilisers (*i.e.* to shorten the turnaround time of P in soil). For instance, farmers can use symbiotic fungi (mycorrhiza) with fertiliser that multiply along the plant roots and improve availability of the phosphate bound in the soil. This allows more phosphorus to reach the roots, increasing yields while using up to 40% less phosphate (Novozymes, 2011). Moreover, farmers can add enzymes to their feed mix which allow the animal to take up more phosphate from the plants and reduce the need for added phosphate by 25-100% (Novozymes, 2011).

Manufactured water-soluble P fertilisers, such as superphosphates, are commonly recommended to correct P deficiencies. Most developing countries import these fertilisers, which are often in limited supply locally and represent a major outlay for resource-poor farmers. In addition, intensification of agricultural production in these countries necessitates the addition of P not only to increase crop production, but also to improve soil P status in order to avoid further soil degradation. Alternative P sources have been explored (Box 2.8).

The *OECD Environmental Outlook* (OECD, 2012a) assesses the impact of measures to reuse nutrients in agriculture and reduce both domestic and agricultural discharges of nitrogen (N) and phosphorus (P). New measures that could bring about these improvements would include an increase in efficiency of fertiliser use, higher nutrient efficiencies in livestock production, and using animal manure instead of synthetic N and P fertilisers in countries with a fertiliser-dominated arable system. Under this scenario (OECD 2012a), by 2050 the global N and P surpluses in agriculture could be almost 20% less than in the business-as-usual scenario and the effluent of nutrients in wastewater could fall by nearly 35%. Total nutrient loads to rivers would be reduced by nearly 40% for nitrogen and 15% for phosphorus, which could help prevent further biodiversity loss in rivers, lakes and wetlands in the long term, and even allow some recovery locally.

Box 2.8. **Increasing fertiliser use in Africa**

From a development perspective, current approaches to increase fertiliser use in Africa are an important component of enhanced agricultural productivity and sustainability and, therefore, comprise issues related to institutional risks, policy environments, financial and economic constraints, market opportunities and agronomic matters, among others. Integrated Soil Fertility Management is part of the solution, along with timely and affordable access to agricultural inputs (fertilisers, improved seeds, crop protection products, etc.), credit facilities, infrastructure development and market opportunities. Successful approaches have entailed building farmer capacity by forming co-operatives and teaching business skills; credit insurance schemes for farmers, retailers and wholesalers; vouchers to purchase inputs from private-sectors dealers; providing market information through the internet and mobile phones; and multi-stakeholder public-private partnerships around specific agricultural development corridors. Significant growth could also take place in Latin America and Eastern Europe where many countries have considerable agricultural potential but where farmers have little access to inputs, knowledge and technologies.

The appropriate utilisation of locally sourced Phosphate Rocks (PR) as P sources can make a limited contribution to sustainable agricultural intensification. On soils low in P and calcium, and having also a low pH, the direct application of PR has proved to be an agronomically and economically sound alternative to the more expensive superphosphates. This only applies to high quality "reactive" PR (*e.g.* Tanzania, Tunisia). Unfortunately, most African PR deposits are of medium to low quality.

Source: FAO (2004), *Use of Phosphate Rocks for Sustainable Agriculture.*

More efficient use of crop protection

Between 26% and 40% of the world's potential crop production is lost annually because of weeds, pests and diseases, and these losses could double without the use of crop protection practices. Studies estimate that the food produced today with the yield levels of 1961 would require an additional 970 Mha, or more than the total land area of the United States. Crop protection products also play a major role in water conservation by efficiently controlling invading alien plants that threaten scarce water resources. They need, however, to be carefully managed, to avoid negative externalities linked to run-off of chemicals into freshwater courses.

Herbicides have enabled conservation practices that require less soil cultivation, keeping topsoil and soil moisture in the field, instead of being lost to evaporation, wind or water erosion. No-till farming can reduce soil erosion by 90% to 95% compared to conventional tillage practices. In addition, conservation or no-till practices are beneficial to enhancing biodiversity in crop fields. In Canada, 64% of farmers planting herbicide tolerant canola are using zero and minimal tillage practices. Using herbicides to control weeds in the United States is estimated to reduce soil erosion by 161 Mt each year.

Natural selection is a constant challenge for the crop protection industry, as pests and disease adapt resulting in decreased susceptibility to that chemical. The crop protection industry claims that restrictive approval rules have added costs and time lags to the development of new products. Researching, developing and registering each new active ingredient can take ten years or more, involving more than 100 scientific studies into the health and environmental impact, with costs upwards of USD 256 million.

A specific success story is the control of viral diseases in tomatoes (FAO, 2011b). Over the past 10 to 15 years, epidemics of viral diseases associated with high populations of whiteflies have plagued tomato production in West Africa, severely reducing yields. In some cases, tomato growing is no longer economically viable. A multi-partner international public-private research collaboration helped establish in Mali an integrated pest management (IPM) programme which included an area-wide campaign to eliminate infected host plants, followed by planting of high-yielding early maturing varieties and extensive sanitation efforts that removed and destroyed tomato and pepper plants after harvest. The programme evaluated early maturing disease tolerant varieties, and used monthly monitoring of whitefly populations and virus incidence to assess the impact of control practices. Recent tomato production was the highest recorded in 15 years.

Harnessing plant breeding and biotechnology

Plant breeding and the use of high quality seed will continue to enhance crop productivity gains in the future, provided that seed markets are properly functioning and farmers have access to high quality seed. During the past decades, crop yields have steadily increased due to a combination of new crop varieties and better cultivation techniques. In the long run, research has shown that crop yields increased by about 2% per annum, of which, about half has been attributed to plant breeding and the adoption of new crop varieties.

Not all seed can be produced in the countries where it will be consumed. In 2010, the export of seeds for sowing amounted to USD 8.2 billion.[15] To facilitate trade of high quality seed of improved varieties it is necessary for governments to provide an enabling environment with the right set of regulations for the plant breeding and seed industry (public and private) to create those new varieties and trade their seeds internationally. Governments need to implement globally harmonised regulations for seed certification, seed testing, variety protection and seed trade. The OECD Schemes for the Varietal Certification of Seed Moving in International Trade promote the use of agriculture seed of consistently high quality.[16]

Some 90 million people per year are affected by drought, 106 million people per year are affected by flooding, and 900 million hectares of soil are affected by salinity (FAO-data). Drought tolerance has been a breeding target for many years in maize, wheat, rice, beans, millet, canola and grasses, among others. Initial trials with drought tolerant wheat in Australia indicate a yield increase of 20%. Drought tolerance will have a major impact on more sustainable cropping systems worldwide, particularly in developing countries where drought is more prevalent. Plant breeders are also working on salt, flood, cold, heat or aluminium tolerance.

The changing climate will also lead to increased disease pressure levels. The annual global loss due to pathogens and insects are estimated at USD 85 billion and due to insects at USD 46 billion, respectively. About half of all the investments into plant breeding are targeted towards breeding for resistant varieties and each year thousands of new varieties are released. This has lead to a significant decrease in the use of pesticides. Disease resistant crops save millions in crop protection products. Worldwide, close to 400 million kg of active ingredient has been saved over the last 15 years. Less spraying also means reduced tractor use with associated reductions in fuel use and CO_2 emissions.

The "first generation" of biotech crops primarily offered one or a combination of three traits: disease resistance, pest resistance, and/or herbicide tolerance. In 1996, when biotech crops were first grown commercially, only 1.7 Mha were planted worldwide. In 2010, more than 15.4 million farmers in 29 countries grew biotech crops on 148 Mha. Plant biotechnology is perhaps the most rapidly adopted crop technology in agriculture today.

Biotech crops can on the one side help farmers reduce the use of other inputs, thereby reduce input costs, and through increased productivity and predictability, improve farmers' output and incomes. On the other side, they can increase the cost of seeds and reduce the seed capital value of farmers. Since plant biotechnology is generally scale-neutral, the benefits may be more accessible to developing countries and smallholders in general. Subramanian and Qaim (2010) showed that cultivation of biotech cotton increased Indian women farmer's wages by 55% In South Africa, it is estimated that biotechnology increased farm income by USD 383 million between 1998 and 2007 due to crop yield increases of 40% and reduced hand-weeding labour.

Companies are working today on biotech crops that can resist abiotic stresses such as drought, extreme temperatures, saline soils, and flooding. Researchers are developing drought-resistant crops that allow plants to survive with less water, or recover and re-grow after dry conditions. Research is also developing biotech crops that can handle the stress of high temperatures and salt-tolerant crops that can grow in saline soils that were traditionally viewed as unproductive. In all these areas, work is aimed at developing biotech crops that cannot only survive in extreme climates, but can maintain and even increase crop yields. The next generation of biotech crops is also expected to provide direct consumer benefits, such as nutrient-enhanced varieties (*e.g.* vitamin-enriched foods, healthier cooking oils).

Continued innovation and investment in new technologies depends on intellectual property protection. Intellectual property rights, such as patents, help ensure inventors receive compensation for their time and resource investments. Agricultural advances are no different from other inventions and should be afforded the same intellectual property protections that other industries enjoy. Intellectual property protection supports continued agricultural improvements and innovations. Most plant biotechnology companies reinvest a significant portion of their sales in order to continue their research and development into new products.

Policy challenges

Throughout this chapter a number of policy issues have been identified, emphasising the need for appropriate incentives and greater "knowledge intensity" to make increased productivity and sustainability into complements rather than trade-offs. They can be grouped into three broad areas: improving agronomic practices, creating the right policy environment, and strengthening the agriculture innovation system. A brief summary is presented here. These and other policy recommendations are discussed in more detail in the 2012 Interagency Report[17] to the Mexican G20 Presidency entitled *Sustainable Agricultural Productivity Growth and Bridging the Gap for Small Family Farms.*

Encouraging better agronomic practices

Given the increasing resource constraints and growing environmental pressures, the challenge is to increase agricultural productivity in a more sustainable manner. Farmers recognise the importance of conserving natural resources and ecosystems but lack incentives to do so at socially optimal levels. There have been improvements in environmental performance but progress has been uneven. There is growing pressure on natural resources, including land, water, forests, and biodiversity which are fundamental to sustainable production.

Sustainable intensification of production will require widespread adoption of sustainable land management practices, more efficient use of water (especially for irrigation), better plant nutrient management, increased control of disease and pests through modern crop protection (herbicides, pesticides, integrated pest management systems) and the harnessing of plant breeding and biotechnology. Achieving sustainable agricultural productivity growth will largely depend on farmers' ability to adopt the latest technologies.

Adoption of agronomic practices such as minimum or zero till cultivation, drip irrigation, optimum fertiliser use, higher nutrient efficiencies in livestock production and better use of organic fertilisers would all contribute to more sustainable agriculture. New biotech developments and plant breeding techniques may introduce crops that are more drought and heat resistant, or be able to survive saline soils or excessive rainfall. In many developing countries, the shift to larger, more modern production systems will improve productivity and sustainability.

Many sustainability initiatives have come from the private downstream sector. Industry associations such as the Sustainable Agricultural Initiative (SAI) support the development and implementation of sustainable agricultural practices. Industry and producers work together to promote good agricultural practices through such partnerships as Global GAP. Still, some farm groups have expressed concern about the plethora of standards and their coherence with government initiatives. More recently there has been greater emphasis on public–private partnerships, such as the FAO-UNEP task force and the European Union's European Food and Sustainable Consumption and Production Round Table (SCP).

Creating the right policy environment

The key to increased efficiency on a global scale is the establishment of a commercial, technical and regulatory environment that encourages agronomic improvements at farm level, particularly in developing countries. To improve production efficiency in a sustainable manner, well-functioning markets must provide clear price signals that reflect the scarcity value of natural resources, and property rights must be defined so as to encourage optimal use of resources, both individually and collectively. The over-arching policy challenge is to create the right incentives that would optimise resource use from an economic, environmental and social perspective.

Actions include not just technical options to promote sustainable intensification and reduce production risks, they also comprise a set of conditions to remove constraints and build flexibility. Creating the right enabling environment involves creating a business climate that is conducive to investments (domestic and foreign), therefore, limiting trade restrictions as well as those domestic support schemes that distort production and investment in agriculture. There is a need to develop national investment schemes and increased development assistance to agriculture for R&D, innovation adoption and infrastructure development. The importance of securing fair and efficient institutions for natural resource tenure and management is a critical precondition to reduce risks and enhance incentives.

For example, when pursuing productivity growth under water availability and quality constraints, it is important to set up the right incentives for technological change to take place, such as through providing the right incentives to use less water (*e.g.* government water quota policies). Regulatory reforms could reduce the time and costs of getting new technologies to the market. Effective intellectual property rights, such as patents, are essential to stimulate innovation and investment in new technologies. There are a variety of options available that may improve the system of international property rights (IPR) protection to provide further incentives for private investment in innovation, without compromising the sharing of knowledge. Some of these issues can be addressed by use of best practices in regulation and innovation policy frameworks such as with respect to collaborative approaches, public-private partnerships, or licensing of genetic inventions (*e.g.* OECD, 2011b and 2006).

For developing countries, investments in physical capital (infrastructure) in rural areas, is needed in the areas of storage, transportation and irrigation systems, and in electrification and information/communication systems. Such investments can provide access to markets for smallholders and incentives to produce more. Investment in human capital (production and management skills) is equally important and depends on public spending for health care, education and training. Human capital directly affects when and how technology is adapted to a particular situation and how efficiently inputs are used.

Strengthening the agriculture innovation system (AIS)

With respect to agricultural innovation systems, the focus is on improving institutional design, the regulatory environment for innovation systems, and the relevance of R&D and innovation for small family farms. Most developing countries do not yet have an innovation policy. Specific areas that could be addressed include the need for better policy coherence for agricultural innovation, for more demand-driven research system, for rejuvenated agricultural education and training programmes, and for greater private sector engagement.

Strengthening the AIS system requires effective coordination to bring together the interests and skills of the many different stakeholders in the innovation process. Enhanced public-private partnerships would be a benefit in all these areas. However, the private sector will tend to focus on high value and market oriented production systems. While the private sector is increasingly engaged in resource management and maintaining ecosystems, the provision of such public goods is primarily a responsibility of governments.

Future productivity gains will largely depend on investments in research and development. Agricultural R&D is the main source of innovation and is needed to sustain agricultural productivity growth in the long-term. Public R&D expenditures on agriculture in developing countries is generally low, with funding often dependent on foreign aid, and is granted for time-limited projects which may hamper the development of national R&D institutions and capacity building. An important challenge is to better adapt research to local conditions.

Of course developing innovative tools and approaches will not succeed without on the ground adaptation. Improving the ability of farmers to innovate requires education and extension, which has been neglected in many countries. Low levels of general education of the agricultural labour force in developing counties can be a serious constraint. Building an effective education and extension system is a long term investment that requires sustained political support.

Summary

The focus of this chapter is on the need to increase agricultural productivity growth in a sustainable manner. There are several key messages.

- The demand for food, feed, fuel and fibre is steadily increasing with a growing population, increasing incomes in developing countries, an urbanising population, higher protein diets and expanding biofuel production.
- The potential to meet this rising demand by increasing the amount of land in agriculture is limited due to the investment costs involved, environmental considerations, and competition from other sectors.
- With limited area expansion, increased agricultural production will need to come from increased productivity in the same way as it has in the last 50 years. Globally, there is no conclusive evidence of declining total factor productivity (TFP) growth, but large differences exist across countries and commodities.
- A recent study predicts that TFP growth over the next 30 years will exceed that of the previous decades, primarily due to the expansion of the production possibilities frontier rather than through a reduction of the productivity gap.
- This *Outlook* suggests annual growth in global agricultural production over the next ten years will be lower than the previous ten years, although it will remain ahead of population growth such that output *per capita* will continue to increase at the global level (though experiences will vary widely across countries).
- Actual yields for the main food crops are well below potential yields in many regions with yield gaps in many developing countries in excess of 50% (data from 2005). The yield gaps, at 76%, were greatest in Sub-Saharan Africa.
- A stylised scenario indicates that reducing the yield gap in developing countries by a fifth from 2012 to 2021 would increase global production by only 5.1% in 2021, with an estimated 33% of increased coarse grain production projected to go into biofuels. Nevertheless, it would exert downward pressure on commodity prices, from the current historically high levels and would have a positive effect on food consumption in developing countries.
- Given the resource constraints and environmental pressures, exacerbated by climate change and increased volatility, the major challenge for the agri-food sector is to increase agricultural productivity in a more sustainable manner.
- Sustainability of ecosystems is fundamental to maintaining the resilience in productivity growth. There is growing pressure on natural resources, including land, water, marine ecosystems, fish stocks, forests, and biodiversity, which are fundamental to sustainable production. Environmental pressures are also caused by agriculture, which accounts for about 14% of total anthropogenic GHG emissions.
- Increasing productivity and improving sustainability of agriculture are not mutually exclusive objectives. Much can be done that contributes to both these objectives in a complementary fashion. However, the possible farm practices to undertake generally

involve a long lead time before realising the benefits, suggesting that actions are needed now.

- Encouraging better agronomic practices, creating the right commercial, technical and regulatory environment and strengthening the agriculture innovation system (*e.g.* research, education, extension) are all essential steps. As the agri-food chain is essentially a business activity, the private sector will need to play a lead role in developing and adopting innovation, putting more emphasis on public-private partnerships, in such areas as research and extension services, that can facilitate progress. Moreover, reducing food loss and food waste could significantly ease the pressure to increase productivity.
- Progress is being made. Governments have started to orient their policy priorities to take account of the environmental consequences of food and agriculture production and consumption, and to improve incentives for optimal resource use by farmers.
- The specific approach varies by agro-ecology, farming system and market conditions but consistently will involve increasing the conservation and sustainable use of natural resources in agricultural production systems, as well as the reduction of waste and pollution associated with inefficient input use and degraded ecosystems.

Notes

1. Please note that "productivity growth" refers to gains in production efficiency (*i.e.* technical and efficiency change) and not to "production growth". In this chapter we refer to productivity growth through both partial (*e.g.* yields) and overall (*e.g.* total factor productivity) indicators.
2. Productivity gaps are further analysed in section 2.5 as yield gaps, basically the difference between the economically attainable yields and average farm yields in different world regions.
3. Least-squares growth rate (see Glossary).
4. It does not include the EU12.
5. It is important to note the limitations of the TFP indicator since it does not take account non-market outputs and inputs, especially long-term resource degradation such as soil loss or nutrient mining, and environmental externalities. There are other alternative indicators in the literature that measure TFP growth in agricultural production incorporating environmental effects, such as the Total Factor Nutrient-Orientated Productivity Index (Hoang 2009) and the Total Social Factor Productivity indices (Byerlee and Murgai, 2000). Nevertheless, they present problems of measurement (*e.g.* non-marketed goods), data availability (*e.g.* resource quality) and time and geographical coverage.
6. Caution should be applied to assessing global yield estimates in terms of productivity change. Land expansion to more marginal land may reduce average yield estimates. Similarly, production shifts to areas/regions of lower yields will reduce average global yield, even if yields from these respective areas are increasing.
7. Wheat, coarse grains (maize, barley, oats and sorghum), rice, roots and tubers, oilseeds (soybean, rapeseed and sunflower), sugar beet, and sugarcane.
8. Suboptimal from a social welfare perspective but possibly optimal for individual farmers given the constraints they face (rational behaviour).
9. For instance, economic policies, barriers to trade and poor market infrastructure could play a significant role in stifling economic incentives to farmers to raise yields. In fact, differences in relative input/output prices (Peterson 1979, 1988) and fertiliser application (Tilman *et al.* 2011) have been found as major determinants of global yield gaps. Overcoming these constraints might involve substantial costs, which are not addressed in this chapter.
10. This share of the gap has been arbitrarily chosen for the scenario analysis presented.
11. Non-negligible costs are to be faced in R&D and improvement of knowledge information systems, but they are not directly attributed to the agricultural sector in a partial analysis framework.

12. It is important to acknowledge that increasing yields might involve higher rates of fertiliser use, especially in regions currently experiencing low yields (Tilman *et al.* 2011). Nevertheless, this level of detail is beyond the current capabilities of the Aglink-Cosimo model and will be subject of further study.
13. While this chapter focuses on crops, in Chapters 7 (Meat) and 9 (Dairy), additional information on livestock productivity is provided.
14. This includes a broad range of technologies, such as genetic indicators, accelerated breeding processes and micropropagation.
15. Information provided by the International Seed Federation.
16. For more information on the OECD Seeds Scheme see: *www.oecd.org/tad/seed*.
17. Bioversity, CGIAR Consortium, FAO, IFAD, IFPRI, IICA, OECD, UNCTAD, UN High Level Task Force on the Food Security Crisis, WFP, World Bank, and WTO.

References

Alloway, B.J. (2008), Zinc in Soils and Crop Nutrition, International Zinc Association, Brussels and Paris, URL: *www. fertiliser.org/ifa/HomePage/LIBRARY/Publication-database.html/Zinc-in-Soils-and-Crop-Nutrition.html*.

Alston, J. (2010), "The Benefits from Agricultural Research and Development, Innovation, and Productivity Growth", *OECD Food, Agriculture and Fisheries Working Papers*, No. 31, doi: *10.1787/5km91nfsnkwg-en*.

Alston, J.M., Beddow, J.M. and P.G. Pardey (2010), "Global Patterns of Crop Yields and Other Partial Productivity Measures and Prices", Chapter 3 in Alston, J.M., B.A. Babcock, and P.G. Pardey (eds.) (2010), *The Shifting Patterns of Agricultural Productivity Worldwide*, CARD-MATRIC Electronic Book, Center for Agricultural and Rural Development, The Midwest Agribusiness Trade Research and Information Center, Iowa State University, Ames, Iowa. Available at: *www.matric.iastate.edu/shifting_patterns*.

Beintema N. and Stads G.J (2008), Measuring Agricultural Research Investments: A Revised Global Picture. ASTI Background Note, Agricultural Science and Technology Indicators, Rome, October.

Bruinsma J. (2011), "The Resource Outlook: By How Much do Land, Water and Crop Yields Need to Increase by 2050?", Chapter 6 in *Looking Ahead in World Food and Agriculture: Perspectives to 2050*, edited by Piero Conforti. Rome: FAO (*www.fao.org/docrep/014/i2280e/i2280e.pdf*).

Byerlee D. and Murgai R. (2000), Sense and Sustainability Revisited: The Limits of Total Factor Productivity Measures of System Sustainability, Agricultural Economics, 26(3), 227-236

Cakmak, I (2005), Identification and correction of widespread zinc deficiency problem in Central Anatolia, 73rd IFA Annual Conference, Kuala Lumpur, Malaysia, 6-8 June, URL: *www. fertiliser.org/ifa/Media/Files-Publications/2005_IFA_KL_cakmak.pdf*.

Cakmak, I. (2006), *Enriching Grain with Micronutrients: Benefits for Crop Plants and Human Health* , IFA Agriculture Conference on "Optimizing Resource Use Efficiency for Sustainable Intensification of Agriculture", Workshop on Micronutrients, Kunming, China, 27 February – 2 March, URL: *www. fertiliser.org/ifacontent/download/6891/109156/version/1/file/2006_ag_kunming_cakmak.pdf*.

Calderini, D.F. and G.A. Slafer (1998), "Changes in yield and yield stability in wheat during the 20th century", *Field Crops Research*, Vol. 57, pp. 335-347.

Cassman, K.G. (2001), "Crop science to secure food security", in J. Nösberger, H.H. Geiger, P.C. Struik (eds.), *Crop Science – Prospects and Progress*, IV International Crop Science Congress, Hamburg, Germany, CABI International, Wallingford, UK.

CEEP (2011), *La production agricole mondiale à l'horizon 2050: comparaison de quatre prospectives*, Centre d'études et de prospective.

CIMMYT (2005), Solving the Zinc Problem from Field to Food, International Maize and Wheat Improvement Center, URL: *www.cimmyt.org/en/newsletter/86-2005/274-solving-the-zinc-problem-from-field-to-food*.

Cline, W (2007), *Global Warming and Agriculture: Impact Estimates by Country*. Washington, DC: Center for Global Development and Peterson Institute for International Economics.

Davis, K., E. Nkonya, E. Kato, D.A. Mekonnen, M. Odendo, R. Miiro and J. Nkuba (2010), *Impact of farmer field schools on agricultural productivity and poverty in East Asia*, IFPRI Discussion Paper 00992, Washington, DC: International Food Policy Research Institute.

Easterling, W., P. Aggarwal, P. Batima, K. Brander, L. Erda, M. Howden, A. Kirilenko, J. Morton, J.-F. Soussana, S. Schmidhuber, and F. Tubiello (2007), *Food, fibre and forest products. In Climate Change 2007: Impacts, Adaptation and Vulnerability*. Contribution of Working Group II to the Fourth Assessment Report of the Intergovernmental Panel on Climate Change. M.L. Parry, O.F. Canziani, J.P. Palutikof, P.J. van der Linden, and C.E. Hanson, Eds. Cambridge University Press, pp. 273-313.

European Commission (2011), *Foresight report on agriculture*, Brussels, URL: *ec.europa.eu/research/agriculture/scar/pdf/scar_feg3_final_report_01_02_2011.pdf*.

Evans, L.T. and R.A. Fischer (1999), "Yield potential: its definition, measurement and significance", *Crop Science*, Vol. 34, pp. 1544-1551.

FAO (2004), *Use of Phosphate Rocks for Sustainable Agriculture*, A joint publication of the FAO Land and Water Development Division and the International Atomic Energy Agency, URL: *www.fao.org/docrep/007/y5053e/y5053e00.htm*.

FAO (2009), *How to Feed the World in 2050*, FAO High-level Expert Forum, Rome, 12-13 October.

FAO (2011a), Global Food Losses and Food Wastes: Extent, Causes and Prevention, URL: *www.fao.org/fileadmin/user_upload/ags/publications/GFL_web.pdf*.

FAO (2011b), *Save and Grow. A policymaker's Guide to the Sustainable Intensification of Smallholder Crop Production*, URL: *www.fao.org/ag/save-and-grow/*.

FAO (2011c). *The State of the World's Land and Water Resources for Food and Agriculture. Managing systems at risk*, FAO Conference document C2011/32. Thirty-seventh Session. Rome, 25 June – 2 July, URL: *www.fao.org/nr/solaw/solaw-home/en/*.

FAO (2011d), *State of Food and Agriculture 2010-11: Women in* Agriculture – Closing the gender gap for development. Rome, 2011.

FAO (2012), *Stability of food security in a green economy environment*. GEA Rio+20 Working Paper 3. Rome.

FAO (2012 forthcoming), World Agriculture Towards 2030/50: the 2012 Revision, Rome.

FAO *et al.* (2011), *Price Volatility in Food and Agricultural Markets: Policy Responses*, Policy report to the G20 including contributions by FAO, IFAD, IMF,OECD, UNCTAD, WFP, the World Bank, the WTO, IFPRI and the UN HLTF.

Fischer, G., E. Hizsnyik, S. Prieler, and D. Wiberg (2010), *Scarcity and abundance of land resources: competing uses and the shrinking land resource base. SOLAW Background Thematic Report TR02*, Rome, *fao.org/nr/solaw/*.

Fischer G., H. van Velthuizen, and F. Nachtergaele (2009), *Global Agro-ecological Assessment*, 2009 revised edition, IIASA, Laxenburg, Austria.

Fischer R.T., D Byerlee, and G.O. Edmeades (2011), "Can technology deliver on the yield challenge to 2050?", in P. Conforti (ed.), *Looking ahead in world food and agriculture: Perspectives to 2050*, FAO, Rome.

Foley, J.A., N. Ramankutty *et al.* (2011), Solutions for a cultivated planet. Nature 478: 337-342.

Fuglie K.O. (2008), "Is a slowdown in agricultural productivity growth contributing to the rise in commodity prices?", *Agricultural Economics*, Vol. 39 supplement, pp. 431-441.

Fuglie K.O. (2010a), Indonesia: From Food Security to Market-Led Agricultural Growth, in Alston, J., B.A. Babcock and G. Pardey (eds.) The Shifting Patterns of Agricultural Production and Productivity Worldwide, the Midwest Agribusiness Trade Research and Information Center, Iowa State University, pp. 343-381.

Fuglie K.O. (2010b), Total Factor Productivity in the Global Agricultural Economy: Evidence from FAO Data, in Alston, J.M., B.A. Babcock, and P.G. Pardey (eds.) (2010), *The Shifting Patterns of Agricultural Productivity Worldwide*, CARD-MATRIC Electronic Book, Center for Agricultural and Rural Development, The Midwest Agribusiness Trade Research and Information Center, Iowa State University, Ames, Iowa. Available at: *www.matric.iastate.edu/shifting_patterns*.

Fuglie K.O. (2012), "Productivity Growth and Technology Capital in the Global Agricultural Economy", in Fuglie, K.O., S.L. Wang, and V.E. Ball (eds.) (2012), *Productivity Growth in Agriculture: An International Perspective*, CAB International, Oxfordshire, UK.

Fuglie, K.O., Heisey P., King J., Pray C.E., Day-Rubenstein K., Schimmelpfennig D., Wang S.L. and Karmarkar-Deshmukh R. (2011), Research Investments and Market Structure in the Food Processing, Agriculture Input and Biofuel Industries Worldwide. Economic Research Report 130, Economic Research Service, US Department of Agriculture, Washington, DC.

Hoang, V (2009), Measurement of agriculture total factor productivity growth incorporating environmental factors: a nutrients balance approach, Australian Agricultural and Resource Economics Society>2009 Conference (53rd), February 11-13, 2009, Cairns, Australia.

IAASTD (2009), Synthesis Report: A synthesis of the Global and Sub-Global IAASTD Reports, International Assessment of Agricultural Knowledge, Science and Technology for Development, Island Press, Washington DC, URL: *www.agassessment.org*.

IFA(2012), *The Role of Fertilizers in increasing agricultural productivity sustainably*, International Fertilizer Industry Association.

IFDC (2011), *Fertilizer Deep Placement*, International Fertilizer Development Center, *www.ifdc.org/Expertise/Fertilizer/Fertilizer_Deep_Placement_(UDP)*.

IPCC (2007), Synthesis report, Contribution of Working Groups I, II and III to the Fourth Assessment Report of the Intergovernmental Panel on Climate Change, URL: *www.ipcc.ch/pdf/assessment-report/ar4/syr/ar4_syr.pdf*.

Lobell, D.B., K.G. Cassman, and C.B. Field (2009), Crop yield gaps: their importance, magnitudes, and causes, *Annual Review of Environmental Resources*, Vol. 34, pp. 179-204.

Ludena C.E., Hertel T.W., Preckel P.V., Foster K. and A. Nin (2007), Productivity growth and convergence in crop, ruminant and nonruminant production: measurement and forecasts, *Agricultural Economics*, No. 37, pp. 1-17.

Millennium Ecosystem Assessment (2005), Ecosystems and human well-being: synthesis. Washington, DC, Island Press.

Novozymes (2011), *Facts about Phosphates*, *www.novozymes.com*.

OECD (1995), *Technological Change and Structural Adjustment in OECD Agriculture,* OECD Paris.

OECD (2006), *Guidelines for Licensing of Genetic Inventions*. OECD publishing.

OECD (2008), *Environmental performance of agricultural in OECD countries since 1990*. OECD Paris.

OECD (2010a), *Sustainable Management of Water Resources in Agriculture*, OECD, Paris, URL: *www.oecd.org/dataoecd/2/50/49040929.pdf*.

OECD (2010b), "The Environmental Performance of Agriculture", in OECD *Review of Agricultural Policies: Israel 2010*, URL: *www.oecd.org/document/56/0,3343,en_2649_33773_45174776_1_1_1_37401,00.html*.

OECD (2011a), *A Green Growth Strategy for Food and Agriculture*, OECD, Paris.

OECD (2011b), iCollaborative Mechanisms for Intellectual Property Management in the Life Sciences, OECD publishing pp. 22-23.

OECD (2011c), *Fostering productivity and competitiveness in agriculture*, OECD, Paris. URL: *dx.doi.org/10.1787/9789264166820-en*.

OECD (2012a), *Environmental Outlook to 2050*, Water Chapter, OECD, Paris.

OECD (2012b), *OECD Review of Agricultural Policies in Indonesia*, forthcoming.

OECD (2012c), *Agricultural Policies: Monitoring and Evaluation 2012*, OECD Countries, forthcoming.

Pardey, P.G., Pingali, P.L (2010), Reassessing International Agricultural Research for Food and Agriculture, report prepared for the Global Conference on Agricultural Research for Development (GCARD), Montpellier, France, 28-31 March.

Peterson W. (1979), International Farm Prices and the Social Cost of Cheap Food Policies, *American Journal of Agricultural Economics* 61, 1: 12-21.

Peterson W. (1986), International Supply Response. Agricultural Economics 2, 4: 365-374.

Piesse, J. and C. Thirtle (2010), "Agricultural R&D, technology and productivity", *Philosophical Transactions of the Royal Society B*, No. 365, pp. 3015-3047.

Rabbinge, R., S.W. Kazibwe, and M.S. Swaminathan (2004), *Realizing the promise and potential of African agriculture*, InterAcademy Council, Netherlands.

Sheng Y. (2011), Productivity pathways: Climate adjusted production frontiers for the Australian broadacre cropping industry, presentation for the 8th Meeting of the OECD Network for Farm-level Analysis, Paris, 3-4 November 2011.

Sinclair, T. (1998), Options for sustaining and increasing the limiting yield-plateaus of grain crops, paper presented at the NAS Colloquium Plants and Population: Is there Time?, Irvine, California, 5-6 December 1998.

SREX (2012), Managing the Risks of Extreme Events and Disasters to Advance Climate Change Adaptation (SREX). Special Report of the Intergovernmental Panel on Climate Change (IPCC).

Subramanian, A., M. Qaim (2010), The Impact of Bt Cotton on Poor Households in Rural India. *Journal of Development Studies* 46: 295-311.

The Government Office for Science (2011), Foresight, The Future of Food and Farming, *Final Project Report*, London.

Tilman, D. (1999), Global environmental impacts of agricultural expansion: the need for sustainable and efficient practices, Proceedings of the National Academy of Science, Vol. 96, No. 11, pp. 5995-6000.

Tilman D., Balzer C., Hill J. and Befort B. (2011), Global food demand and the sustainable intensification of agriculture. Proceedings of the National Academy of Sciences (PNAS), October 12.

OECD-FAO Agricultural Outlook 2012

Chapter 3

Biofuels

Biofuels were added to the *Outlook* in 2008 as an emerging sector that would increasingly affect agricultural markets. This has certainly turned out to be the case with currently some 65% of EU vegetable oil, 50% of Brazilian sugarcane, and about 40% of US corn production being used as feedstock for biofuel production. Today, it would be inconceivable to prepare an agricultural projection without taking biofuels into account. The biofuels chapter has been expanded this year to provide a more detailed description of the very complex US biofuel policy and an analysis of the policy options facing the US Environmental Protection Agency over the medium term.

Market situation

World ethanol prices (Figure 3.1) increased strongly in 2011 well above the levels of the 2007/08 highs in a context of strong energy prices, although the commodity prices of ethanol feedstock, mainly sugar and maize, decreased from their peaks in 2010. The two major factors behind this increase were the stagnating ethanol supply in the United States and a drop in Brazilian sugarcane production. Additionally, ethanol production was also significantly below expectations in developing countries having implemented mandates or ambitious targets for the use of biofuels.

World biodiesel prices (Figure 3.1) also increased in 2011. Contrary to the global ethanol market, production did not stagnate in 2011; the four major biodiesel producing regions (the European Union, the United States, Argentina, and Brazil) increased their supply compared to 2010. This increase was moderated by a decreasing biodiesel production in Malaysia (from about 1 Bnl in 2010 to almost nothing in 2011).

Projection highlights

- Over the projection period, ethanol and biodiesel prices are expected to remain supported by high crude oil prices and by the implementation and continuation of policies promoting biofuel use. Changes in the implementation of biofuel policies can strongly affect biofuel markets.
- Global ethanol and biodiesel production are projected to expand but at a slower pace than in the past. Ethanol markets are dominated by the United States, Brazil and to a smaller extent the European Union. Biodiesel markets will likely remain dominated by the European Union and followed by the United States, Argentina and Brazil.
- Biofuel production in many developing countries is projected to remain below expressed targets as the cultivation of non-edible crops to produce biofuels remains, in most cases, on a project or small-scale level and high prices of agricultural commodities do not encourage their use as biofuel feedstock.

- Biofuel trade is anticipated to grow significantly, driven by differential policies among major producing and consuming countries. The United States, Brazil and the European Union policies all "score" fuels differently for meeting their respective policies. This differentiation is likely to lead to additional renewable fuel trade as product is moved to its highest value market, resulting in potential cross trade of ethanol and biodiesel.

Figure 3.1. **Strong ethanol and biodiesel prices over the outlook period**

Evolution of prices expressed in nominal terms (left) and in real terms (right)

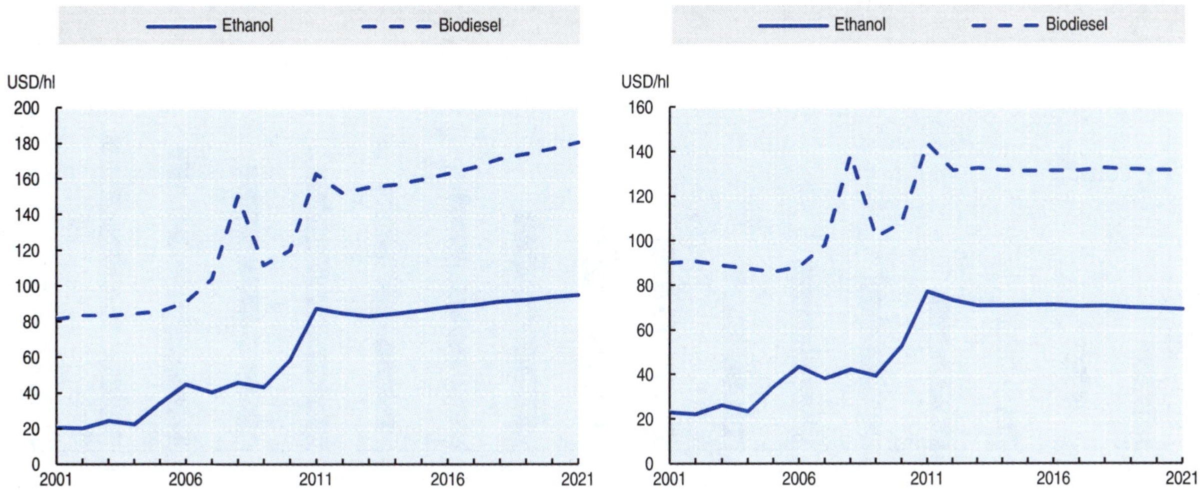

Notes: Ethanol: Brazil, Sao Paulo (ex-distillery), Biodiesel: Producer price Germany net of biodiesel tariff.

Source: OECD and FAO Secretariats.

StatLink http://dx.doi.org/10.1787/888932639362

Market trends and prospects

Prices

World ethanol prices[1] increased strongly in 2011, well above the levels of the previous 2007/08 highs. In 2012, a slight drop is projected but the price is expected to stay constant in real terms after 2013 following the price paths of the two major feedstocks maize and sugar (Figure 3.1). However, ethanol prices are not expected to increase as much as the crude oil price is assumed to over the projection period to reflect recent trends of the ethanol to crude oil price ratio.

World biodiesel prices[2] have increased in 2011 as well in a context of rising vegetable oil prices and high crude oil prices. This increase was smaller than for the world ethanol price because biodiesel production did not stagnate in 2011. Comparable to ethanol prices, biodiesel prices are projected to decrease slightly until 2013 and stay constant in real terms thereafter; this is in line with major biofuel feedstock prices.

Production and use of biofuels

Global ethanol production is projected to almost double over the projection period when compared to the 2009-11 base period and to reach some 180 Bnl by 2021 (Figure 3.2). The three major producers are expected to remain the United States, Brazil and the European Union. Production and use in the United States and the European Union are mainly driven by the policies in place, namely the US Renewable Fuel Standard (RFS2) final rule and the EU Renewable Energy Directive (RED). The growing use of ethanol in Brazil is linked to the development of the flex-fuel vehicle industry and the import demand of the United States to fill the advanced biofuel mandate. In the developing world, China should remain the main producer and user of ethanol with a production of 8 Bnl in 2011, projected to increase to 10 Bnl by 2021 (most of it is projected to be used for non-fuel applications), followed by India (4.2 Bnl in 2021).

Figure 3.2. **Development of the world ethanol market**

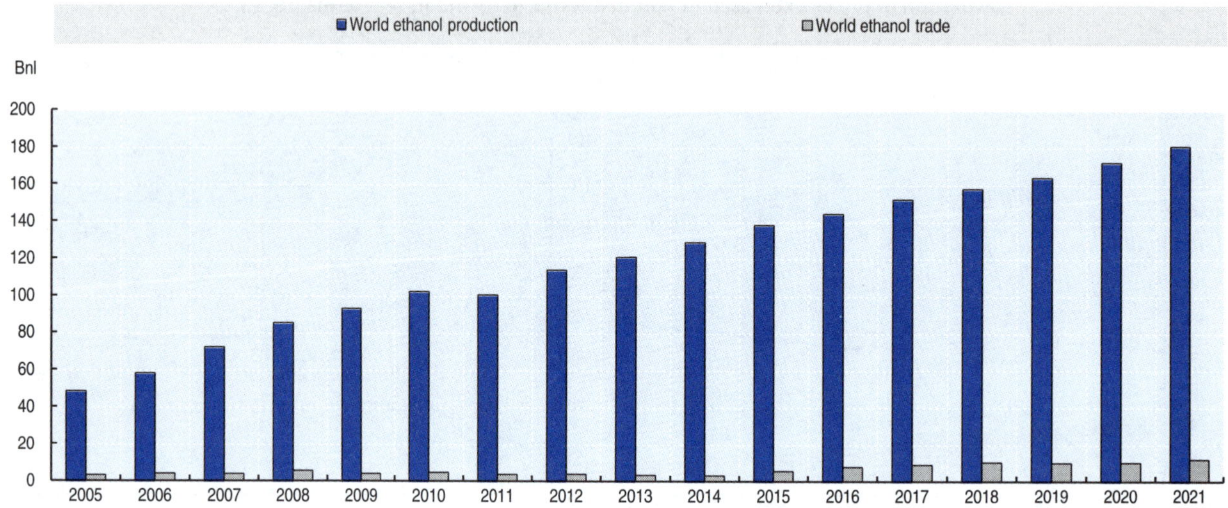

Source: OECD and FAO Secretariats.

StatLink http://dx.doi.org/10.1787/888932639381

Global biodiesel production is expected to increase to above 42 Bnl by 2021 (Figure 3.3). The European Union is expected to be by far the largest producer and user of biodiesel. Other significant players are Argentina, the United States, Brazil, as well as Thailand and Indonesia.

Figure 3.3. **Development of the world biodiesel market**

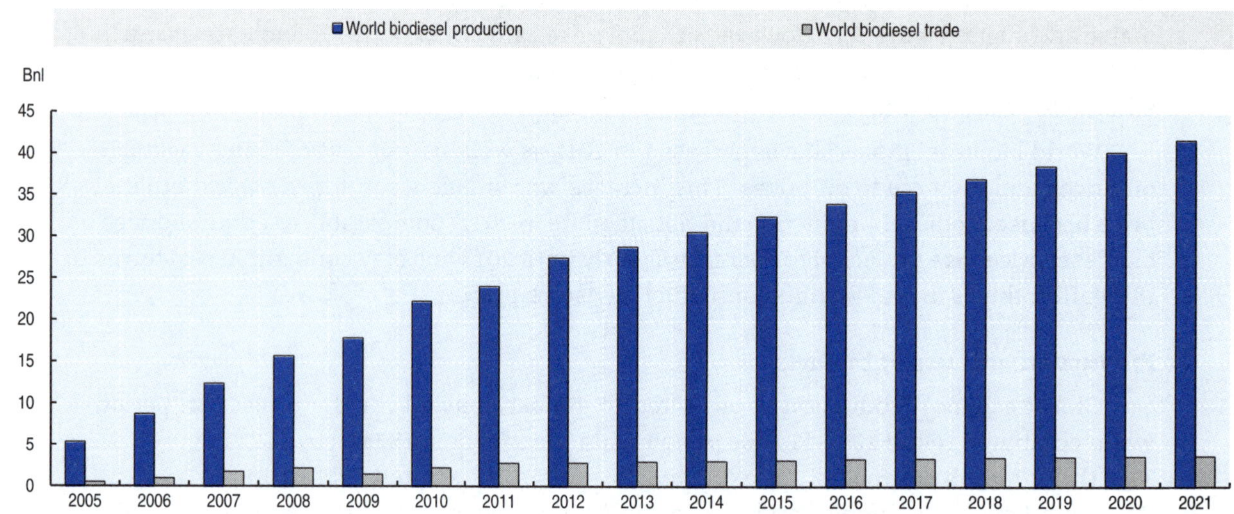

Source: OECD and FAO Secretariats.

StatLink http://dx.doi.org/10.1787/888932639400

To put in perspective the use of biofuel in total transport fuel use, Table 3.1 presents the projections for total transport and biofuel use both in energy and volume terms for a certain number of countries.

Table 3.1. **Transport fuel use in major biofuel producing countries**

		2009-2011			2021		
		Total	*Of which:* biofuel	Share of biofuel %	Total	*Of which:* biofuel	Share of biofuel %
Energy basis (1000toe)	Argentina						
	Gasoline type	3.5	0.1	2.7	4.1	0.1	3.4
	Diesel type	9	0.3	3.2	11	0.4	4.0
	Australia						
	Gasoline type	15	0.2	1.3	947	0.3	1.5
	Diesel type	16	0.5	3.1	18	0.5	3.1
	Brazil						
	Gasoline type	23	11.0	47.0	29	18.9	64.2
	Diesel type	40	1.6	4.0	54	2.4	4.6
	Canada						
	Gasoline type	30	0.8	2.6	32	1.1	3.4
	Diesel type	26	0.1	0.7	28	0.4	1.6
	China						
	Gasoline type	61	1.1	1.8	104	1.4	1.3
	EU						
	Gasoline type	103	2.8	2.7	103	8.6	8.3
	Diesel type	189	9.4	5.1	200	16.7	8.5
	USA						
	Gasoline type	409	21.9	5.4	412	45.0	10.9
	Diesel type	215	1.9	0.9	249	3.8	1.5
Volume basis (bnl)	Argentina						
	Gasoline type	4.7	0.2	4.0	5.4	0.3	5.0
	Diesel type	11	0.4	4.0	13	0.6	5.0
	Australia						
	Gasoline type	20	0.4	1.9	23	0.5	0.0
	Diesel type	19	0.6	3.9	22	0.7	3.8
	Brazil						
	Gasoline type	31	21.7	57.0	39	37.4	72.9
	Diesel type	48	2.1	5.0	64	3.2	5.7
	Canada						
	Gasoline type	40	1.6	3.8	42	2.1	5.0
	Diesel type	31	0.2	0.8	33	0.6	2.0
	China						
	Gasoline type	81	2.2	2.7	137	2.7	2.0
	EU						
	Gasoline type	137	5.5	4.0	136	16.9	12.0
	Diesel type	225	12.5	6.3	239	22.0	10.4
	USA						
	Gasoline type	541	43.4	7.8	545	89.1	15.5
	Diesel type	257	2.5	1.1	298	5.0	1.9

Source: OECD and FAO Secretariats.

StatLink http://dx.doi.org/10.1787/888932640540

Developed countries

With a global production share of about 50% in 2011, the United States is currently the biggest ethanol producer. The development of US biofuel markets has taken off since the enactment of the Energy Independence and Security Act of 2007 (EISA).[3] The implementation of this policy is made by the Environmental Protection Agency (EPA) through annual rules setting the levels for different fuel types. The Annex of the biofuel chapter provides a detailed description of US biofuel policies and, in particular, of the nested structure of quantitative minimums in place. An analysis of different implementation options is provided in the last section of the chapter. Current technological developments seem to suggest that the cellulosic biofuel mandate as it is currently regulated by the EPA is unlikely to be met by 2022.

It was assumed in the baseline that the production of cellulosic ethanol would rise steadily over the course of the outlook period to reach 16 Bnl by 2021, *i.e.* only about 30% of the cellulosic biofuel mandate.[4] EPA announcements for 2012 are incorporated in the baseline projections. For 2013 and remaining years of the projection period, the assumptions were made that the conventional ethanol gap would stay at the quantities in the legislation and that the other advanced gap could not shrink from year to year following the shortfall in cellulosic biofuels, *i.e.* that the total and advanced mandates would be reduced in parallel.[5]

This adjusted total US biofuel mandate would amount to 96 Bnl in 2021. As the total biofuel mandate is projected to be binding throughout the projection period, ethanol use in the US is projected to follow the path of this mandate when subtracting the biodiesel mandate and reaches almost 90 Bnl (Figure 3.4). However, because of the high crude oil price, conventional ethanol production mostly based on coarse grains would be above the conventional gap.[6] Concerning the blend wall,[7] the EPA provided a decision in January 2011 to expand the ethanol blending percentage in regular gasoline from 10% to 15% expressed in a volume share for cars built in 2001 or later. At present, gasoline retailers are not ready to propose different types of gasoline to their customers because of logistics, warranties on motors as well as liability issues. It is assumed in the baseline projection that this issue will be resolved allowing cars built before 2001 to gradually disappear from the roads so that the full use of the 15% blend fuel would be reached at the end of the projection period. The assumed effective blend wall would be reached by 2017.[8] To meet the mandates, a slight expansion of the fleet of flex fuel vehicles is expected towards the end of the projection period.

The mandate for biodiesel defined in the RFS2 is extended from 3.8 Bnl to 4.8 Bnl to be used by 2012, driving the initial growth in US biodiesel use. Biodiesel production from tallow or other animal fat is expected to represent an important share of US biodiesel production. Because of relatively high ethanol Renewable Identification Numbers (RIN) prices, biodiesel production is expected to surpass the biodiesel mandate to reach 5 Bnl in 2021.

The RED[9] of the European Union requires that renewable fuels should increase to 10% of total transport fuel use by 2020. The RED allows for substitution with other renewable sources including electric cars. In that context, when adding together the energy content of ethanol and biodiesel, the *Outlook* assumes that only a 9.5%[10] share of renewable fuels can be reached by 2021.

In that context, fuel ethanol production mainly from wheat, coarse grains and sugar beet is projected to reach 16 Bnl in 2021 and ethanol fuel consumption amounts to an

Figure 3.4. **Projected development of the US ethanol market**

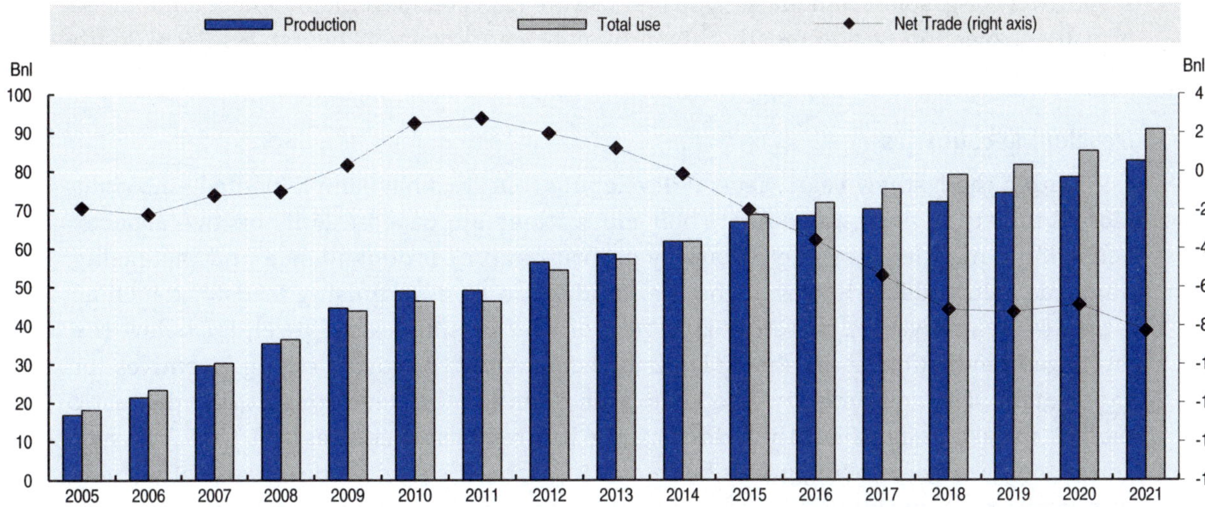

Source: OECD and FAO Secretariats.

StatLink http://dx.doi.org/10.1787/888932639419

average share of 8.3% in gasoline type transport fuels. Second generation ethanol is not assumed to play a major role throughout the projection period. Stimulated by mandates and tax reductions in European Member States, total biodiesel use is projected to reach 22 Bnl by 2021 (Figure 3.5) representing an average share of biodiesel in diesel type fuels of 8.5%. Domestic biodiesel production should increase to keep pace with demand. Second generation biodiesel production is assumed to reach about 4 Bnl in 2021.

Figure 3.5. **Projected development of the European biodiesel market**

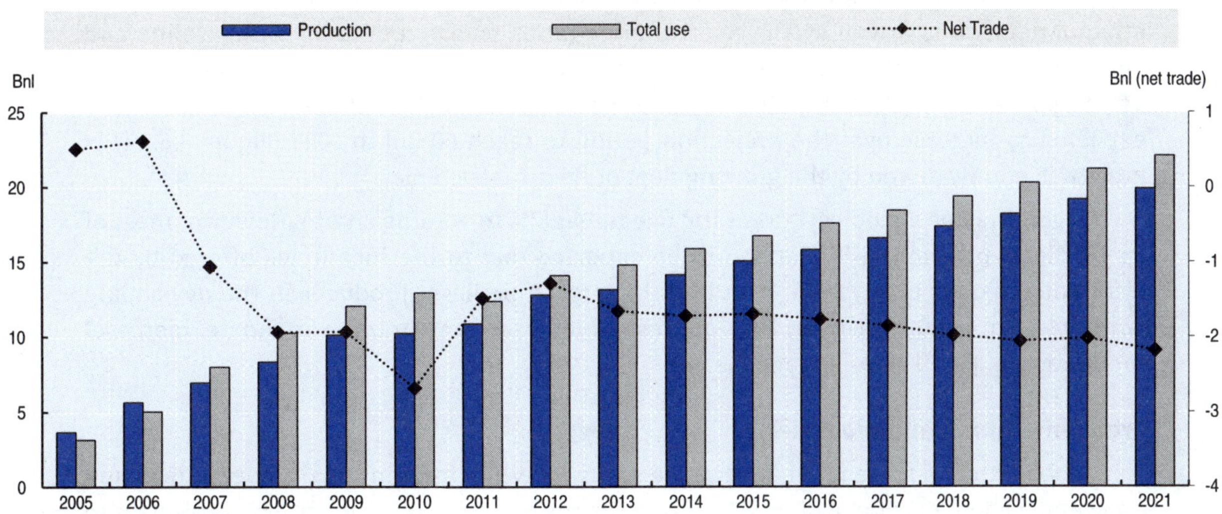

Source: OECD and FAO Secretariats.

StatLink http://dx.doi.org/10.1787/888932639438

Canadian mandates require an ethanol share of 5% in gasoline type fuel use and a biodiesel share of 2% in diesel type fuel and heating oil use, both expressed in volume terms. Both mandates are projected to be filled; ethanol and biodiesel uses should grow in line with gasoline and diesel consumption. In Australia, the ethanol and biodiesel shares respectively in gasoline and diesel type fuel use are expected to remain almost unchanged over the projection period mostly driven by policies in place in two states (New South Wales and Queensland).

Developing countries

Within the last few years, several developing countries have implemented ambitious biofuel targets or even mandates. Their motivations are based mainly on two aspects: achieving a high level of energy supply security and/or independence and increasing domestic value added. However, the fuel production from promising feedstock such as jatropha or cassava are currently still on a project or small-scale level, far below the envisaged production levels. Rising biofuel feedstock prices provide strong incentives for exportation of agricultural raw products. This hampers the development of a domestic biofuel industry significantly; additionally, limited resources restrict the ability of governments to implement policies by supporting domestic production and use of biofuels through financial incentives. Subsequently the fill-rates of mandates and targets in several developing countries remain low.

Countries which already have a high potential for sugarcane and molasses production, such as India, Thailand, Colombia and the Philippines, or vegetable oil production such as Malaysia, Indonesia and Thailand, are expected to produce and use more ethanol and biodiesel over the projection period. However, it is very likely that, except for Brazil and Argentina, biofuel use in developing countries remains significantly below the targets/mandates and an export oriented biofuel industry does not develop anywhere.

Brazil is projected to be the second largest ethanol producer. Brazilian ethanol derived from sugarcane should reach 51 Bnl and represent 28% of global ethanol production in 2021. One characteristic of the Brazilian ethanol industry is that it is very flexible. The sugarcane industry can quickly switch between sugar and ethanol production. Domestic ethanol demand is driven by the relative price ratios between ethanol and gasoline and between sugar and ethanol. It shifts with the growth of the flex-fuel vehicles fleet as well as the percentage of ethanol blended into gasoline. Brazilian ethanol domestic use is expected to increase over the projection period to reach 40 Bnl in 2021 (Figure 3.6). This growth is mainly driven by the growing fleet of flexi-fuel vehicles.[11]

Argentina has a biodiesel domestic use target (7% in volume share). However, most of its biodiesel production is planned to be exported due to the incentives offered by the differential export tax system. It will be the largest biodiesel producer in the developing world (4.2 Bnl in 2021). Driven by a domestic biodiesel consumption mandate, biodiesel production in Brazil should reach 3.2 Bnl.

Trade in ethanol and biodiesel

Global ethanol trade is set to increase strongly. While international trade represented on average about 4% of global production in the previous decade, the outlook projects it to increase to about 7% by 2021 (4.5 Bnl to 12 Bnl). Most of this increase is due to ethanol trade between Brazil and the United States. In 2021, the United States is expected to import about 16 Bnl of sugarcane based ethanol from Brazil which is assumed to be the cheapest alternative to fill the advanced biofuel mandate.[12] At the same time Brazil is projected to import 7.5 Bnl corn based ethanol from the United States to satisfy the flexfuel demand. Despite some tariffs, the European Union should increase imports by 2 Bnl of ethanol over

Figure 3.6. **Projected development of the Brazilian ethanol market**

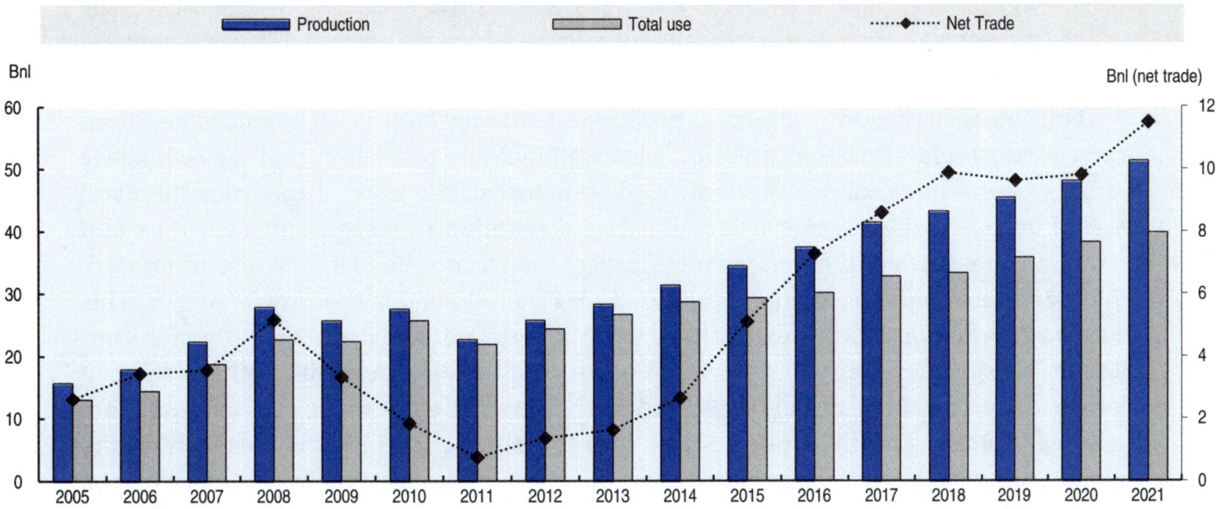

Source: OECD and FAO Secretariats.

StatLink http://dx.doi.org/10.1787/888932639457

the projection period while some countries like Thailand, Pakistan or South Africa increase their export supply only marginally. Recently, the two major palm oil producers, Indonesia and Malaysia have developed flexible refining capacities that enable them to quickly switch to biodiesel production for export once the relative prices become favourable. Yet given the expected price ratio in the coming decade, biodiesel trade is projected to increase only slightly with Argentina remaining the major exporter due to its differential export tax system.

Feedstocks used to produce biofuels

Coarse grains are projected to remain the dominating ethanol feedstock but the share of coarse grains based ethanol production in global ethanol production is projected to 44% by 2021. By then, 14% of global coarse grain production should be used to produce ethanol by 2021. The sugarcane based ethanol share in global ethanol production should increase from 23% in 2009-11 to 28% in 2021. By 2021, 34% of global sugarcane production is expected to be used for ethanol production. While the share of ethanol produced from wheat and molasses should decrease, cellulosic ethanol is projected to take a global share of almost 9.5% – almost all stemming from production in the United States.

The share of biodiesel produced from vegetable oil in global biodiesel production is expected to decrease by 10% over the projection period down to 70%. Sixteen per cent of global vegetable oil production should be used to produce biodiesel by 2021. Second generation biodiesel production is projected to increase slightly over the projection period, mainly coming from the European Union.

Main issues and uncertainties

Global issues

The development of biofuel markets over the past few years has been strongly related to the level of crude oil prices, biofuel policy packages in place, and the macroeconomic environment. This *Outlook* is marked by the assumption of strong energy prices which favour the development of biofuels. A scenario on the effect of a lower crude oil price is presented in the Overview. It shows that if the crude oil price was lower by 25% on average over the projection period, the world ethanol price would be on average 12% lower and the world biodiesel price would be 5% lower on average.

The first generation of biofuels produced from agricultural feedstocks could be progressively replaced in the future by advanced biofuels produced from lignocellulosic biomass, waste material or other non-food feedstocks. The pace of this transition will depend on profitability expectations determining industry investment decisions and private R&D research and development efforts as well as on the biofuel policy framework which determines public spending and provides guidelines for the private sector. This *Outlook* remains very cautious on the medium-term potential of second generation biofuels. No specific assumptions have been made on the development of other advanced biofuels including drop-in fuels[13] such as bio-butanol. The conversion of some ethanol facilities in Brazil and the United States into bio-butanol facilities is currently in the pipeline, although potential associated environmental and safety problems still need to be resolved. Important investments are currently being made on these advanced biofuels, especially in the defence sector. Advancements should be monitored as they could displace many of the projected paths presented in this *Outlook*.

The sustainability criteria embedded in the US and European biofuel policies are expected to increasingly affect biofuel markets. In the coming years, biofuel producers will have to comply with GHG emission targets. This could limit the availability of imported biofuels or biofuel feedstock. Given the steadily increasing amount of agricultural commodities used as biofuel feedstocks it is expected that regulations set forth by biofuel policies will shape not only biofuel markets but all agricultural commodity markets.

The rest of this section presents a quantitative analysis of the uncertainties around the implementation of US biofuel policies. It is complemented by a description of US biofuel policies presented in the Annex of the chapter.

Implementation of US biofuel policies

Baseline assumptions concerning the implementation of US biofuel policies can be challenged as implementation possibilities open to the EPA are numerous. Until now, the yearly decisions taken by EPA did not have important impacts on agricultural and biofuel markets because the level of the cellulosic ethanol shortfall was small. But by 2021, the end of this *Outlook*, the amounts will be much larger and EPA's decision will likely have impacts on agricultural markets. This section identifies the effect of three alternative implementation options (as described in Annex 3.A1):

- *Option 1:* Lower the total and advanced mandates by the shortfall in the cellulosic mandate; EPA has not so far chosen this option which could seem to be the "simplistic" one.
- *Option 2:* Maintain both the advanced and total mandates, *i.e.* increase the other advanced gap. This is the option that has been chosen by the EPA. This scenario provides some insights regarding the sustainability of such an implementation option, especially when focusing on the interactions between US and Brazilian ethanol markets.
- *Option 3:* Maintain the total mandate and lower the advanced mandate by the shortfall in cellulosic production, *i.e.* increase the conventional gap. Maize based ethanol production is expected to exceed the conventional ethanol gap in baseline projections especially in

the latter years of the projection period when the conventional gap cannot exceed 56.8 Bnl. This scenario highlights the effects on international markets of the nested structure of US biofuel mandates.

The assumptions regarding the implementation of US biofuel policy in the baseline and in the three envisaged scenarios for 2021 are summarised in Figure 3.7. Scenarios were conducted after the completion of the revision of the US biofuel module of the AGLINK-COSIMO model, which captures the complex interplay of the different mandates, a simplified market of Renewable Identification Numbers (RINs) as well as the possibility to transfer these RINS between two years (*i.e.* roll-over). Scenario results are presented in Table 3.A2.1.

The decision taken by EPA will not be reflected fully by any of the scenario options. Those scenarios have been produced to illustrate the policy space, not to promote any particular policy option. This analysis focuses in different sub-sections on the impacts of the scenarios in comparison to baseline projections on ethanol markets (United States, Brazilian, European and global), on biodiesel markets and on agricultural markets. The last section provides key conclusions.

Figure 3.7. **Structure of US biofuel mandates in the law (RFS2), the baseline and the 3 options for 2021**

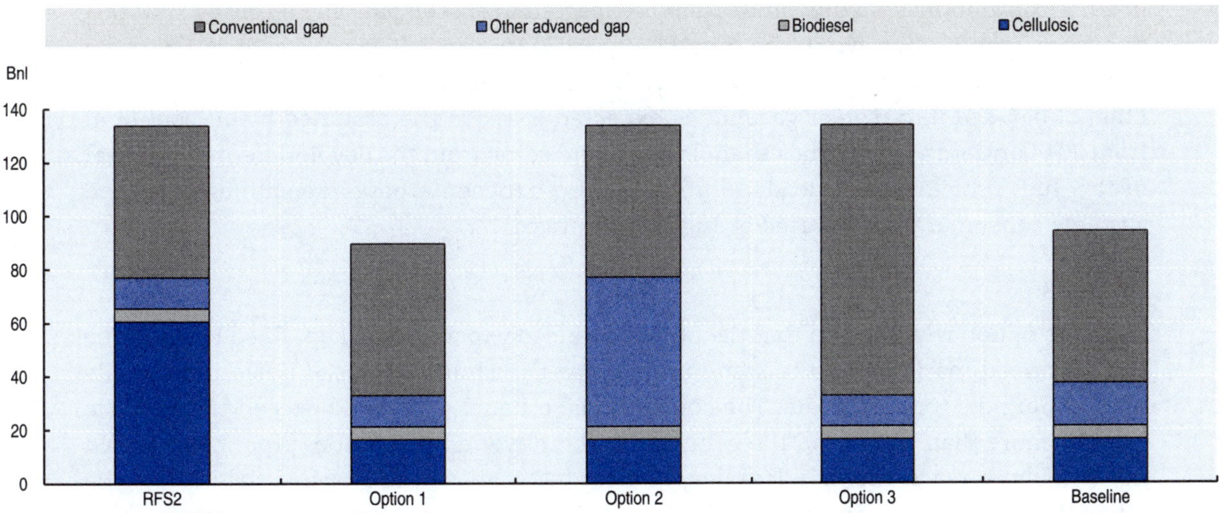

Source: OECD and FAO Secretariats.

StatLink http://dx.doi.org/10.1787/888932639476

Impacts on US ethanol market

This section illustrates the key impacts in terms of supply, use, net trade and prices of the three implementation options on the US ethanol market. Results are summarised in Figure 3.A2.1. The three scenario options underline the fact that the US ethanol market – on the supply side as well as on the demand side – can adjust relatively easily to policy changes and to world price variations. On the demand side, the blend wall issue[14] is a major constraint for further expansion in ethanol use. An increase in the size of the flex-fuel vehicles is expected to be the most plausible outcome if the total mandate was to remain at the level defined in EISA towards the end of the projection period.

Option 1

With this implementation option, the total and advanced mandates are lowered by the shortfall in meeting the cellulosic ethanol mandate which keeps the conventional ethanol and other advanced fuel gaps unchanged from original levels. In 2021 the need for ethanol imports from Brazil to meet the other advanced gap is 30% lower than in the baseline, which leads to a 2% decrease of the world ethanol price. United States conventional ethanol production is projected to still exceed the conventional gap, but to be reduced by 1% in 2021 when compared to the baseline, in line with the reduction of the ethanol producer price. Option 1 leads to lower percentages of ethanol blended into regular gasoline: the blend wall is not achieved in any year of the projection period and consequently there is no need to expand the fleet of flex-fuel vehicles.

Option 2

In this case, EPA would maintain both the advanced and total mandate. This would result in the widening of the other advanced gap and in an important increase of advanced ethanol imports, *i.e.* imports of sugarcane based ethanol from Brazil. Those would reach 51 Bnl in 2021, compared to 16 Bnl in the baseline. This additional demand for advanced biofuels on world markets triggers a 17% higher world ethanol price in 2021 when compared to the baseline which is transmitted in part to the US ethanol producer price. In 2021, conventional ethanol production is expected to exceed baseline levels by 10%; this additional production would be largely exported to Brazil (see next section). On the demand side, Option 2 leads to ethanol use being 40% higher in 2021 than in the baseline. Ethanol blended into regular gasoline is expected to reach the assumed blend wall limit from 2014 onwards. Additional ethanol use should come from the development of the fleet of flex fuel vehicles which leads to a lower ratio between ethanol consumer price and gasoline consumer price induced by higher RIN prices.

Option 3

This option would mean that the other advanced gap would be kept fixed by reducing the advanced mandate by the same amount as the shortfall in cellulosic fuels while maintaining the total mandate. The conventional ethanol gap would exceed the baseline level by more than 70% in 2021, reaching 97 Bnl. Conventional ethanol production would not be able to reach the mandate despite being 40% above the baseline in 2021[15] – the ethanol producer price exceeds baseline levels by 40% – and US ethanol exports outside North America would be close to zero. To meet the global mandate, the United States would have to import ethanol. The world ethanol price in 2021 is projected to be 6% above the baseline level. This disparity in the movement of the Brazilian and US ethanol price is caused by the passage of the US price from the export floor (world price minus transport cost) to the import ceiling (world price plus transport cost plus a small *ad valorem* tariff) basis.[16] On the demand side, Option 3 leads to a situation very similar to Option 2 because the total mandate that has to be consumed is the same: ethanol blended into regular gasoline is expected to reach the assumed blend wall limit from 2014 onwards and additional ethanol use should come from the development of the flex fuel vehicle fleet. However, a stronger increase in biodiesel production leads to an ethanol consumption increase of only 38% compared to 40% in Option 2.

Interactions between the US and Brazilian ethanol markets

The different EPA implementation options analysed in this section have major implications for US import demand of ethanol able to qualify for the advanced biofuel mandate. Currently, the only ethanol type qualifying and being produced on a large scale is from sugarcane. In the outlook period, Brazil is the sole country that has the capacity and the flexibility to respond to strong additional demand from non domestic markets.[17] This means that the three implementation options have direct effects on Brazilian ethanol and sugar sectors.

Figure 3.A2.2 illustrates the most important interactions between the US and Brazilian ethanol markets. US ethanol imports directly impact Brazilian ethanol exports. In Brazil, the expansion/contraction of ethanol exports are due to several inter-related factors on the domestic market: expansion/contraction of domestic ethanol production and thus of sugarcane and sugar production, but also shifts in domestic ethanol demand through the adjustment of the car fleet as well as possibilities of ethanol re-imports from the United States.

Option 1

In the case of Option 1, US ethanol import demand is reduced. It is interesting to note that Option 1 has hardly any effects on the Brazilian and the world sugar markets when compared to baseline levels. Although ethanol exports to the United States are 30% lower in 2021, ethanol production in Brazil is only reduced by 3%, reducing sugarcane area by 2% while domestic consumption with a rising flex-fuel fleet increases by 3%. However, the lower sugarcane production does not have a visible impact on sugar production given the flexibility of the Brazilian sugar industry.

Option 2

Option 2 is associated with the strongest increase in US ethanol import demand when compared to baseline levels in 2021. This additional demand of about 35 Bnl induces larger Brazilian ethanol production by only about 10 Bnl. The rest will become available because of lower Brazilian consumption and higher imports from the United States.

Impact on Brazilian sugar markets: To produce more ethanol, the Brazilian sugarcane area is extended by 9% when compared to the baseline and the share of sugarcane used for biofuel production is increasing at the expense of sugar production. On the domestic Brazilian sugar market, lower sugar production implies higher domestic sugar prices, a lower sugar demand and a significant decrease of sugar exports. As a consequence, world sugar prices in Option 2 are 6% above baseline levels in 2021.

Impact on Brazilian ethanol use: Brazilian ethanol demand in a context of higher prices is expected to decrease considerably when compared to baseline levels in 2021. This decrease can be decomposed into two components:

- Low blend demand is reduced to the minimum blending requirement (18% of total fuel consumption on an energy equivalent basis).
- Ethanol used by flex-fuel vehicles is reduced to 21% of total fuel consumption – the 2011 level – compared to 41% in the baseline.

Ethanol imports from the United States: To meet domestic demand – even if it is much lower than in the baseline – in a context of tremendous increase[18] of Brazilian ethanol exports, Brazil needs to import some ethanol. Imports are projected to reach 18 Bnl, to a large extent originating from the United States where, in turn, the maize based ethanol production is stimulated by high ethanol prices. So Option 2 would create a large policy driven two-way trade in ethanol.

Option 3

The same argumentation can be built for Option 3. However, impacts on Brazilian ethanol and sugar markets are lower as US import demand is only 11% higher than in the baseline case in 2021. With much higher requirement for other conventional ethanol, the price of ethanol in the United States increases to levels eliminating the possibilities of exporting any ethanol outside North America. Brazil replaces this amount (close to 7 Bnl in the baseline) by domestic production and increases exports to the United States.

Implications on global ethanol production

The impacts of the scenarios on the European Union are only visible on the supply side, because consumption is bound by the EU mandate. In Option 2, with high world ethanol prices and a lot of competition on the world market, EU ethanol production is increasing by 9% (Figure 3.8). In the rest of the world, the supply and demand responses follow the world price incentives. In Option 2, China, India, Thailand and Canada make more than 50% of the production increase and even more in Option 3, where Canada shows the strongest supply increase given the tight connection to the US ethanol market. Consumption changes mainly take place in China, Thailand and Ukraine.

Figure 3.8. **Global ethanol market effects**

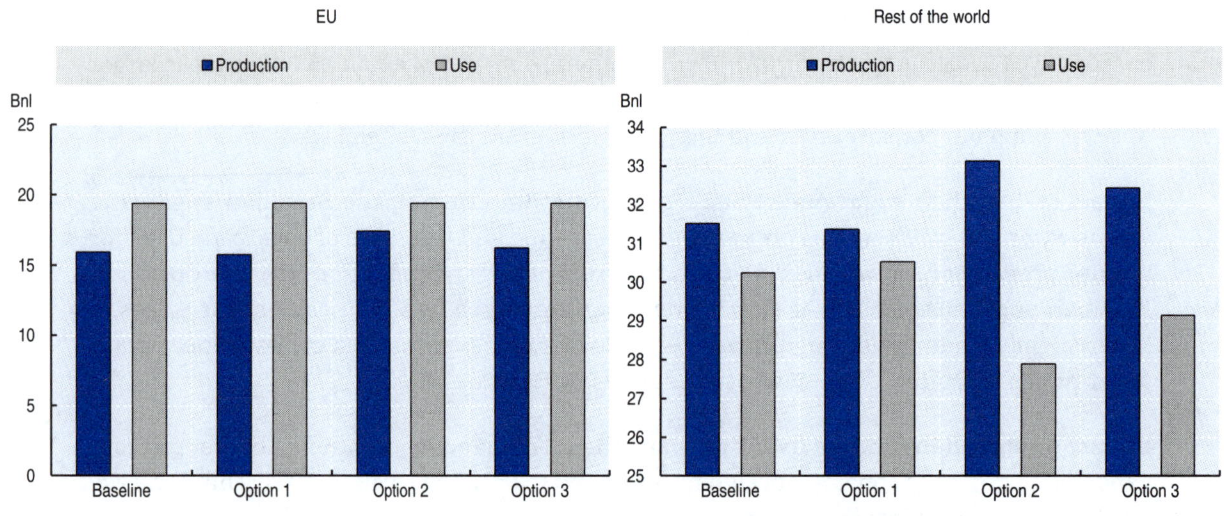

Source: OECD and FAO Secretariats.

StatLink http://dx.doi.org/10.1787/888932639495

Implications on biodiesel markets

Given the implicitly strong increases in RIN prices for ethanol in Options 2 and 3, biodiesel is likely to become more competitive against ethanol to meet the advanced mandate. In Option 2, US biodiesel production and use are increasing by about 50% to

7.5 Bnl when compared to the baseline. They increase even more in Option 3 where they reach 8 Bnl. Effects on global biodiesel markets are quite low, as the US biodiesel net trade position does not change considerably in the scenarios when compared to the baseline. In that context, the world biodiesel price does only increase slightly.

Implications on other agricultural sectors

The increasing production of ethanol from sugarcane and from coarse grains in Options 2[19] and 3 is sufficient to generate significant impacts on the other sectors, which is not the case for Option 1. Therefore, only Options 2 and 3 are reflected in this section. The impacts are summarised in Figure 3.A2.3.

Impacts on biofuel feedstock sectors

The starting point is obviously an increase in the demand for coarse grains and for sugarcane by the ethanol producers by 11% and 20% respectively in Option 2 and by 35% and 3%, respectively, under Option 3. This leads to an increase in the world price of coarse grains and sugar of 5% and 6%, respectively, in Option 2 and of 16% and 4% in Option 3. Many factors are mitigating the price impact and in particular the strong reduction in consumption of ethanol by flex fuel cars in Brazil and an increase in coarse grains and sugarcane production by 1% and 6% in Option 2 and by 2.5% and 0.5% in Option 3.

Overall, the larger amount of coarse grains consumed by ethanol producers (20 Mt and 64 Mt respectively in Option 2 and 3) is accounted for in the model by a larger production, increase in distiller's dry grain (DDG) production (5 Mt and 20 Mt) and by a reduction in the amount consumed by human either directly or indirectly through non-ruminant meats. Basically, the reduction in human consumption represents less than 50% of the additional demand by ethanol producers in Option 2 and Option 3. In the case of sugarcane, 80% of the additional amount used by ethanol producers is accounted for by larger production and 20% by lower sugar consumption in Option 2. In Option 3, these percentages are 41 and 59, respectively.

Impact on other sectors

The increase in the world coarse grains price affects many other sectors. First, through demand and supply substitution, it leads to a higher price of wheat and oilseeds by 2% in Option 2 and by 5% and 4% in the case of Option 3. The higher oilseed price reduces crush demand leading to lower supply of protein meal and vegetable oil. This combined with substitution on the feed demand side lead to a significant increase in the price of protein meal by 2% and 5% in Options 2 and 3 respectively.

The increasing price of feed generates a reduction in supply and production of non-ruminant meats. World pigmeat and poultry production falls respectively by 0.1% and 0.2 % in Option 2 and by 0.2% and 0.7% in Option 3. This leads to higher price and lower consumption of these meats. Taking the Pacific market as an example, the price of pork is 2% higher in Option 2 and 7% higher in Option 3. The US price of poultry increases by about the same percentage.

Considering the smaller share of feed in the variable cost of producing beef and the longer production cycle, the impact on the beef sector is different. In fact, the increasing demand for beef generated by the higher price of pork and poultry crosses the lower supply generated by the higher feed prices at a point leading to higher price and to a small increase in world production by 0.1% and 0.3% in Options 2 and 3.

The impact on the fish sector is also different since capture and raised molluscs, the largest share of supply, are not directly influenced by feed prices. On the other hand, demand for fish as food is entirely influenced by the movement in meat prices. Another important point is that China, which counts for 61% of world aquaculture production, is not strongly tied to the movement in the world price of coarse grains. Chinese coarse grain price is only 3% higher in Option 3 compared to a 16% increase for the world price. The combination of all these elements and world capture being mostly controlled by production quotas, leads to a small impact on production. For aquaculture production, the increasing price caused by the larger demand generated by higher meat prices compensates for the increasing feed cost.

Key conclusions of the scenarios

Option 1 (the total and advanced mandates are lowered by the shortfall in the cellulosic mandate), does not differ much from the baseline except from the fact that low blend ethanol use in the United States would not reach the blend wall in any years and that the United States would be less dependent on advanced ethanol imports.

Option 2 analysed in this section corresponds to maintenance of the actual policy of the EPA: both the advanced and total mandates are kept at the EISA level. The main conclusions of Option 2 compared to baseline projections are the following:

- Important policy driven two-way ethanol trade emerges between Brazil and the United States.
- Spill-over effects are expected in the coarse grains market as ethanol trade is completely free between the United States and Brazil, but the impact on the world price of coarse grains is not expected to be large.
- The largest adjustment will come from a severe reduction in consumption of ethanol by flex fuel cars in Brazil, *i.e.* the improvement in the US energy independence would be partly achieved through a reduction in Brazil's energy independence.
- The potential increase in sugarcane production is sufficient to prevent a large increase in the sugar price.

If, on the contrary, the EPA decides to reduce as well the advanced mandate without changing the total mandate as is the case in Option 3, then the impact on the coarse grains markets will be much larger. This is due to the fact that the US ethanol price will be much higher because it will go from an export floor price basis to an import ceiling. Not surprisingly, this will put even more upward pressure on the price of coarse grains. The main conclusions of this scenario are the following:

- US ethanol exports outside North America disappear and imports from Brazil driven by price advantage increase significantly.
- World coarse grains price is almost 16% higher in 2021, compared to the baseline.
- About half of the coarse grains or sugarcane used to produce the additional ethanol is derived from lower human consumption, taking into account additional production and the greater availability and use of DDGs.

- Quantities of food consumed around the world are somehow similar but at higher prices. Option 3 would put even more pressure on countries where food expenditure already accounts for a large share of income.
- The reduction in feed demand comes entirely from the non-ruminant meat sectors.

Finally, the impacts of the decisions to be taken by the EPA concerning the implementation of the US biofuel policy in the coming years are not fully reflected by the scenario options presented. However, it is clear from this analysis that the impacts will vary according to the decisions taken, that they are likely to be important, and that they will affect not only the biofuel sector in the United States but more broadly the global biofuel and agricultural markets. The implementation decision will have an impact on world ethanol and agricultural commodity prices. It will require some adjustment in terms of ethanol production and consumption patterns, as well as in terms of ethanol feedstocks use around the world.

Notes

1. Brazil, Sao Paolo (ex-distillery).
2. Producer price Germany net of biodiesel tariff.
3. Energy Independence and Security Act of 2007, Public Law 110–140 (2007) *www.gpo.gov/fdsys/pkg/PLAW-110publ140/pdf/PLAW-110publ140.pdf.*
4. Cellulosic ethanol production is an exogenous model component.
5. The total and advanced mandates are reduced by about 90% of the difference between the assumed applied and the legislated cellulosic biofuel mandate at the end of the projection period.
6. The conventional gap is the difference between the total mandate and the advanced mandate, see Annex 3.A1 for more explanations.
7. For more information on the blend wall, see Annex 3.A1.
8. In baseline assumptions, the blend wall is gradually extended from 10% to 15% over the projection period (accounting for the disappearance of older vehicles and for the resolution of logistic problems by blenders). These assumptions result in an assumed effective blend wall slightly lower than E15 in all years of the projection period except 2021. For example, it is assumed that the maximum ethanol blending percentage in regular gasoline would be of 13% in 2017.
9. *eur-lex.europa.eu/LexUriServ/LexUriServ.do?uri=OJ:L:2009:140:0016:0062:EN:PDF.*
10. This percentage takes into account the fact that the contribution of second generation biofuels will be counted twice toward the EU RED mitigation targets.
11. Currently, gasoline prices in Brazil are not allowed to exceed a certain cap value. The *Outlook* assumes that this cap will be adjusted upwards given rising energy prices so that the driving ethanol/gasoline price ratio remains slightly in favour of ethanol.
12. According to the RFS2, sugarcane based ethanol is classified to be an advanced biofuel, while maize based ethanol is not.
13. Drop-in fuels are defined as renewable fuels that can be blended with petroleum products, such a gasoline, and utilised in the current infrastructure of petroleum refining, storage, pipeline and distribution.
14. Vehicles produced in 2001 or later are allowed since 2011 to use blends up to 15% ethanol. Annex 3.A1 contains a specific section on the blend wall and associated constraints on US biofuel demand.
15. In Option 3, in 2021, 53% of US coarse grains production would be consumed by ethanol producers.
16. US imports in Option 2 occur even if Brazilian ethanol prices are high because of the classification of sugarcane based ethanol as advanced biofuel. The US ethanol price, which can be interpreted as the conventional ethanol price, is therefore tight to the marginal quantity of US ethanol exported.

In Option 3, exports completely disappear and Brazilian sugar-cane ethanol exports now compete inside the conventional gap.

17. Other producers in the world are also reacting to a smaller extent to the higher ethanol price and mitigate some of the shortfall on the world market created by the US policy.
18. In 2021, Brazilian exports that qualify for the US advanced mandate are projected to be more than 260% higher than in the baseline.
19. All impacts reported are with respect to the baseline for the last year of the *Outlook* period, *i.e.* 2021.

ANNEX 3.A1

US biofuel policy

Biofuel policies in the United States are entering a new phase as the long standing blenders credits on ethanol and biodiesel and the tariff on imported ethanol expired at the end of 2011 and mandated quantities of biofuels continue to expand.

The expiration of the ethanol blenders credit of USD 0.45 per gallon (USD 0.12 per litre) with an offsetting USD 0.54 per gallon (USD 0.14 per litre) import tariff and the USD 1.00 per gallon (USD 0.26 per litre) blenders credit on biodiesel ends a decade's long policy of subsidisation to mix the renewable fuels into general motor fuel use.[1] The unique producers' credit for cellulosic biofuels of USD 1.01 per gallon (USD 0.27 per litre) is set to expire at the end of 2012. While there are calls for renewal of the credits, and it has happened in the past (even retroactively), as of the writing of this text the credit paid for by US taxpayers has expired. What remains is a system of mandates on blenders for inclusion of four classes of renewable fuels, total, advanced, bio-based diesel and cellulosic biofuels, into broader petrol and distillate use.

US biofuel mandates

The mandates on blenders represent their share of the calendar year quantitative national mandates laid out in the Energy Independence and Security Act of 2007 (EISA).[2] The mandates are segmented into four classes presented in Figure 3.A1.1 based on the fuel's feedstock and its estimated greenhouse gas (GHG) reduction score relative to the 2005 base level as specified in EISA but are not independent of each other; they are a nested structure of quantitative minimums.

The overarching total mandate (T) requires fuels to achieve at least a 20% GHG reduction. Advanced fuels (A), as specifically defined in the legislation, are fuels which achieve a 50% greenhouse gas reduction score, ethanol derived from sugar is explicitly defined as an advanced fuel. Of that advanced mandate, a minimum quantity must come from bio-based diesel fuels (B), a distillate replacement with a 50% GHG reduction score, and cellulosic renewable fuels (S), either petrol or distillate replacement fuels, with a 60% green house gas reduction score.

The biodiesel and cellulosic minimums leave another advanced gap (O), the difference between the advanced mandate and the minimum that must come from cellulosic fuels and biodiesel, which can be met with fuels such as sugar based ethanol or excess biodiesel (B) and cellulosic fuel (S) consumption.

The conventional gap (C), the difference between the total mandate and the minimum that must come from advanced fuels, is then the portion of the total mandate that could

Figure 3.A1.1. **Mandated quantities and implied gaps**

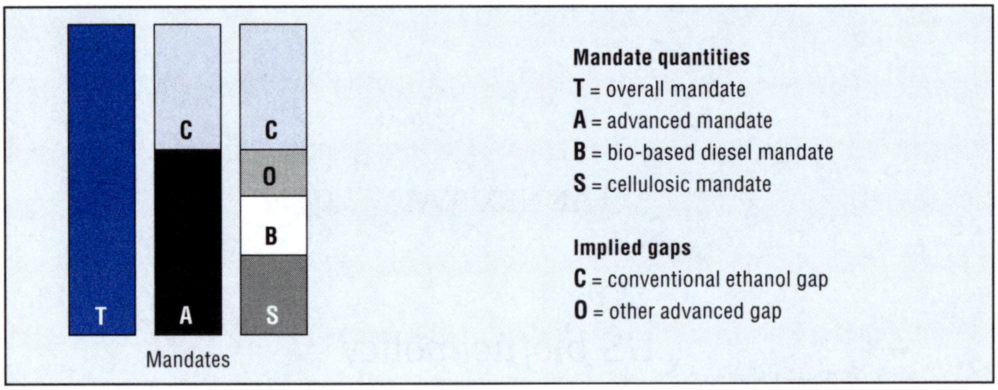

Source: OECD and FAO Secretariats.

potentially come from conventional biofuels such as maize starch based ethanol and therefore only needs to meet the 20% GHG reduction criteria. It is worth noting here that there is no explicit mandate for maize based (specifically maize starch) ethanol in the system, only that it may compete with both other conventional biofuels[3] and advanced biofuels which may be consumed in excess of its mandate, in filling the conventional gap (C).

The mandates only restrict minimum quantities and are nested within each other, creating a hierarchy of biofuel types. Any overproduction in a sub-category can be used to fulfill the next broader mandate. Under varying conditions all, some or none of the four mandates may be binding at any given time.

RIN markets and prices

Blenders are the obligated party in the system of mandates and show compliance in all four mandate categories, total, advanced, bio-based diesel and cellulosic biofuels, through the submission of Renewable Identification Numbers (RINs). A RIN is a 38-digit number which indicates the year, volume and highest mandate classification the renewable fuel is capable of meeting and is obtained from the US Environmental Protection Agency (EPA) by the biofuel producer upon production and registration of the fuel. Conveyed along with the fuel, for example maize starch based ethanol, is the associated RIN (in this case a conventional RIN) where the blender can detach and use the RIN for compliance or sell the RIN to another blender to help satisfy their obligation. The RIN price may be very low if the market demands quantities in excess of the mandate, such as when oil prices are high relative to biofuel prices, or the RIN may be very costly if the mandate quantity is well in excess of true market demand.

When the market (P_M) demands more than the mandated quantity (frame A in Figure 3.A1.2) the price paid for the renewable fuel from producer (P_P), blended and sold into the retail supply chain (P_R) will be equivalent when adjusted for taxes and margins. However, when the mandate is in excess of that the market would otherwise demand the wholesale price of the renewable fuel will rise relative to its value to consumers (frame B). In this context, blenders must pay a price to producers high enough to obtain the quantities they need to meet the mandate (P_P). The blenders cannot impose the cost directly on the ethanol share of the retail fuel or risk reducing demand for renewable, making the mandate even harder to achieve. They therefore must sell it at a lower price (P_R)

based on consumers preferences. Blenders must spread the cost of RINs out over the entire motor fuel sales, both petrol and distillates, maintaining relative renewable and conventional fuel prices; which in turn raises costs to motor fuel consumers. This difference between what the blenders pay (P_P) and what they impose on the retail market (P_R) is reflected in the RIN price. With four separate mandates there are potentially four separate RIN prices each of which reflects the per gallon cost born by motor fuel consumers of imposition of that mandate.

Figure 3.A1.2. **Determination of a binding mandate and RIN price evaluation**

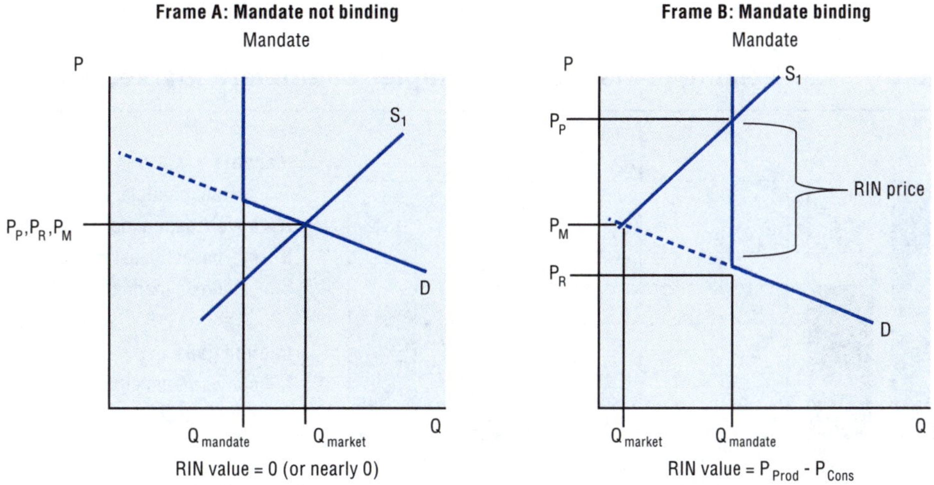

Source: OECD and FAO Secretariats.

The hierarchical nature of the mandates will be reflected in the RIN prices. A biodiesel RIN can be priced no lower than an advanced RIN as any lower priced biodiesel RINs would be diverted to satisfy the advanced mandate equalising prices. If the biodiesel mandate is highly binding, biodiesel RIN prices would rise, but advanced RINs which, conversely, cannot be used for biodiesel compliance may lag behind.

Examples illustrating the nested nature of the biofuels mandates

A number of examples not intended to be exhaustive, can highlight some of the possible outcomes and clarify the hierarchical nature of the mandates (Figure 3.A1.3).

Market outcome 1 shows the situation where, perhaps due to high petroleum prices and low agricultural commodity prices, maize ethanol consumption exceeds the conventional mandate gap (C) and therefore total ethanol RIN supplies exceed the total mandate. The total mandate would then be non-binding, conventional RIN prices would approach zero.

Market outcome 2 highlights the point that no specific mandate for conventional ethanol exists within EISA, but only a conventional biofuel gap. This case may be reflected in a situation where the total biofuel mandate may be binding, but imports of sugarcane ethanol, perhaps from high maize prices as a result of a short-crop, could enter and displace maize starch based ethanol in meeting the total mandate. In this instance the total mandate may be binding while the advance mandate is not and conventional and advanced RIN prices will be close in value.

Finally, market outcome 3 further highlights the hypothetical situation where there is a technological breakthrough in cellulosic ethanol production which reduces the cost of production, while the overall mandate remains binding, perhaps in the context of a low petroleum price. In this instance, cellulosic production may far exceed its mandate, but it cannot displace bio-based diesel production which has its own category specific mandate. Together, biodiesel and cellulosic ethanol may provide sufficient quantities to meet and exceed the advanced biofuel mandate and even displace some of the corn starch based ethanol being used to meet the total mandate. The biodiesel mandate and the total mandate may be binding but the cellulosic and advanced mandates would not be. In this situation, the prices for cellulosic and conventional RINs would be very close.

Figure 3.A1.3. **Nesting of mandates, examples of different market outcomes**

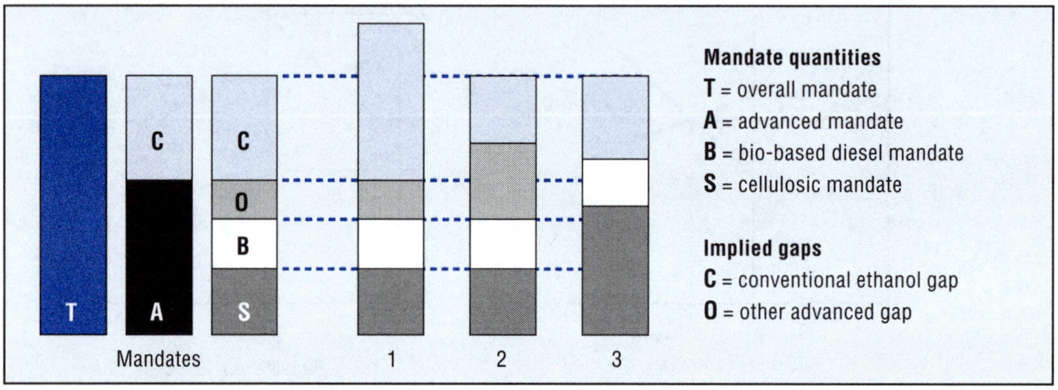

Source: OECD and FAO Secretariats.

Mandate flexibilities

Additional flexibility and complexity is added to the mandate system with provisions allowing blenders to "rollover" or run a "deficit" of RINs into the following year. Up to 20% of a given mandate may be met with RINs produced in the previous year. This allows for limited "stock holding" of obligations which can be drawn down in years where RIN prices rise. The blender can hold an additional stock of RINs as a hedge against rising biofuel and RIN costs or other compliance issues. This allows for some moderation of feedstock prices when a transient shock, such as below average crop yields, push RIN prices higher.

On an individual basis, blenders may fall short of the mandate in a particular year if in the following year they make up the "deficit" from the previous year and fully comply with the mandate in the current year. Running a deficit in the current year introduces considerable rigidity in the following year for blenders, as failure to comply with mandates can result in a fine of USD 37 500 per day plus any economic benefit derived from non-compliance.[4] Such flexibility in the mandate should mitigate swings in feedstock and biofuel prices from transient shocks in energy prices and crop production.

Mandate waivers and the implication of EPA implementation

The OECD-FAO baseline maintains current US biofuel policy with respect to mandates;[5] however, implementation of the policy by the Environmental Protection Agency (EPA) remains a significant source of uncertainty and could have significant effects on commodity markets.

Each year, the EPA puts forth the minimum quantities for each of the four classes of biofuels required (total, advanced, bio-based diesel and cellulosic biofuels), taking into account what can be viably produced or imported. Thus far, the production capacity for cellulosic ethanol has lagged well behind the quantities mandated in 2010, 2011 and 2012. For 2012 the EISA legislation calls for 500 Mn gallons (1.893 Bnl), but has been reduced by the EPA to just 8.65 Mn gallons (32.7 Mnl) or just 1.7% of the targeted quantity. The cellulosic mandate also grows at an increasing rate for the remainder of the projection period. While this shortfall has its own implications for biofuel markets in terms of potential feedstock use and production, there is concern that meeting the cellulosic mandate faces considerable hurdles.[6, 7]

This leaves the EPA with an important decision each year regarding the other mandates. It is within their power to adjust each of the other mandate levels or leave them as legislated in EISA. The EPA may choose Option 1 in Figure 3.A1.4, in this case they lower the total and advanced mandate by the shortfall in cellulosic ethanol which keeps the conventional ethanol gap and other advanced fuel gap consistent with EISA. This policy maintains the maximum quantity of maize based ethanol that can be used to meet the mandate as well as the need for advanced fuels to meet the "other advanced gap". This choice is likely to lead to the lowest commodity and food prices while also resulting in the lowest GHG savings.

Alternatively the EPA could choose Option 2 in Figure 3.A1.4 and maintain both the advanced and total mandate which results in the widening of the other advanced gap and potentially drawing in additional imports such as sugarcane ethanol from Brazil. This option is likely to have a larger impact on commodity and food prices and mandate compliance costs than Option 1.

The EPA could alternatively choose to keep the other advanced gap fixed by reducing the advanced mandate by the same amount as the shortfall in cellulosic fuels while maintaining the total mandate. This would result in a growth in the conventional ethanol gap and a larger potential market for maize ethanol (Option 3 in Figure 3.A1.4). The EPA could also choose to do a partial adjustment on either the advanced mandate or total mandate or any combination of the two.

Figure 3.A1.4. **EPA mandate implementation options**

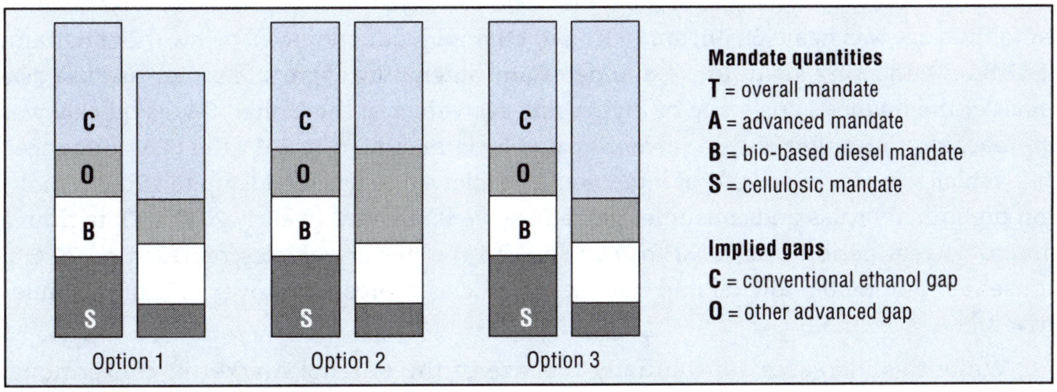

Source: OECD and FAO Secretariats.

Thus far, with the cellulosic mandate at relatively low levels, the EPA has chosen to keep the total and advanced mandate at their original levels (*i.e.* Option 2 in Figure 3.A1.4). This has led to the opening up of the "other advanced gap" of undefined advanced fuels needed to meet the mandate, such as imports of sugarcane ethanol from Brazil, a gap which will grow rapidly in the future if EPA maintains this option (Table 3.1).

Under legislated quantities, in 2020 the advanced gap would require 2.58 Bn gallons (9.76 Bnl) of other advanced fuel. Under our projected cellulosic biofuel production path, the continuation of current EPA implementation would result in the need for 10.731 Bn gallons (40.624 Bnl) of other advanced fuels in 2020. In developing the baseline for the *OECD-FAO Agricultural Outlook 2012-2021*, this was deemed an unlikely outcome; the most viable fuels to fill this gap, under current projections, would appear to be significant additional imports of sugarcane ethanol with possible additional production of biodiesel beyond its mandated minimum. This volume of imports would represent more than the total ethanol production for Brazil in 2011.

In the *OECD-FAO Outlook 2012-2021*, it was therefore decided to reduce both the total and advanced mandate by a proportion of the shortfall in cellulosic biofuels such that the other advanced gap did not shrink from year to year and the conventional ethanol gap was held to the quantities in the legislation. Changes in this assumption would have significant impact on commodity prices and consumer fuel costs as well as biofuel prices and trade. The production of cellulosic biofuels is an exogenous component in the model; all other categories of biofuels as defined in the nested structure of mandates are modeled endogenously.

The blend wall and constraints on biofuel demand

While the system of mandates in US policy specify quantities of biofuels which must be domestically consumed it provides no direction on *how* such fuels should be consumed. Petrol dominates US fuel consumption, representing 62% of consumption, with diesel fuels representing another 28%.[8] Short run technical constraints, referred to as "the blend wall" in the petrol market, act as an impediment to increased ethanol consumption. Biodiesel use could face similar constraints in the future.

Prior to 2011, conventional petrol vehicles in the United States were limited, by EPA rules, to a maximum blend of 10% ethanol by volume with a small number of flex fuel vehicles (FFV) able to take up to 85% blends.[9] The 10% constraint posed little problem when motor fuel use was near 568 Bnl annually and ethanol production well below the constraint of 57 Bnl. With rising quantitative mandates and stagnating aggregate motor fuel use as a result of the financial crisis and of higher mileage vehicles, the United States quickly was approaching saturation of the conventional vehicle market.[10] In 2011 the EPA announced that vehicles produced in 2001 or later would be allowed to use blends up to 15% ethanol[11] and preliminary rules and consumer guidelines were released in early 2012.[12] Data from a similar 11 year period from 1998 to 2009 showed the newer vehicles represented 70% of household automobile ownership but these vehicles represented over 77% of the miles driven.[13]

While this increases substantially the size of the ethanol market in conventional vehicles, many obstacles remain along the distribution chain. These constraints can have significant impact on the costs to consumers of the mandate system and the competition between renewable fuels, primarily ethanol and biodiesel, to fill the undefined advanced fuel quantities (O) within the EISA mandate. While EPA rules allow the dispensing of E15, retailers may be hesitant to offer it to consumers until the issue of liability is resolved. Earlier car warrantees may limit ethanol content to the previous 10% limit and would expose retailers to other consumer complaints. In addition, with a bifurcated market of newer and older vehicles, retailers must take action to minimise the mis-fuelling of vehicles by consumers who may be unaware of the restrictions. There may also simply be no "room" at the pump to add yet another handle dispensing an additional fuel type (different octane and ethanol inclusion rate combinations). Furthermore, the installation of additional underground tanks is very costly.

While even modest growth in E15 dispensing would allow for full absorption of maize ethanol that could be used to fulfill the conventional ethanol mandate gap (C), any significant growth in cellulosic ethanol production[14] or imports of sugarcane ethanol to meet the advanced mandate gap (O) could put pressure on the distribution system. This pressure will be reflected in increased RIN prices, ultimately born by consumers, and increase the incentives for blenders to expand the availability of E15 and E85 fuels and to price them competitively. This pressure also increases the motor fuel costs to consumers who may consume less in aggregate and thus make the ethanol blend-wall even more constraining. As an alternative, the constraint of the blend-wall also increases the potential for biodiesel consumption to exceed its own mandate to fulfill the larger advanced mandate if consumption of renewable diesel is less constrained.

It is assumed in baseline projections that the blend wall is gradually extended from 10% to 15% over the projection period and that the assumed effective blend wall would be reached by 2016.

Further reading

The discussion of US biofuel policy and its implementation are drawn from the following works where additional detail may be found.

Meyer, Seth and Wyatt Thompson. "EPA Mandate Waivers Create New Uncertainties in Biodiesel Markets", *Choices*, Vol. 26 (2), 2011.

Thompson, Wyatt, Seth Meyer and Patrick Westhoff. "Renewable Identification Numbers are the tracking Instrument and Bellwether of US Biofuel Mandates", *EuroChoices*, Vol. 8 (3), pp 43-50, 2009.

Notes

1. The vast majority of cars in the US have gasoline engines while the trucking fleet is dominated by diesel engine trucks.
2. Energy Independence and Security Act of 2007, Public Law 110–140 (2007) *www.gpo.gov/fdsys/pkg/PLAW-110publ140/pdf/PLAW-110publ140.pdf*.
3. Ethanol derived from corn starch is explicitly named as a conventional biofuel but it is not the only conventional biofuel. Other grains could be used to produce ethanol and if a 50% GHG reduction is not achieved the derived ethanol would be considered as a conventional biofuel.
4. EPA clams this authority under sections 205 and 211 of the Clean Air Act *www.epa.gov/air/caa/title2.html*.
5. Including the assumption that the cellulosic mandate will continue to be set by EPA at a reduced volume relative to that legislated in EISA.
6. *www.fas.org/sgp/crs/misc/R41106.pdf*.

7. The *Outlook* baseline for cellulosic biofuel production in the United States is exogenous and dependent on a fixed technology path.
8. Jet fuel consumption represents the remaining 10%, *www.eia.gov/forecasts/steo/report/us_oil.cfm*.
9. In October of 2010, the EPA granted a partial waiver for the use of E15 in model year 2007 and newer vehicles.
10. The mandates are quantitative and do not respond to aggregate motor fuel use. Factors which increase or decrease aggregate motor fuel use, change the effective share of biofuels required in consumption.
11. *http://edocket.access.gpo.gov/2011/2011-1646.htm*.
12. *www.gpo.gov/fdsys/pkg/FR-2011-07-25/pdf/2011-16459.pdf*.
13. National Travel Household Survey (*http://nhts.ornl.gov/download.shtml*) Author's query from data set using NTHS estimates of miles driven by age, self reported miles driven would increase the share of newer vehicle miles to over 81%. The results do not correct for potential differences in miles per gallon based on age of vehicle.
14. Cellulosic biodiesel also qualifies as a cellulosic fuel.

ANNEX 3.A2

Uncertainties around the implementation options of US biofuel policies: Results of the scenarios

Table 3.A2.1. **Results of the three options scenarios**

		Baseline		Option 1	Option 2	Option 3
		Average 2009-2011	2021	2021	2021	2021
Ethanol production						
USA	MN L	47 617	82 610	81 860	89 553	108 960
Brazil	MN L	25 331	51 300	49 625	61 048	52 627
European Union	MN L	6 424	15 748	15 572	17 145	15 986
Canada	MN L	1 565	1 992	1 978	2 135	2 550
China	MN L	8 094	10 058	10 016	10 507	10 146
India	MN L	1 976	4 194	4 174	4 376	4 237
Rest of World	MN L	7 213	14 673	14 598	15 337	14 776
Ethanol use						
USA	MN L	45 582	90 757	86 217	126 462	125 778
Brazil	MN L	23 347	39 805	41 287	25 902	34 467
European Union	MN L	7 877	19 388	19 388	19 388	19 388
Canada	MN L	1 759	2 356	2 356	2 356	2 356
China	MN L	7 994	10 242	10 433	8 905	9 646
India	MN L	2 254	4 384	4 385	4 381	4 383
Rest of World	MN L	8 406	13 460	13 573	12 524	13 076
Energy share in Gasoline type fuels						
USA	%	5.4	10.9	10.4	15.3	15.2
Brazil	%	47.1	64.3	66.8	40.4	55.1
European Union	%	2.7	8.3	8.3	8.3	8.3
Canada	%	2.6	3.4	3.4	3.4	3.4
China	%	1.8	1.3	1.4	0.7	1.0
Ethanol trade						
USA	MN L	1 864	–8 268	–4 479	–37 030	–16 943
Brazil	MN L	1 984	11 495	8 338	35 146	18 160
European Union	MN L	–1 453	–3 640	–3 816	–2 243	–3 402
Canada	MN L	–195	–364	–378	–221	194
China	MN L	100	–183	–416	1 602	500
India	MN L	–278	–190	–211	–5	–146
Rest of World	MN L	–1 205	1 214	1 025	2 813	1 700
Biodiesel						
USA production	MN L	2 834	5 083	5 083	7 571	8 006
USA consumption	MN L	2 546	4 979	4 979	7 515	7 956
USA net trade	MN L	288	104	104	56	50
Prices						
World						
Ethanol	USD/hl	64	96	94	113	102
Biodiesel	USD/hl	132	181	181	184	185
Coarse grains	USD/t	228	246	245	259	286
Raw sugar	USD/t	533	483	482	516	503
Wheat	USD/t	267	279	279	286	294
Oilseeds	USD/t	503	550	549	562	572
Vegetable oils	USD/t	1 067	1 232	1 232	1 256	1 265
Beef and veal (USA)	USD/t	3 477	4 718	4 711	4 780	4 900
Pigmeat (USA)	USD/t	1 658	2 380	2 375	2 434	2 542
Poultry (USA)	USD/t	1 074	1 121	1 119	1 148	1 204
Fish	USD/t	2 500	3 445	3 441	3 484	3 532
USA						
Ethanol	USD/hl	61	77	76	85	108

Note: For the definition of world prices, please refer to footnotes of Table 1.A.2. 30 and 31.
Source: OECD and FAO Secretariats.

Figure 3.A2.1. **Implications of the three options on the US ethanol market**

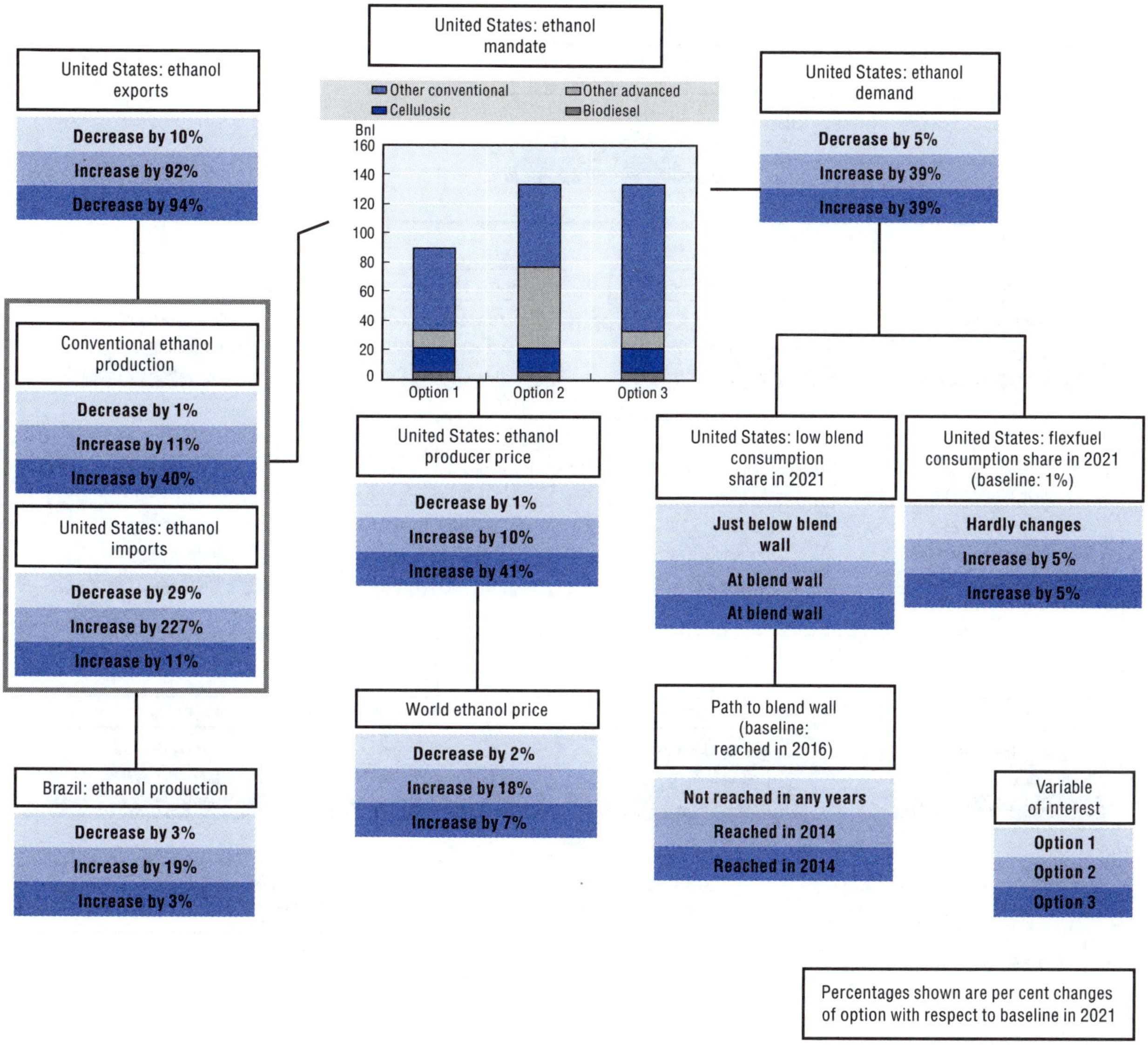

Source: OECD and FAO Secretariats.

Figure 3.A2.2. **Interactions between US and Brazilian ethanol markets**

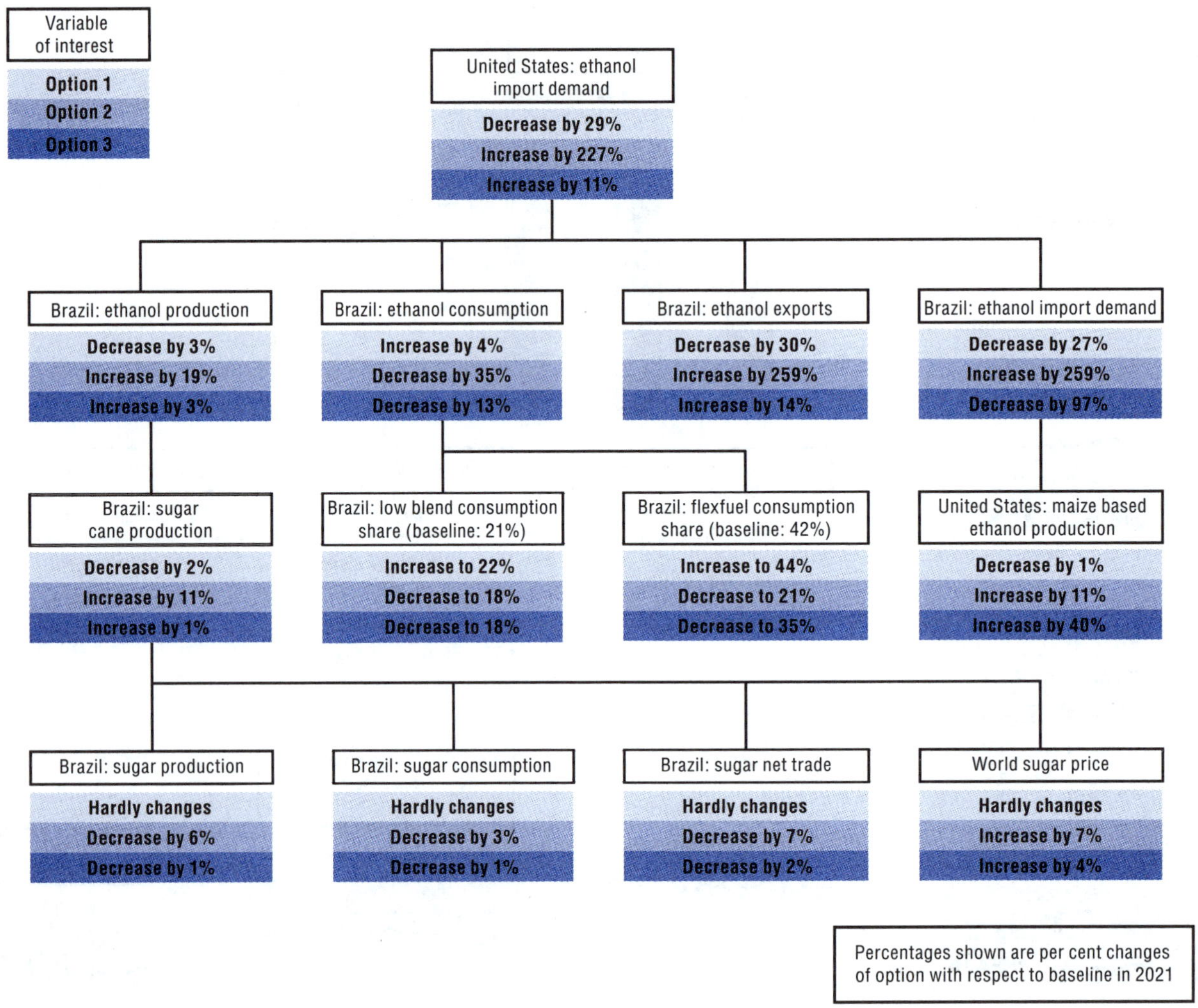

Source: OECD-FAO Secretariats.

Figure 3.A2.3. Impacts on the other agricultural sectors

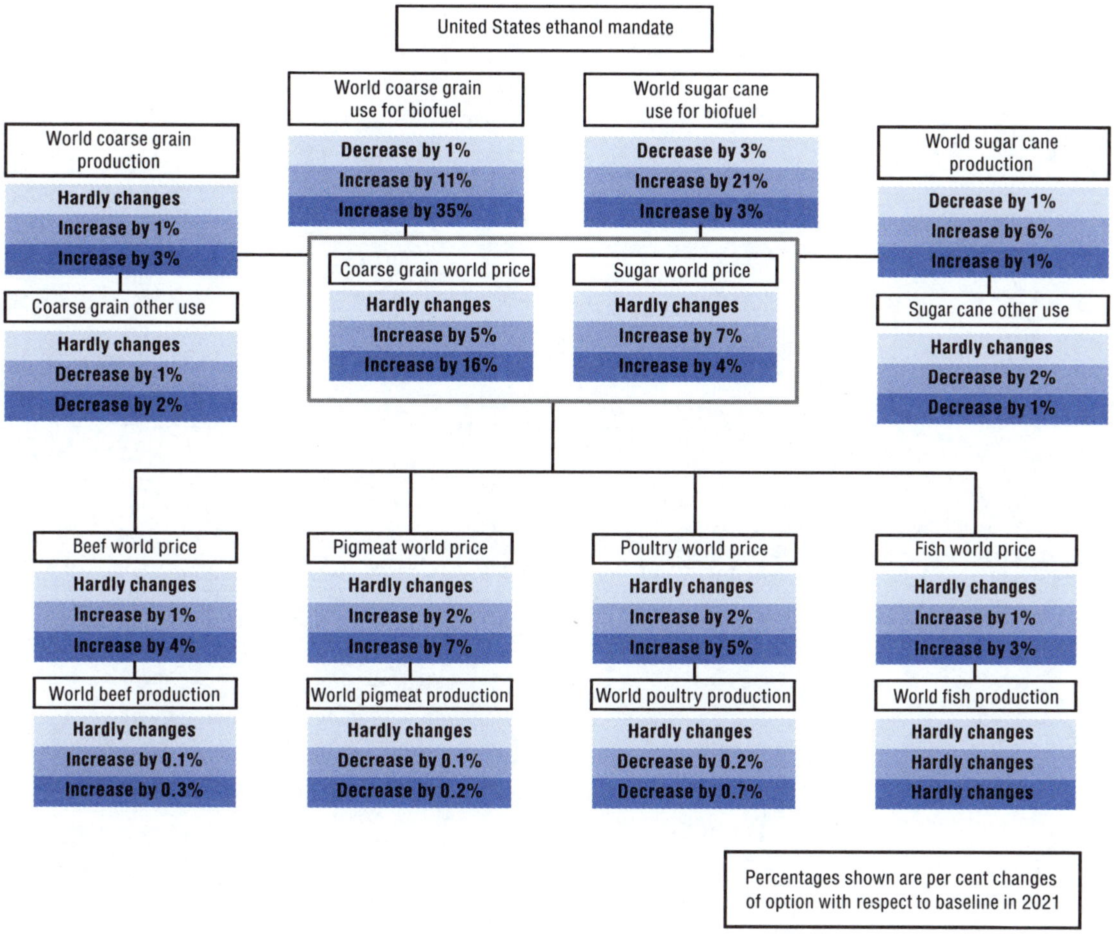

Source: OECD and FAO Secretariats.

OECD-FAO Agricultural Outlook 2012

Chapter 4

Cereals

Market situation

Despite record cereals production in 2011, international prices remained elevated. The bumper crop helped to replenish stocks and moderate prices in the second half of 2011, but a weakening US dollar and low freight rates bolstered export prices in early 2012.

Grain prices were further supported by excessive cold across much of Europe and the Commonwealth of Independent States (CIS), and the excessive dry conditions that prevailed in the southwest United States and portions of South America. A tight maize supply situation, following last year's below trend yields in the United States, contributed to the increase in coarse grain prices while deteriorating conditions of the South American soybean crop put additional upward pressure on markets. Early reports revealed the intention of United States farmers to plant 38.8 Mha of maize, a level not seen since 1937, but price movements during planting showed the market bidding for more oilseeds (soybeans) area which could pull plantings down from the initial intentions. Given the low level of maize inventories, markets are expected to remain sensitive to the eventual size and progress of the maize crop in the United States, the world's largest producer. Despite significant losses due to floods in the second half of 2011, especially in Thailand, outstanding yields in India and several other major producers lifted world rice production in 2011. Rice prices over the year were very much influenced by the implementation of new polices in Thailand and India.

Projection highlights

- By 2021, wheat prices in nominal terms are projected above the previous decade, but below those prices seen during the last two years. Prices in real terms are expected to remain flat to moderately declining from 2012. Coarse grain and rice prices follow a similar pattern with nominal prices dipping in 2012, and then modest growth for the remainder of the period (with declining real prices).
- World cereal production is expected to grow in 2012 as a response to higher returns, increasing gradually during the rest of the projection period. Stocks-to-use ratios will remain below historical averages, potentially adding to price volatility.
- Harvested area for cereals continues expanding, but at a slower pace than in the past. Yield growth prospects are less optimistic and mostly concentrated on maize and rice.
- Trade of wheat and coarse grain increases at a slightly slower pace than in the past. The United States keeps its leading position as maize exporter. The CIS becomes an even more important source of wheat exports by 2021 than in the base period.[1] Trade of rice is expected to increase faster than in the past driven by growing shipments from LDC Asia, in particular Cambodia and Myanmar, and by increasing imports by African countries. Nevertheless, trade in rice remains small compared with other grains.

Market trends and prospects

Prices

The benchmark US wheat export price is expected to average slightly lower in 2011 (USD 279/t for June/May marketing season) from the high 2010 level (USD 301/t) following a rebound in world wheat production in 2011. The benchmark US maize price is also likely to decline in 2011 (USD 246/t), though more modestly compared to wheat. For 2012, wheat prices are projected to further decrease (USD 248/t), remaining at a higher level than in the past, mainly due to additional pressure on input prices, with crude oil prices at USD 110/barrel of Brent. This is also the case for coarse grain prices, with a 2012 price forecasted at USD 245/t. The current situation reveals historical low wheat to maize price differential between 1.01-1.02.

By 2021, wheat prices are projected to approach USD 279/t in nominal terms[2], increasing over the projection period at 1.5% per year, supported by strong energy prices and spillover effects in coarse grain markets. They consolidate at about 15% above the historical average. In real terms, however, wheat prices remain at around the same levels as in the past decade (Figure 4.1).

Similar to wheat, coarse grain prices are projected to increase only in nominal terms over the projection period, to reach USD 246/t by 2021. In real terms, they are projected to stagnate. An expected trend in world markets is a continued narrowing of the price differential between wheat and maize, with the wheat to maize price ratio falling below 1.2 by 2021, compared to 1.4 in the previous decade. The primary driver is an anticipated tighter supply and demand balance for maize relative to wheat. The main reason for this development is the fact that the overall demand for wheat is for food, which tends to be less elastic than demand for maize which is largely for feed and more recently biofuels.

Figure 4.1. **Tight stocks and high input costs underpin prices throughout the projection period**

Evolution of prices expressed in nominal and real terms

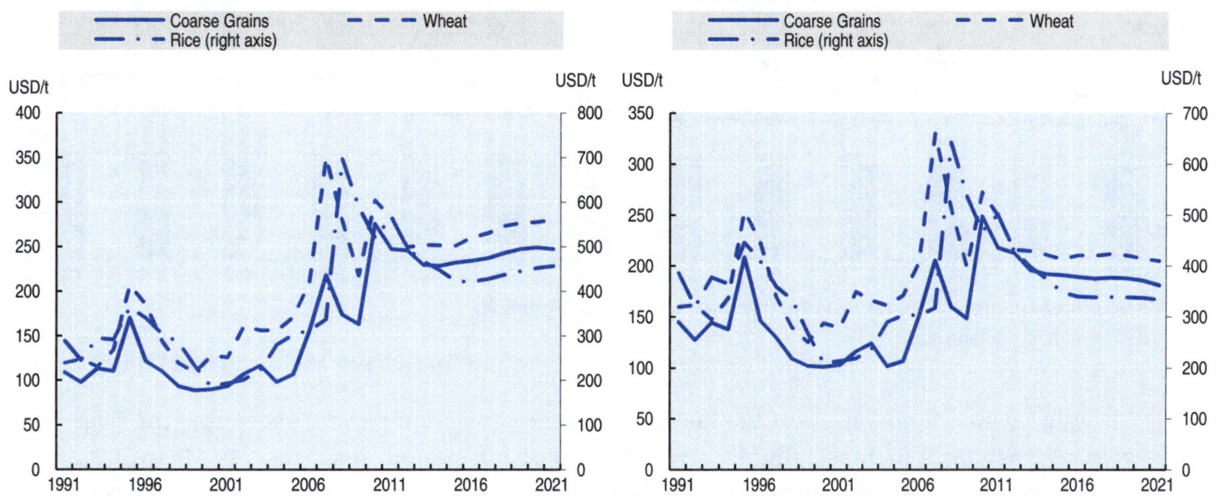

Note: The left figure shows nominal prices and the right figure shows real prices. The world reference price for wheat is the No. 2 Hard Red Winter, USA f.o.b. Gulf Ports. For coarse grains, it is the US maize price No. 2 Yellow, f.o.b. Gulf Ports and for rice, it is the Thai white 100% B, milled, f.o.b.

Source: OECD and FAO Secretariats.

StatLink http://dx.doi.org/10.1787/888932639514

World rice prices started to increase strongly in the summer of 2011 due to the Thai government's new support policy, and then dropped rapidly pressured by the return of cheap Indian rice on the market (Figure 4.1). Rice prices are projected to decline gradually in real terms, reflecting the ample supply of a few rice exporting countries in Southeast Asia combined with slowing import demand, especially by those countries that are targeting rice self-sufficiency. The ratio of rice to coarse grains prices is projected to fall to 1.8, lower than the 2.5 observed in the last ten years, while the rice to wheat ratio is expected hover around 1.6, closer to the 1.8 value seen in the past decade. In nominal terms, the benchmark rice price looks set to fall to USD 454/t in 2021.

Production of cereals

World wheat production is projected to reach 761 Mt by 2021, about 12% higher than in the base period 2009-11, but with slower annual growth relative to the previous decade (Figure 4.3). By 2021, the wheat area is projected to be 3% higher than the base period. The largest area expansions are projected for the Russian Federation, Ukraine and Kazakhstan. Average global yield growth for wheat is projected at only 0.7% p.a., a slowing from historical trends and influenced by lower average yields in regions of area expansion (Figure 4.2). As highlighted in Chapter 2, how to improve productivity is a critical issue to be addressed in the coming years.

Figure 4.2. **Less expansion in area while yield growth also slows**

Evolution of global cereal harvested area and yields over the projection period

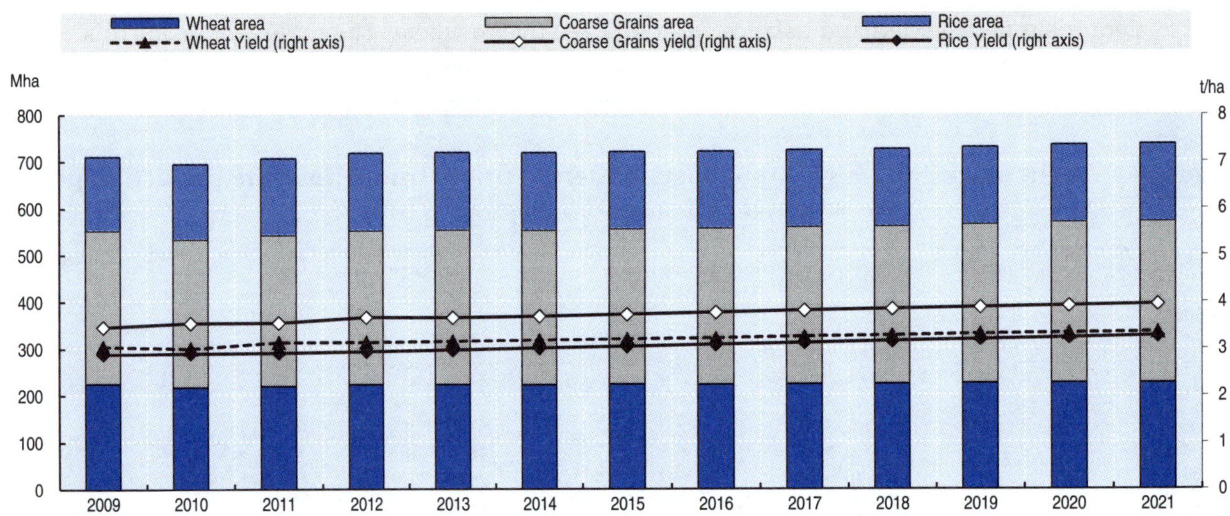

Note: The first three columns (2009, 2010 and 2011) include historical information.

Source: OECD and FAO Secretariats.

StatLink http://dx.doi.org/10.1787/888932639533

Wheat stocks are ample at the beginning of the projection period, exceeding the five-year average, and increase slightly over the projection period, reaching 219 Mt in 2021. Most of the build-up with respect to the base period is expected to occur in CIS countries. At this level, the ratio of wheat stocks of major exporters to its utilisation will approach 32% in 2021, which is slightly below the base period but 10 percentage points higher than in the 2007/08 food crisis period. Similarly, the ratio of major exporters' wheat stocks-to-disappearance (*i.e.* defined as the domestic utilisation plus exports in major exporting countries[3]) is projected to approach 21% (Figure 4.3).

Figure 4.3. **World wheat production and stock ratios**

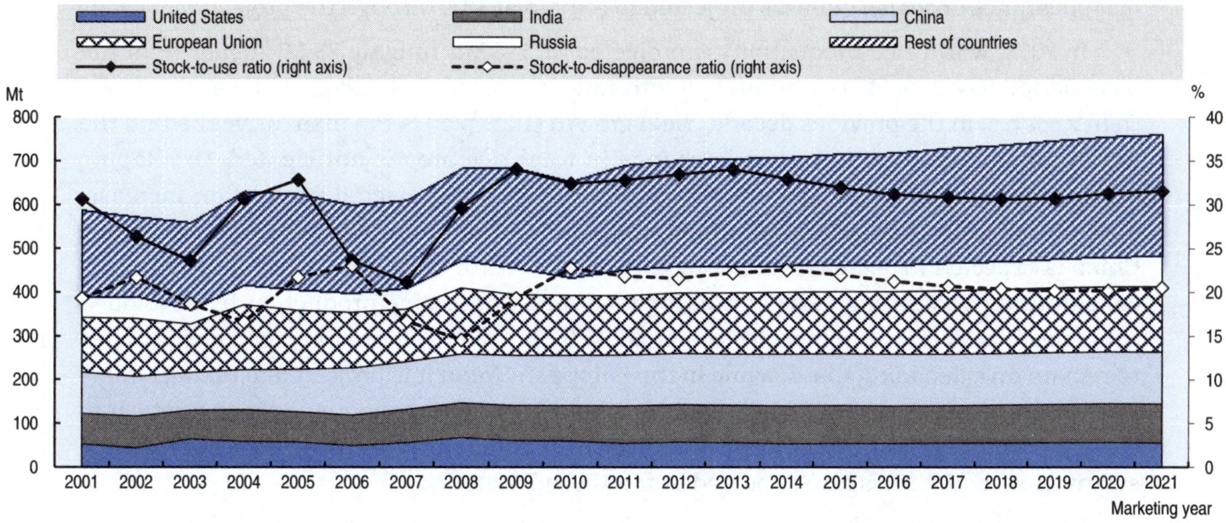

Source: OECD and FAO Secretariats.

StatLink http://dx.doi.org/10.1787/888932639552

World coarse grain production is expected to reach 1 359 Mt by 2021, up 20% from the base period (Figure 4.4), with significant increases projected for Argentina, Brazil, China, the Russian Federation, Ukraine, and the United States. The increase in the total coarse grain area is projected to be more significant than other crops by 2021, up 7% from the base period, with notable increases in Brazil, Argentina and Canada, as well as several Sub-Saharan African countries. Coarse grain yields are projected to increase by 0.9% p.a., well below historical trends (Figure 4.2).

World stocks of coarse grains are projected to rise by 12% from its critically low level of 205 Mt in the base period. Most of the build-up is projected to occur in the United States and Brazil. The ratio of major exporters' coarse grains stocks to its utilisation is projected

Figure 4.4. **World coarse grain production and stock ratios**

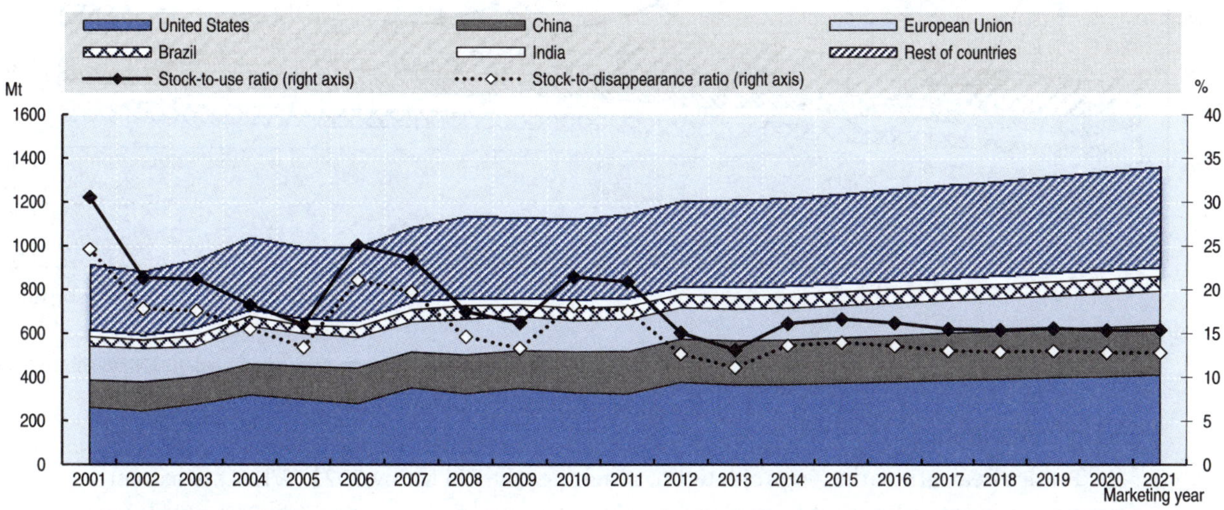

Source: OECD and FAO Secretariats.

StatLink http://dx.doi.org/10.1787/888932639571

to slightly recover to 16%. More importantly, the ratio of major exporters' stocks-to-disappearance of coarse grains is projected to remain at 13% through the projection period.

In 2021, world rice production[4] is projected at 542 Mt, roughly 75 Mt higher than the base period (Figure 4.5). The annual growth rate is projected at 1.2%, significantly slower than 2.5% p.a. in the previous decade. Yield growth (1.2% p.a.) is the main driver behind the global production increase, as little change in total rice area is anticipated. Developing countries are expected to account for virtually all of the projected production increase, particularly India, Cambodia, Myanmar and African countries. Among large producers, China is expected to decrease output by 6 Mt, as the sector responds to declining domestic consumption and strong competition for land and water. Rice production by developed countries is anticipated to increase by a modest 0.3% p.a. Rice output in Japan and Korea is to remain on a declining trend, while in the European Union it is projected to remain stable around the base year level (1.7 Mt) over the whole projection period. Assuming a return to a more favourable rainfall pattern, production is anticipated to recover in Australia. A steady growth of 1.3% is also expected in the United States.

World rice stocks have been increasing strongly since 2008, boosted by positive production outcomes and the need by a few governments to maintain increased rice reserves for their public distribution systems. Rice stocks peak early in the projection period at just under 157 Mt in 2013, then drift moderately lower through the remainder of the projection, reaching 145 Mt in 2021, mainly driven by a contraction in China and India from the extraordinarily high inventory levels held in recent years. The major exporters' stock-to-use and stock-to-disappearance ratios continue to decrease all through the projection period.

Figure 4.5. **World rice production and stock ratios**

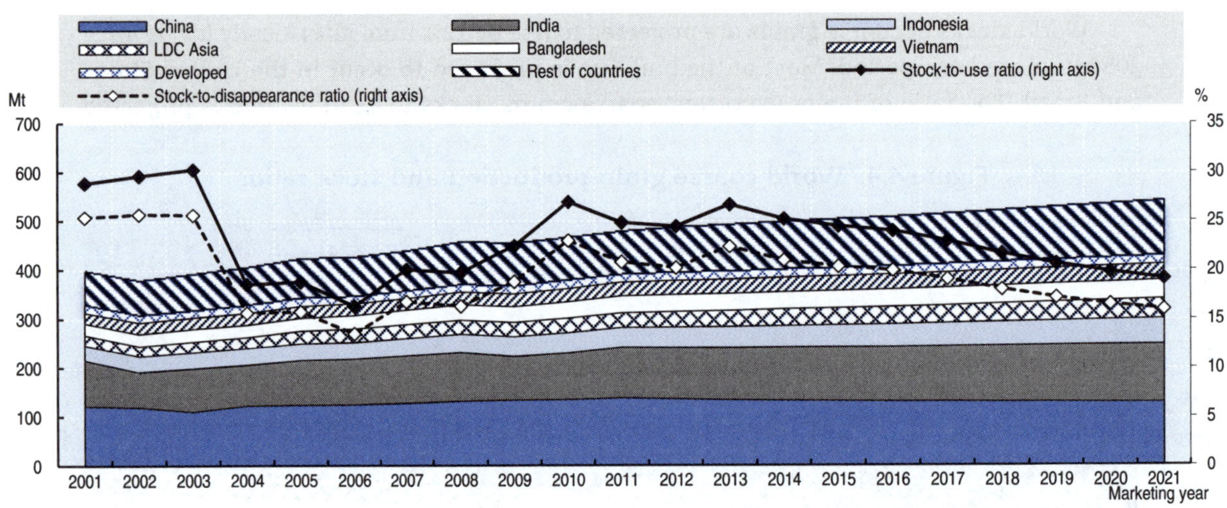

Source: OECD and FAO Secretariats.

StatLink http://dx.doi.org/10.1787/888932639590

Use of cereals

Total wheat utilisation is projected to reach nearly 755 Mt by 2021. Wheat is expected to remain a commodity that is predominantly consumed for food, roughly 67% of total use by 2021, 2% below its current share (Figure 4.6). *Per capita* food consumption is projected to

Figure 4.6. **Wheat consumption in developed and developing countries**

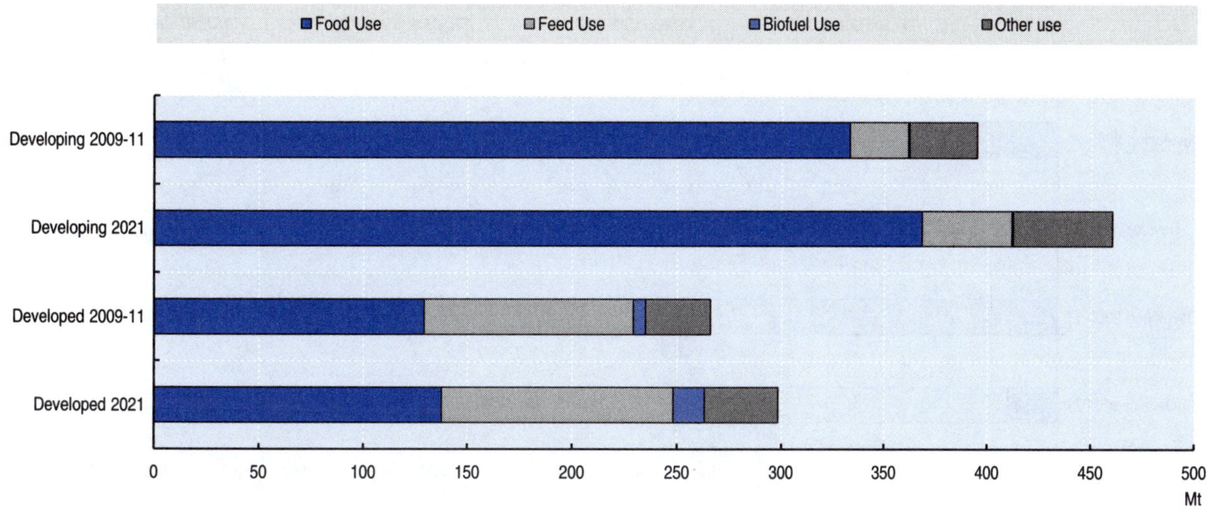

Note: In "other use", we include other non-disaggregated industrial demand sources (*e.g.* processing of starch or straw).
Source: OECD and FAO Secretariats.

StatLink http://dx.doi.org/10.1787/888932639609

remain steady at around 65 kg per person p.a. World wheat feed utilisation is expected to reach 154 Mt by 2021, growing at a slightly slower pace than during the historical period, though still representing around 20% of total use (37% in developed countries and 10% in developing countries). Wheat use for biofuels in developed countries increases from 0.09% in the base period to 2.1% of world wheat utilisation by 2021. The projected increase of 7% p.a. will be driven largely by growth in EU wheat-based ethanol production which, by 2021, may account for almost 83% of global wheat use for biofuel production (compared to 68% in the base period).

World utilisation of coarse grains is projected to increase by 19% by 2021, compared to the base period and reach 1 350 Mt, driven largely by expansions in demand for feed and biofuel. The projected annual growth (1.5%) is less than observed over the previous decade (2.7%) (Figure 4.7). Food use is projected to reach 239 Mt, up 18% from the base period and resulting in a small increase in *per capita* level to around 31 kg. Total coarse grains feed use is projected at 722 Mt, up 20% from the base period, mostly driven by strong growth in the Russian Federation (58%), China (47%), and the United States (24%). Among its industrial usages, maize-based ethanol production in the United States is projected to continue expanding until 2015, following the ethanol mandate, and then decreasing slightly until the end of the projection period. This demand path is dependent on the assumptions on policy implementation discussed in Chapter 3 on biofuels and remains a source of uncertainty. World use of coarse grains for production of biofuels is projected to reach 185 Mt, 34% more than in the base period and representing 13.6% of total world coarse grains production (within the United States, the share of maize used for ethanol production goes up to 38.6% of total maize domestic production).

World rice utilisation, consumed chiefly as food (85%), is set to reach 542 Mt in 2021, up from 460 Mt in the base period. The increase is driven mainly by population growth, while *per capita* food consumption is also anticipated to rise slightly from 56.7 kg in 2009-11 to 59.9 kg in 2021. The annual growth rate in aggregate consumption over the projection period is 1.2%, down from 1.7% in the past ten years.

Figure 4.7. **Coarse grains consumption in developed and developing countries**

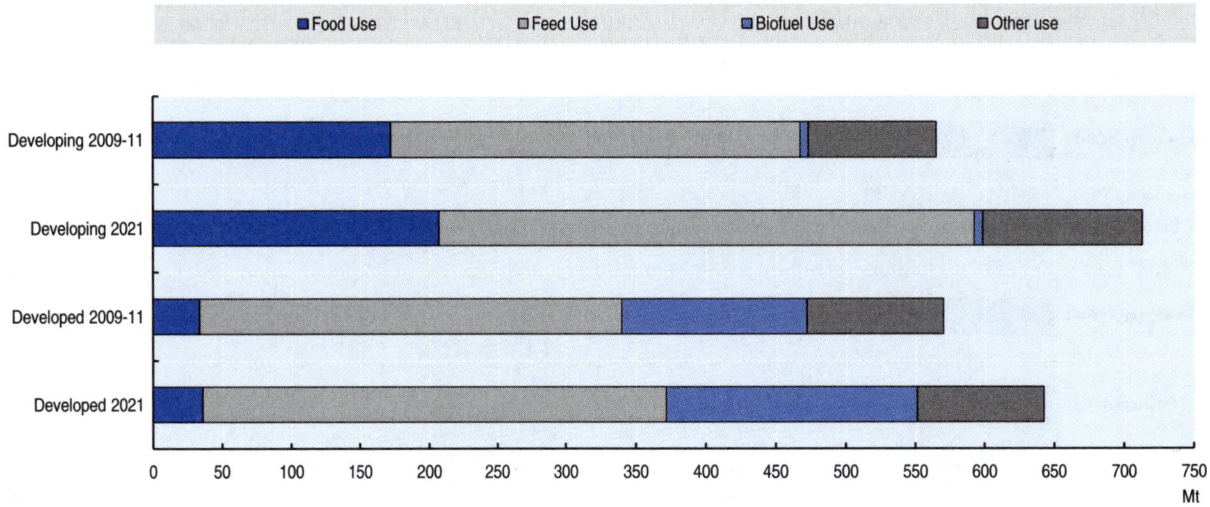

Note: In "other use", we include other non-disaggregated industrial demand sources (*e.g.* production of high fructose corn syrup).
Source: OECD and FAO Secretariats.

StatLink http://dx.doi.org/10.1787/888932639628

Trends of rice consumption vary across regions. In Southeast Asia, where rice is a major food staple, consumption is projected to expand by 1.1% to 2.5% p.a, often spurred by vigorous population and income growth. Rice consumption in Africa is anticipated to show a significant overall increase of 52% by 2021 from the base period, equivalent to 3.7% growth p.a, due to relatively strong population growth and continued changes in diet towards rice. The reverse is foreseen in China, where total and *per capita* consumption are projected to decline. Consumption by developed countries looks set to rise to 20 Mt of rice by 2021, or 0.5% per year. With only a 4% share of total consumption, the contribution of developed countries to global consumption growth will continue to be small.

Trade of cereals

World trade of wheat and coarse grains in 2021 is projected to expand to 298 Mt, 42 Mt (19%) higher than in the base period. This sharp expansion in global cereal trade is mostly on account of a surge in wheat exports, which is anticipated to approach 152 Mt, 17% higher than in the base period. This is boosted by the recovery of exportable supplies of grains in the CIS and the stronger import demand for feed wheat as a result of tight domestic supplies of coarse grains. The largest increases in wheat imports are expected in China, the European Union, Indonesia, and the Islamic Republic of Iran. World trade in coarse grains is projected to reach 146 Mt, 20% higher than the base period. Whereas the Russian Federation is projected to achieve the highest export share of wheat in 2021 (17%), the United States keeps its leading position in coarse grains markets (43%) (Figure 4.8). Kazakhstan increases its share in world wheat trade but does not fully exploit its potential, primarily for reasons relating to institutional and infrastructure constraints (Box 4.2). Traditionally high production variability in CIS countries due to changing weather conditions may have implications for global trade and world price volatility.

Figure 4.8. **Wheat (left) and coarse grains (right) export shares in 2021**

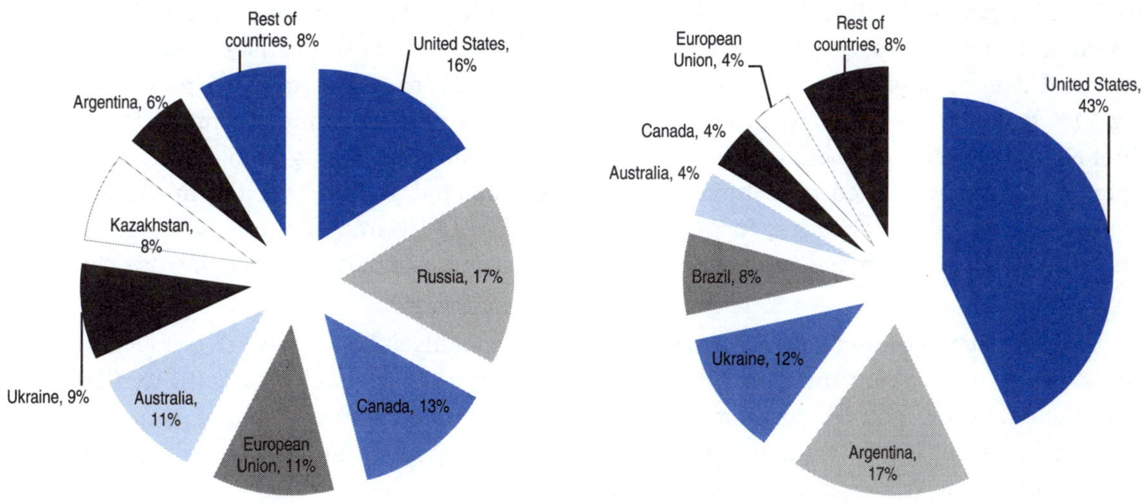

Source: OECD and FAO Secretariats.

StatLink http://dx.doi.org/10.1787/888932639647

A sharp rise in imports by China as well as higher imports by Japan, Mexico and Saudi Arabia are expected to be offset by reductions in imports by the European Union and Korea. China and India together account for 20 Mt of grains[5] imports in 2021 (Figure 4.9), experiencing an increasing gap between supply and demand over the projection period.

Figure 4.9. **Production, demand and net trade projections for coarse grains in China (left) and wheat in India (right)**

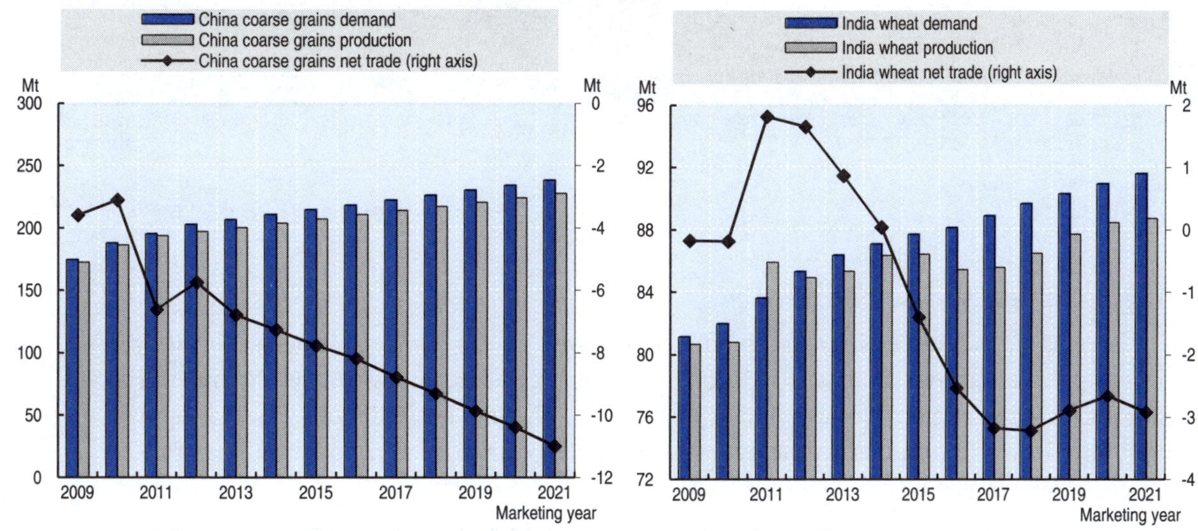

Source: OECD and FAO Secretariats.

StatLink http://dx.doi.org/10.1787/888932639666

Rice trade is projected to increase by 1.9% p.a, going from 33 Mt to 43 Mt between 2011 and 2021 (Figure 4.10). Global rice trade is likely to be fuelled by growing imports by African countries, where the production gains are unlikely to be sufficient to satisfy the growing demand. Among large importing countries, the Philippines is expected to reduce purchases by 21% by 2021, assuming it maintains its policies favouring domestic production. Viet Nam is projected to be the largest rice exporter by 2021, displacing Thailand's higher quality and higher priced rice. India's exports are projected at 5 Mt over the projection period. Other Asian countries, in particular Myanmar and Cambodia, are expected to make major inroads in the international rice market, with exports growing by 12% p.a. over 2012-21.

Exports by the United States are expected to grow steadily at 2.5% p.a. With production stagnating, imports by the European Union are projected to expand, facilitated by preferential access granted to least developed countries under the Everything But Arms (EBA) programme. Little change is foreseen in imports by Japan and Korea, which will largely reflect the WTO minimum quota provisions.

Figure 4.10. **World rice exports projections**

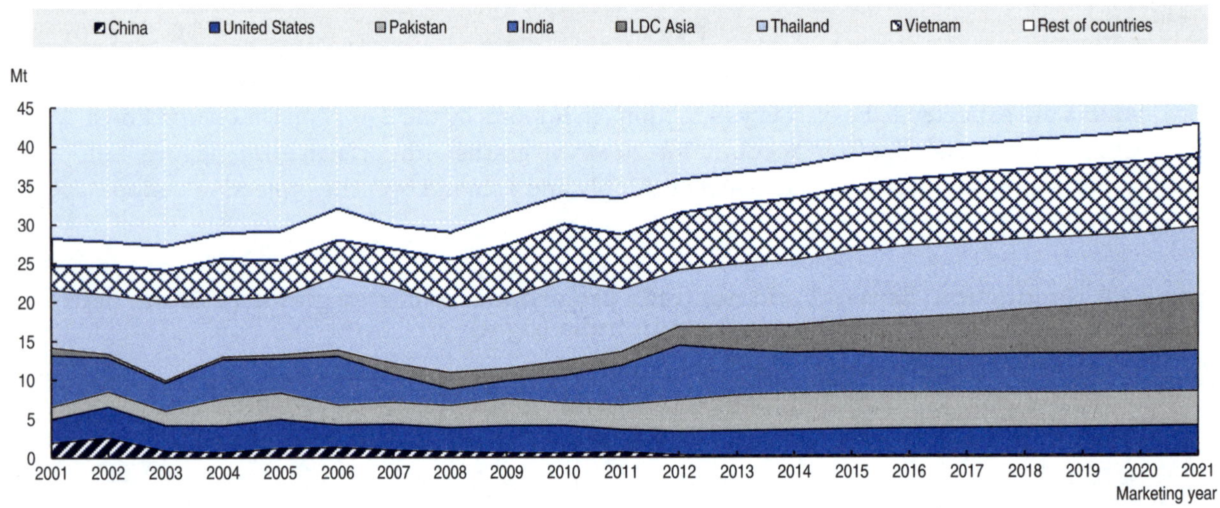

Source: OECD and FAO Secretariats.

StatLink http://dx.doi.org/10.1787/888932639685

Main issues and uncertainties

World cereal stocks increase in the early years of the projection period under the assumption of normal weather conditions and average yields, but overall world inventories are projected to remain low in relation to the past. The low level of stocks increases price volatility in world markets, in the event of sudden or unexpected production shortfalls in major producing countries.

Another source of uncertainty is the level of production at any given year. In view of the fact that most crop prices are projected to remain strong, competition for land is likely to intensify with planting decisions shaped by the inherent inter-seasonal price volatility (*e.g.* maize and soya in the United States), which in turn will contribute to unexpected changes in production levels.

The evolution of crude oil prices, affecting the economic incentives to use biofuels, is an important assumption in the baseline again this year. Oil prices are assumed to increase steadily, rising above USD 140 per barrel by the end of the projection period. This increase in oil prices provides upward support for both maize and sugar prices, which may move more closely together than they have historically.

Policies remain as an important source of uncertainty. Any significant changes in biofuel policies in the United States, the European Union, and Brazil could have significant impact on the global cereal market, in particular on maize. Moreover, trade policy behaviour in exporting countries adds uncertainty to cereal markets in periods of natural yield fluctuations, as experienced recently in wheat and rice. Policy interventions in the rice economy are frequent and play a critical role in influencing international rice trade and prices (Box 4.2). Although several LDC Asian countries are projected to increase their shares in world exports, this is unlikely to translate in a lessening of government interference, which means that changes in policies could significantly alter the rice projections presented in this chapter.

Box 4.1. **Competitiveness of the wheat sector of Kazakhstan and constraints for future productivity growth**

Kazakhstan has a large land area, with comparatively dry conditions and inherent soil fertility for wheat production. Since the Soviet era ended in 1991, the economy of Kazakhstan has undergone rapid structural change and agriculture is experiencing strong inter-sectoral competition. Currently, Kazakhstan is the dominant wheat producer in Central Asia and ranks third, behind the Russian Federation and Ukraine, in wheat production amongst CIS countries. Roughly one-quarter of arable land under cash crops in the Kazakhstani wheat belt belongs to agro-holdings, *i.e.* large-scale operators holding more than 100 000 ha each. Moreover, Kazakhstan is the most important high protein wheat producer in Central Asia.

Figure 4.11. **Wheat export projections from major exporters**

Historical data: 2005-11, baseline projections: 2012-21

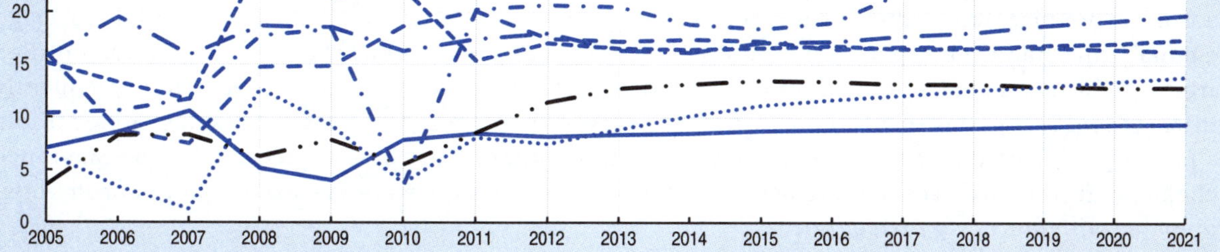

Source: OECD and FAO Secretariats.

StatLink http://dx.doi.org/10.1787/888932639704

Box 4.1. Competitiveness of the wheat sector of Kazakhstan and constraints for future productivity growth *(cont.)*

As shown in Figure 4.11, wheat exports are expected to increase in the short term due to higher expected yields and area expansion, due mainly to conversion of marginal lands, driven by good market returns. However, this trend is not expected to continue after 2015 due to different limiting factors, and its export volume will be surpassed by Ukraine at the end of the projection period. Productivity growth in Kazakhstan is currently hampered by several constraints. First, the lower intensity of agriculture is probably driven largely by the dry climate and risks of drought. At the beginning of the 2000s, the Kazakhstan government adopted a legislative framework that was favourable to the leasing of agricultural machinery and provided privileged access to farmers. This allowed wheat growers to massively renovate their machinery fleet, implement new growing techniques, and consequently significantly improve yields and productivity. Nevertheless, only the biggest growers can actually afford foreign consultants and agronomists to fully utilise new technologies. For medium and small farms, access to knowledge and information remains limited to the recommendations provided by input suppliers and national extension system, which is at the initial development stage. Second, Kazakhstan exporters are constrained by high transportation costs and disadvantages in accessing Black Sea port facilities – the key point for wheat delivery to North Africa and Europe. The Central Asian market is already saturated by Kazakhstan wheat and flour, with future gains limited to the population growth in the region. Exports to the East (South-East Asia, Korea and Japan) look promising, but transportation costs and infrastructure limitations do not allow traders to effectively compete with other world wheat suppliers. In addition, access to finance and proper financial management is still an issue for the small- and medium-sized farms. The lack of turnover capital does not allow them to increase input use and store wheat while waiting for better prices.

In short, there is considerable scope to improve the on-farm productivity of wheat in Kazakhstan. The potential methods for improvement can be categorised as: varietal improvement and related seed management, agronomic and crop management improvements; and institutional and infrastructure improvements.

Box 4.2. The Impact of Thailand's new rice support policy

Thailand is the leading world rice exporter and Thai white Rice, 100% B is a standardised grade and quality which acts as a good reference on which to base or discount prices of other types and grades of rice in international trade. The Thai white Rice, 100% B is also the representative international rice quotation used to equilibrate the rice module of the OECD-FAO medium-term projection model.

Thai prices are influenced by government policies. In 2011, the new government, abiding by its election campaign promise, reintroduced the rice pledging programme (also referred to as the "rice mortgage scheme") that had been replaced by the Price Insurance Scheme in 2010/11. The programme entitles farmers to pledge unlimited quantities of paddy rice as collateral for loans, which they can subsequently either redeem by repaying the loans, or forfeit to the government. In 2011/12, the programme offered prices to paddy producers that were substantially above those guaranteed by the precedent price insurance scheme, with the increases varying between 25-60%. The markups were even more pronounced when compared with prevailing market prices.

Box 4.2. The Impact of Thailand's new rice support policy *(cont.)*

Table 4.1. Thailand: Support prices under Paddy Pledging Programme and Price Insurance Scheme

	Paddy Pledging Programme (2011/12)		Price Insurance Scheme (2010/11)	
	Baht/tonne	USD/tonne	Baht/tonne	USD/tonne
White rice	13 800-15 000	446-484	11 000	355
Glutinous	15 000-16 000	484.517	10 000	323
Pathum Thani	16 000	517	11 500	371
Provincial fragrant	18 000	581	14 300	462
Hom Mali	20 000	646	15 300	494

Source: Thai Department of Internal Trade, Ministry of Commerce/USDA.

StatLink http://dx.doi.org/10.1787/888932640559

Figure 4.12. Thailand: Rice exports in 2010 and 2011

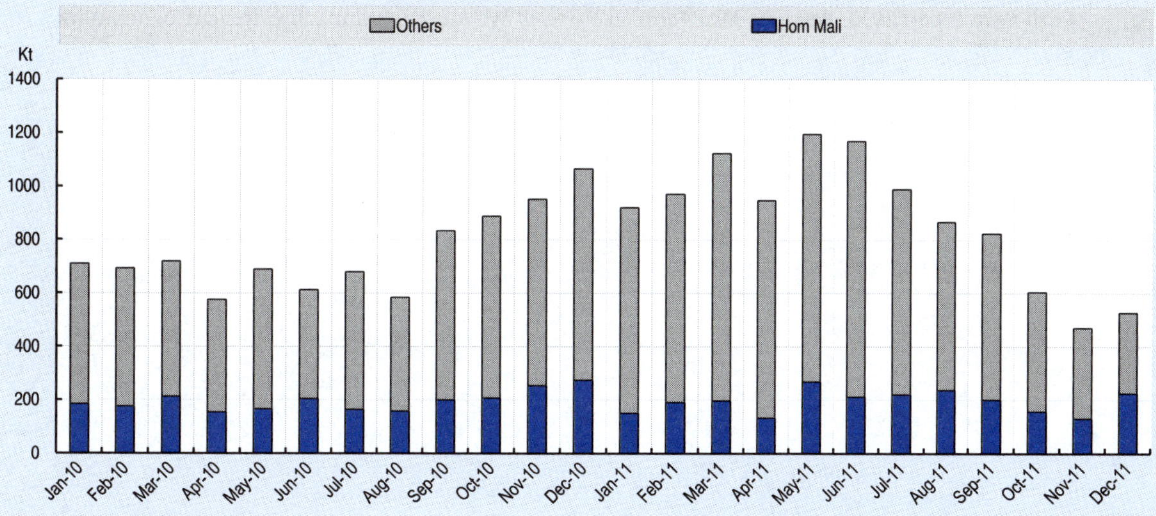

Source: Board of Trade of Thailand.

StatLink http://dx.doi.org/10.1787/888932639723

The announcement of the new support price levels had the effect of supporting Thai export prices in July 2011, even before the implementation of the pledging programme in October. As of September 2011, the Thai white Rice, 100% B was quoted at USD 618 per tonne, up from USD 500 per tonne in May. The measure also had the effect of driving up rice prices from other origins.

The renewed vigour of the rice market began to falter in September when the Indian government suspended the restrictions it had imposed on regular (as opposed to basmati) rice exports since 2008, following the surge in world prices. Deliveries of rice by India surged, taking much of Thailand's market share. Amid growing competition, several other major exporters, in particular Pakistan and Viet Nam, also adjusted down their price quotations closer to the levels offered by Indian traders. This move also eventually depressed the Thai benchmark price, which had retreated 18% by February 2012, compared with its September level. Rice shipments from Thailand fell heavily in the second part of 2011 and beginning of 2012.

Notes

1. Average years, 2009-11.
2. It is important to note that this is an annual average price, and therefore does not reflect intra-annual price variations.
3. The eight major wheat exporters are considered; namely the United States, Argentina, the European Union, Canada, Australia, the Russian Federation, Ukraine and Kazakhstan. In the case of coarse grains, the countries included are the United States, Argentina, the European Union, Canada, Australia, the Russian Federation, Ukraine and Brazil, and for rice Viet Nam, Thailand, India, Pakistan and the United States.
4. All the quantities for rice are on a milled rice basis.
5. This consists of wheat and coarse grains, but does not include rice.

References

Longmire J. and Moldashev A. (1999), *Changing Competitiveness of the Wheat Sector of Kazakhstan and Sources of Future Productivity Growth*, Working Paper 99-06, International Maize and Wheat Improvement Center.

Oshakbayev, Dauren (2010), "Kazakhstan – Agricultural land of opportunity", in Agri benchmark (ed.), *Cash Crop Report 2010. Benchmarking Farming Systems Worldwide*, Braunschweig: agri benchmark, pp. 44-47).

OECD-FAO Agricultural Outlook 2012

Chapter 5

Oilseeds and Oilseed Products

Market situation

In mid-2010, international prices in the oilseed complex embarked on a new upward trend reflecting a progressive tightening of global supplies, combined with steady growth in demand for oils and meals. Spill-over effects from increasingly tight grain markets contributed to this development.

As the 2010 crop season drew to a close, it emerged that supply and demand tightness would continue and possibly intensify during 2011. Although setting out with relatively ample carry-in stocks, the new season began with only a marginal increase in total oilcrop production as increased competition for arable land between oilseeds and grains affected plantings. Adverse weather conditions also impaired the development of several oilcrops, in particular soybeans in Latin America.

With global production not adequate to satisfy demand, a drawdown in global inventories and a drop in global stock-to-use ratios became inevitable. Consequently, after an intermediary period when prices relaxed, quotations in the oilseed complex started firming again at the beginning of 2012. This also resulted because of concerns that competition for land between soybean and maize could be repeated in 2012/13.

Projection highlights

- Strong demand for food, feed and biofuel feedstock combined with high production costs underpin a sustained increase in nominal prices of oilseeds, protein meals and vegetable oils over the projection period.
- Compared to the 2009-11 average, world oilseeds production is expected to expand by only 20% over the coming decade. High costs, environmental constraints and sustained profitability of competing crops limit growth to only about half the rate observed over the previous decade. Production growth is based equally on continued area expansion and yield improvements.
- Oilseeds production and exports continue to be dominated by traditional players, but emerging exporters, such as Ukraine and Paraguay, are expected to increasingly contribute to global export growth. While South American soybean producers continue to dominate global meal exports, Indonesia and Malaysia expand their share of vegetable oil exports to over 60%. Imports of oilseeds and products are less concentrated, yet China and the European Union remain the dominant importers.
- Significant growth in biodiesel use is expected in developed and developing countries. However, food consumption stagnates in the developed world while *per capita* annual food use in developing countries is expected to expand by 2 kg or 12% over the next ten years, leaving it still at only about three-quarters of the level currently found in developed countries.

Market trends and prospects

Prices

Given crude oil price levels, biofuel mandates and sustained demand for food uses of vegetable oil and for oilseed meal, oilseed and oilseed products prices are expected to increase in nominal terms over the projection period. Real prices are expected to weaken, though remaining at a higher plateau in historical terms (Figure 5.1).

Prices for oilseeds are projected to increase in nominal terms by 9% over the outlook period, significantly more than the rise anticipated for coarse grains and wheat, with which oilcrops directly compete for arable land. The price projections suggest a limited supply response in favour of oilseeds relative to competing crops.

Prices for vegetable oils, after having adjusted from their recent peak, are projected to first remain flat and then accelerate, growing faster than in the case of seeds and meals. The biofuel dimension is expected to continue to be an important market driver and the anticipated high crude oil price should contribute to the projected strength in vegetable oil prices.

The oilseed meal price projections are assumed to show a more steady development, driven by constant growth in meat demand. As meals are an important component in intensive livestock rearing, meals and meat prices tend to follow a similar trend, especially in the case of pork.

Figure 5.1. **Oilseeds prices remain at higher plateau**

Evolution of prices expressed in nominal terms (left) and in real terms (right)

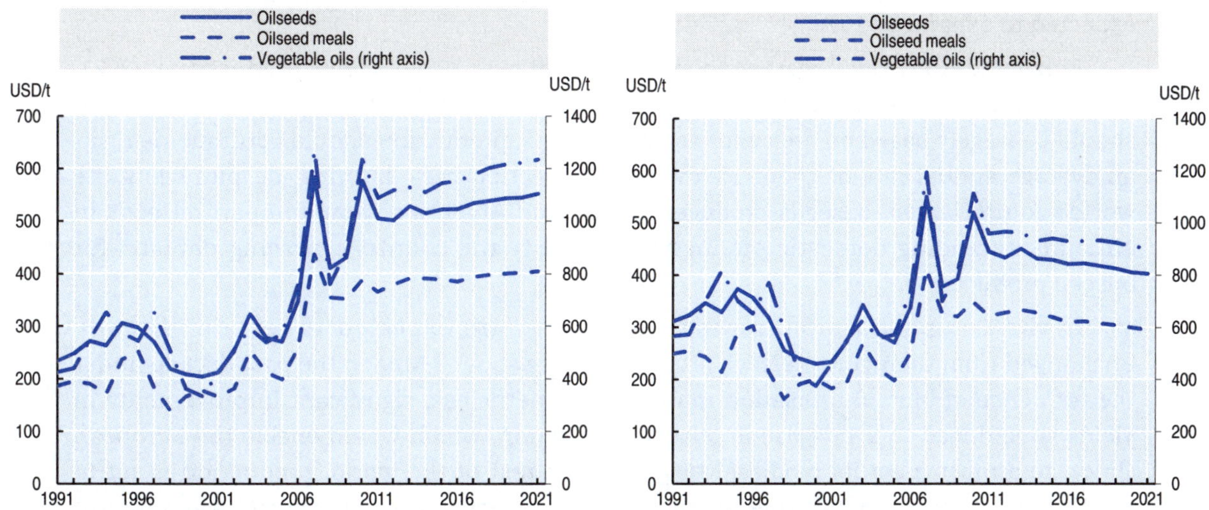

Note: Oilseeds: Weighted average oilseed import price, Europe. Oilseed meals: Weighted average oilseed meal import price, Europe. Vegetable oil: Weighted average export price of oilseed oils and palm oil, Europe

Source: OECD and FAO Secretariats.

StatLink http://dx.doi.org/10.1787/888932639742

Oilseed output and crush

Compared to the 2009-11 average, world oilseeds production is expected to expand by about 21%, exceeding 529 Mt by 2021, thanks to higher area and yield levels and based on sustained demand for oilseeds products. Compared to the past decade, growth is expected to slow down markedly, mostly on account of a deceleration in area expansion due to high marginal costs of planting increases, environmental constraints and sustained profitability of competing crops (Figure 5.2).

Figure 5.2. **Moderate growth in global oilseeds production**

Evolution of global oilseed production over the projection period

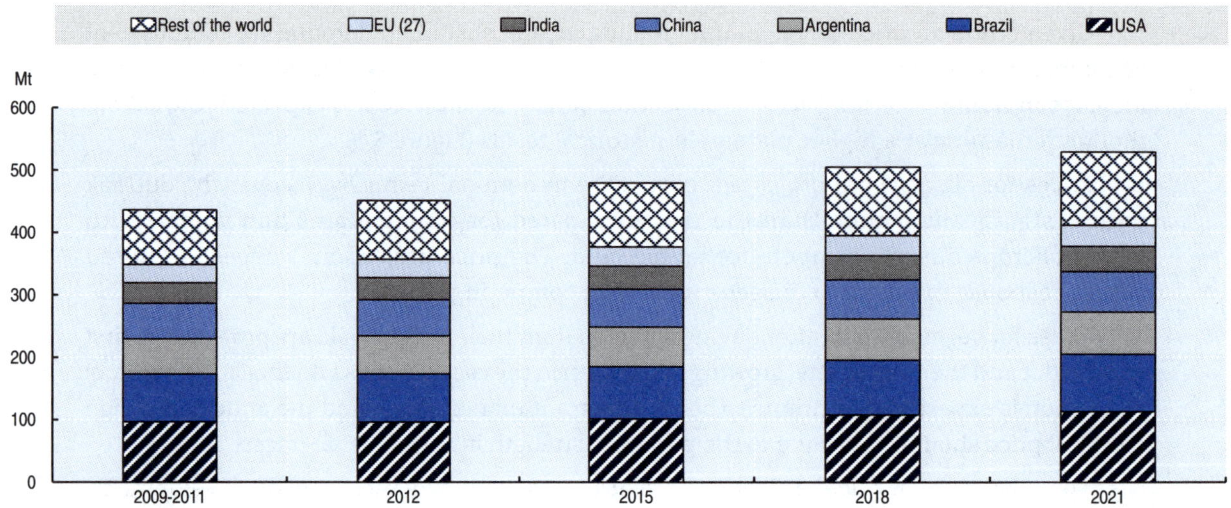

Source: OECD and FAO Secretariats.

StatLink http://dx.doi.org/10.1787/888932639761

The United States is expected to remain the world's leading oilseed producer. Compared to the 2009-11 base period, the area planted to oilseeds in the United States is projected to expand by 7% by 2021, partly at the expense of wheat plantings, and yields are expected to expand by 10%.

Brazil and Argentina together should represent over 30% of global production in 2021. Production in these two countries is expected to rise by 20%. In the case of Brazil, growth should be largely based on area expansion. In China, production should continue rising at an average annual rate of about 1% over the projection period. China, despite being the world's fourth largest oilseeds producer just behind Argentina, would import oilseeds in amounts exceeding domestic production to satisfy the country's growing demand for oilseed products.

In the European Union, rapeseed plantings are expected to remain virtually unchanged from their current level, after increasing about 3 Mha in the past decade under the influence of biofuel utilisation mandates. Nonetheless, significant improvements in yields should lead to a 15% rise in oilseed production, which would enable the European Union to almost meet its biodiesel target. In the rest of the world, production is set to expand by as much as 34% due to traditional producers like Canada and a number of emerging, fast-growing countries, such as Paraguay, Ukraine and the Russian Federation. Overall, global oilseed production is projected to remain fairly concentrated, keeping the world market vulnerable to production shortfalls in major producing areas.

Based on the projected smaller rate of growth in global oilseed production, annual growth in world oilseed crush is expected to be half of what it was in the previous decade. This, in absolute terms, translates into an expansion of 73 Mt over the outlook period (compared to double that amount during the last ten years). The slowdown would involve most locations with the exception of the United States. The largest expansion in absolute terms should occur in Asia. Although China will experience further growth, expansion there is projected to be much lower than previously.

In line with the projected strength in prices, global stock-to-use ratio is expected to decrease from its 2011 level of 8% in the early years of the outlook period to about 7.1% by 2021.

Vegetable oil production and consumption

World vegetable oil production is expected to increase by 35 Mt compared to 2011 or about 28% over the outlook period – a rate exceeding that anticipated for annual oilcrop production, due to the contribution of two perennial oilcrops, palm and coconut oil. Global vegetable oil production should remain geographically concentrated, with a relatively small number of production centres (Indonesia, Malaysia, China, the European Union, United States, Argentina, Brazil and India) accounting for 79% of total output.

Indonesia and Malaysia are set to remain the world's two largest producers accounting for, respectively, 20% and 14% of global oil output in 2021. Over the next ten years, their combined palm oil output is projected to increase by 37% or 12 Mt. As a result, palm oil production is expected to account for one-third of global vegetable oil production in 2021. Compared to the last decade, however, palm oil production would grow considerably less, mainly reflecting possible limitations to area expansion in Indonesia and increasingly binding labour shortages in Malaysia. In Argentina and Brazil, where much of the growth in soybean oil production occurs, output levels are projected to exceed the 2009-11 average by more than 36%, which slightly lifts the share of the two countries in global output. In China, the European Union and the United States, output would expand by between 19 and 24%. The three countries' combined share in global production is expected to remain stable over the projection period. In China, vegetable oil production continues to rely on both domestically grown and imported seed.

Global vegetable oil consumption is expected to grow by about 2% per year, less than half the rate observed in the last decade. Geographically, consumption should remain less concentrated than production; excluding the world's four leading consumers, other countries around the world may account for about half of total utilisation and present strong annual expansion levels, reflecting robust population and income growth rates in many countries. China should remain the single largest user, followed by the European Union, India and the United States.

At the global level, food consumption and biofuel demand are estimated to account for, respectively, around 64% and 33% of the increase in total utilisation when compared to 2009-11. Demand from the biodiesel industry is set to grow less than in the previous decade when biofuel demand accelerated as policies were put in place. The use of edible vegetable oil for biodiesel is still expected to expand to 30 Mt, which corresponds to a 76% increase over the base period and raises the share of vegetable oil consumption used for world biodiesel production from 12% in 2009-11 to 16% in 2021 (Figure 5.3). Although rapeseed and soybean oil are projected to remain the main feedstock, the use of palm oil is expected to more than double over the coming decade, with around 9% of global palm oil production absorbed by the biofuel industry in 2021.

In the developed world, food use and biodiesel demand should account for, respectively, 27% and 73% of total consumption growth. Biodiesel demand growth should continue to be lead by the European Union, where, by 2021, biofuel producers are expected to absorb 51% of domestic vegetable oil up from 40% in 2009-11. In the United States, the absorption rate should rise slightly to 14%. The underlying growth projections for biodiesel output are close

Figure 5.3. **Biodiesel to use a large share of global vegetable oil consumption**

Share of vegetable oil consumption used for biodiesel production in selected countries

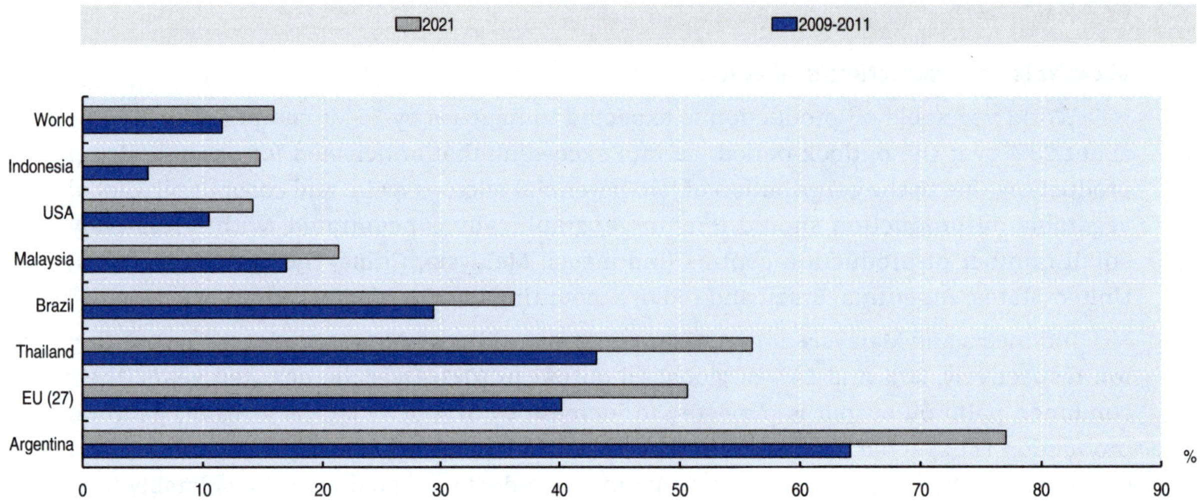

Source: OECD and FAO Secretariats.

StatLink http://dx.doi.org/10.1787/888932639780

to 6% per year for the European Union and less than 2% for the United States. As to developed country food use, it is expected to grow by 10% over the entire outlook period with the average *per capita* consumption about unchanged at 24-25 kg per year.

In developing countries, food demand is projected to continue to drive consumption. Although the average *per capita* intake of vegetable oil is set to rise by about 2 kg to 18 kg per year, the pace of growth would decrease considerably compared to the past. Starting from a relatively small base, demand from the biodiesel industry is expected to almost double in the developing world, with growth in absolute terms not far behind that projected in developed countries. Growth is expected in the traditional producers, Indonesia, Malaysia and Argentina, but also in other parts of Asia (Thailand, India) and South America (Brazil, Colombia). Argentina further expands its export-oriented biodiesel industry, which, by 2021, could absorb 31% of domestic vegetable oil output.

Oilseed meal production and consumption

Global meal output is projected to increase by 23%, reaching almost 315 Mt by 2021. Production remains highly concentrated, with six countries (Argentina, Brazil, China, the European Union, India and the United States) accounting for over 80% of global production. In China and the European Union, meal production would continue to rely on both domestically grown and imported seed. The growth in domestic meal production in the European Union is largely due to the expansion of rapeseed production used to produce biodiesel.

Global meal consumption should rise by 23%, with developing countries accounting for close to three quarters of the increase. Compared to the past decade, annual consumption growth would slow down markedly (Figure 5.4), mostly reflecting the situation in developing countries, where livestock industries are expected to grow at a much slower pace than over the previous decade. Relatively low but stable growth is expected among developed nations, where livestock industries are mature. Consumption growth is expected to accelerate only in the group of least developed countries, in line with expanding domestic meat production.

Figure 5.4. **Oilseed meal use growth rates to slow down**

Comparison of average growth rates of oilseed meal use

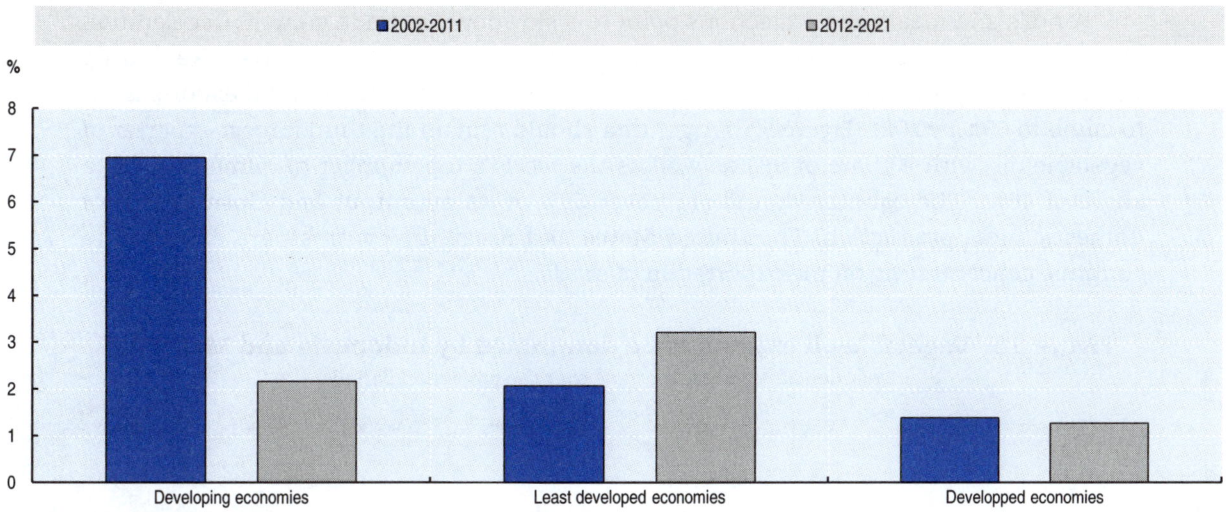

Source: OECD and FAO Secretariats.

StatLink http://dx.doi.org/10.1787/888932639799

The developing countries' share in global consumption rises to 60% in 2021. China should remain the world's single biggest consumer with a share of 26%, although the country's projected annual increase is anticipated to drop markedly as the structural change in the livestock industry should near completion. To meet rising demand, the country would continue to rely strongly on the crushing of imported oilseeds. In the second largest group of consumers, the European Union, annual growth should drop below 1%, whereas in the United States meal use expands again, following a period of decline that was caused by rising availability of dried distillers grains (DDG). Increased growth is also projected for a number of smaller consumers, notably Argentina, Indonesia, Iran and the Russian Federation.

Trade in oilseeds and oilseed products

Growth in world oilseed trade is projected to slow down significantly compared to the last decade. This development is a direct result of the deceleration in the Chinese crushing sector. The country is expected to expand its crush by about 20 Mt less than in the previous decade, which results in an import slowdown of the same amount. Because most other major traders are expected to roughly maintain their growth patterns, global trade growth projections range about 20 Mt lower than observed in the previous decade. Imports by the European Union remain by far the second largest, but should increase only marginally, as increased crush demand is met primarily *via* rising domestic oilseed production. Many smaller importers are expected to expand their imports significantly relative to the base period, but in absolute volumes these additional shipments are small.

As to global oilseed exports, the slowdown in growth is expected to apply equally to developed and developing countries. In the United States, exports should grow by 12% over the projection period. While Brazil's shipments should grow by 9% over the next ten years, Argentina's exports are expected to remain almost unchanged as the country's export tax regime is expected to continue favouring the exportation of processed products. Overall, world trade in oilseeds remains highly concentrated, with the four leading exporters holding a market share of over 80%, and Brazil and the United States alone accounting for two-thirds of the market.

For oils and meals, the projections point to a slowdown in trade growth. Deceleration should be more pronounced in developed than in developing countries. Regarding vegetable oil exports, the combined share of Indonesia and Malaysia in total exports is set to climb to 65% by 2021 (Figure 5.5). Argentina should remain the third largest exporter of vegetable oil (with a share of 9%) as well as the world's top supplier of oilmeals (with a share of 40%). The country should ship over 60% of its oil output and close to 90% of domestic meal production. The United States and Brazil, by contrast, are expected to continue concentrating on the exportation of seeds.

Figure 5.5. **Vegetable oil exports to be dominated by Indonesia and Malaysia**

Evolution of vegetable oil trade over the projection period

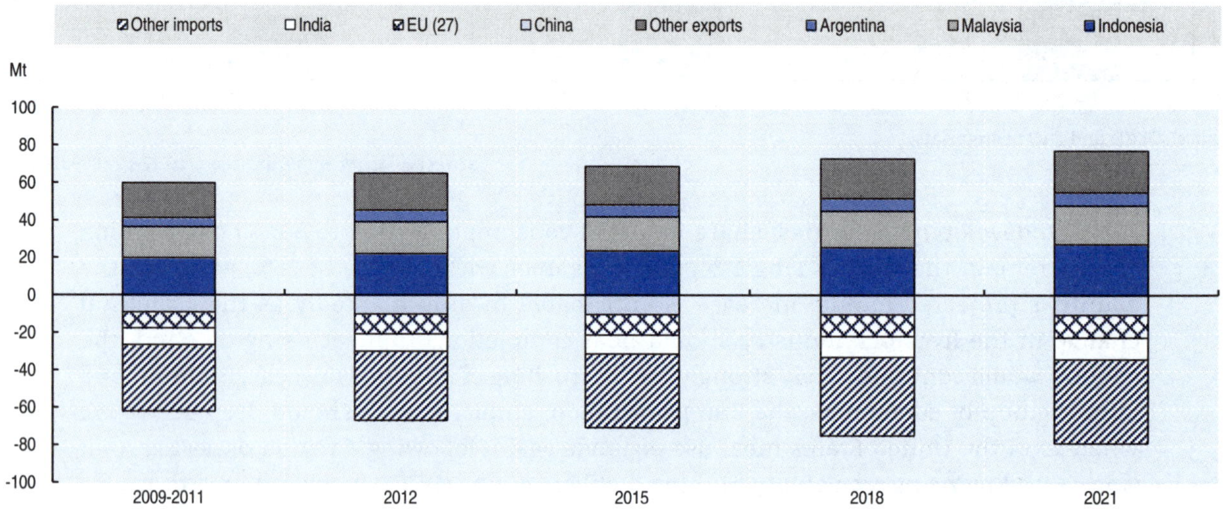

Source: OECD and FAO Secretariats.

StatLink http://dx.doi.org/10.1787/888932639818

As to vegetable oils, the share of developing countries in Asia – led by India and China – in global imports should rise to 54% in 2021. On average, about 50% of consumption in developing Asia would come from imports. In India, where foreign purchases expand by 33%, the rate of import dependence should reach 60%. China, in addition to covering a considerable part of its oil requirements *via* the crushing of imported oilseeds, is set to expand oil imports by 13%, implying that about one-third of total domestic consumption will be covered *via* imports.

In the European Union, the growth pace of vegetable oil imports should slow down compared to the past. To meet both industrial and traditional vegetable oil demand, imports will rise by about 35% over the outlook period. Combined, the European Union, India and China are expected to reach a market share of 44%. During the outlook period, India is set to surpass the European Union as the world's top importer. The group of least developed countries is expected to form a growing deficit region. Vegetable oil production in these countries would not be sufficient to satisfy strong internal demand growth.

With respect to protein meals, when compared to 2009-11, over 80% of the anticipated expansion in global import demand is projected to occur in the developing world, with developing countries in Asia accounting for over 50% of the increase. In China, meal imports should expand but remain insignificant compared to consumption, as growth in domestic meal demand would continue to be primarily met by crushing imported oilseeds. In the European Union, by far the world's largest importer of meals, purchases are projected to grow only marginally, in line with the livestock sector's stable consumption.

Main issues and uncertainties

In addition to the issues and uncertainties common to most commodities (macroeconomic environment, crude oil prices, weather conditions), each sector has its specific supply and demand sensitivities.

Supply side

Area

The *Outlook* projects a significant slowdown in area expansion of oilcrops compared to the previous decade. Relative stagnation in harvested area expansion is expected in the soybean producing countries of South America, rapeseed producing Western Europe and the palm oil producers in South East Asia. These lower growth expectations are based on continuously increasing marginal costs of pasture and forest land conversion into arable land, tightening restrictions on such land use changes and sustained high prices for competing commodities such as cereals and pasture based livestock products. While substitution of crops within the current arable area is mostly guided by economic and agronomic considerations, direct and indirect land use change of pasture and forest areas will be increasingly controlled by government regulations addressing environmental concerns. These policies have to address the complex balance between protecting consumers from high prices, maintaining a viable farm sector and commitments to climate/environment protection goals. The type and magnitude of impact these regulations will have on future area developments is difficult to predict and requires continued monitoring and analysis. Box 5.1 illustrates these issues using the example of the palm oil industry in Indonesia.

Yields

The projected development of global supply patterns is equally driven by yield improvement expectations. As yield growth is based on advances in the production frontier as well as improvements in the commercial realisation of the existing possibilities. The *Outlook* needs to make assumptions about both aspects, as the relative contributions of both factors shape the projections for production, trade and market share development. Depending on the development stage of an agricultural sector, growth potential and main drivers vary. Making adequate assumptions by country and across different oilcrops remains a challenge in this *Outlook*. Chapter 2 provides background information and scenario analysis illustrating this issue.

Demand side

Vegetable oil food uses

Growing food use remains globally the most important driver of vegetable oil consumption. In the developing world, *per capita* consumption is projected to increase by about 12% over the coming decade. In developed countries, on the other hand, no further increases are projected.

Box 5.1. **Palm oil development *vs.* forest conservation in Indonesia-issues and challenges**

Supported by the rapid demand growth for vegetable oils, palm oil production in Indonesia has expanded rapidly over the last two decades. Output grew almost tenfold between 1990 and 2011. In 2005, Indonesia passed Malaysia as the world's largest palm oil producer (OECD, 2012). While the palm oil sector has undoubtedly emerged as an important contributor to export revenue and rural employment, the environmental impacts of these developments (loss of biodiversity, soil loss/degradation, carbon sequestration issues and GHG emission levels) have been controversial. In particular, the conversion of primary forests into plantations has been criticised. The existing palm oil production is concentrated in Sumatra and to a lesser extent in Kalimantan. Further expansion is likely to occur in Sumatra, Kalimantan, Papua, and Sulawesi with a projected production increase from 25 Mt in 2012 to 32.5 Mt in 2021 (OECD, 2012).

With more than 100 Mha of forested land, Indonesia holds the world's third largest area of tropical forests after Brazil and Congo. Due to significant efforts, the annual decline in primary forests has slowed since 1997-98 to reach slightly above 1 Mha in 2009-10 (OECD, 2012). Indonesia, with its fast growing economy, faces strong challenges to find a sustainable balance between economic development and environmental protection. The logging and plantation crop industries are often at the centre of this debate.

The political decentralisation process was launched in 2001. Inconsistent legislation, together with planning and institutional difficulties created uncertainty on the rights of each government level to control forest resources. The multi-layer legal framework can give rise to conflicting interest between different levels of government. As a result, land allocation problems have occurred (World Bank, 2010), because of the primary use of forests for logging and the subsequent conversion of the cleared land into perennial crop plantations, the land administration process has been complicated and cases of license abuse have been reported (Sawit Watch, 2006). Further improvements in the transparency of the allocation of forest use rights are expected to contribute to the implementation of a consistent and sustainable development strategy (World Bank, 2006).

In May 2011, a two-year freeze on new logging, mining and agricultural development permits for more than 44 Mha of primary forest and 21 Mha of peat land came into effect. Based on existing palm oil development permits for 11.4 Mha, slower but continuing oil palm area expansion is, however, projected (GAIN, 2011; OECD, 2012). In addition to the carbon market mechanism REDD (Reducing Emissions from Deforestation and Forest Degradation) launched in September 2010, Indonesia started the Indonesia Sustainable Palm Oil (ISPO) scheme in November 2010 to promote environmental sustainability in palm oil production. It has been implemented voluntarily since February 2011 and will become mandatory for all Indonesian palm oil producers in 2012.

As Indonesia continues its economic and social development, environmental protection issues will remain an important concern for all levels of government, private industry and society in general. As a broad range of interests have to be addressed, well targeted solutions will need to be found in order to provide an enabling environment for sustainable growth in this vast and diverse country.

Interestingly, *per capita* food use seems to have levelled off at about 36 kg per year in the United States, 24 kg in the European Union, and only 17 kg in Japan. Such differences in consumption levels show the challenges involved in anticipating the type of consumption pattern that developing countries are likely to follow. It becomes clear that beyond a certain income level, culture and lifestyle are the crucial factor in determining the consumption level of vegetable oils. As more and more countries approach income levels where food demand becomes virtually inelastic, assessing the cultural aspect becomes an important aspect in food consumption projections. Careful observations of trends in diets and demographic processes, such as urbanisation, are needed to supplement the economic modeling underlying this *Outlook*.

Vegetable oil non-food uses

Increased demand for oils and fats from biodiesel producers has become over the past few years one of the driving forces of the global vegetable oil market. Any changes in biofuel policies in the European Union and in the United States – but also in several other countries, including developing ones – as well as any advances being made on the next generations of biofuels is bound to alter the demand of vegetable oils for non-food purposes. Furthermore, in the coming years, national biofuel policies may also increasingly affect international trade in vegetable oils used as biodiesel feedstock as well as trade in biodiesel itself.

Protein meal use

Meal demand projections have to take into account two offsetting developments. Currently, mostly in developed countries, feed rations are being optimised, reducing the feed need per unit of output. In developing countries, the livestock industry is becoming more industrialised which means commercial feed replacing table scraps and pasture, thus increasing the use of protein meals per unit of output. Once this process is completed, countrywide optimisation will start to shape the national picture. Over the course of a ten-year projection, the relative importance of both of these developments constantly shifts and alters the relationship between feed demand and livestock output. The assessment of such developments presents a source of uncertainty and its evaluation requires diligent market monitoring and adaptation of model specifications.

References

GAIN-ID1127 (2011), *Indonesia Forest Moratorium*, USDA FAS, 8 June.

McCarthy, J.F. and R.A. Cramb (2009), "Policy Narratives, Landholder Engagement, and Oil Palm Expansion on the Malaysian and Indonesian Frontiers", *The Geographical Journal*, Vol. 175, No. 2, pp. 112-123, June 2009.

OECD (2012), *OECD Review of Agricultural Policies in Indonesia*, OECD, Paris, forthcoming.

Sawit Watch (2006), "*Promised Land: Palm Oil and Land Acquisition in Indonesia – Implications for Local Communities and Indigenous Peoples*", by Forest Peoples Programme, Perkumpulan Sawit Watch, HuMA and the World Agroforestry Centre.

World Bank (2006), *Sustaining Indonesia's forests – Strategy for the World Bank, 2006-2009*, Jakarta Indonesia, June 2006.

World Bank (2010), *Environmental, social and economic impacts of oil palm in Indonesia: a synthesis of opportunities and challenges*, Discussion Paper, World Bank, May 2010.

Chapter 6

Sugar

Market situation

World sugar prices continued to experience tremendous volatility in the lead up to this year's *Outlook*. Prices in 2011 bounced between highs and lows as market fundamentals remained obscured under a shroud of conflicting information: early projections of an emerging global surplus and, at the same time, ongoing tight supplies and high prices in several markets. For example, adverse weather and low yields resulted in a large drop in production in Brazil, the leading sugar producing and exporting nation, after consecutive years of growth. Apart from sugar fundamentals, commodity markets in general have seen prices giving way to a generally weaker macroeconomic environment, rising oil and energy prices, and expectations of increasing supplies. World sugar prices started 2012 at lower levels and are expected to continue a downward drift over the remainder of the year as the global sugar market returns to a significant global production surplus in 2011/12 (October to September crop year) in response to recent high prices. This would follow a nearly balanced market in 2010/11 and large global deficits in the previous two years (Figure 6.1). With world sugar production in the current season estimated at a new record level, some stock rebuilding is expected to commence. Overall, global stocks and stocks-to-use ratios will remain at relatively low levels at the beginning of the Outlook period.

Projection highlights

- The raw sugar price (Intercontinental Exchange No. 11 contract nearby futures) is projected in nominal terms at USD 483/t (USD 22 cts/lb) in 2021/22. While slightly lower than at the

Figure 6.1. **World sugar balance moves into production surplus**

World sugar production minus consumption

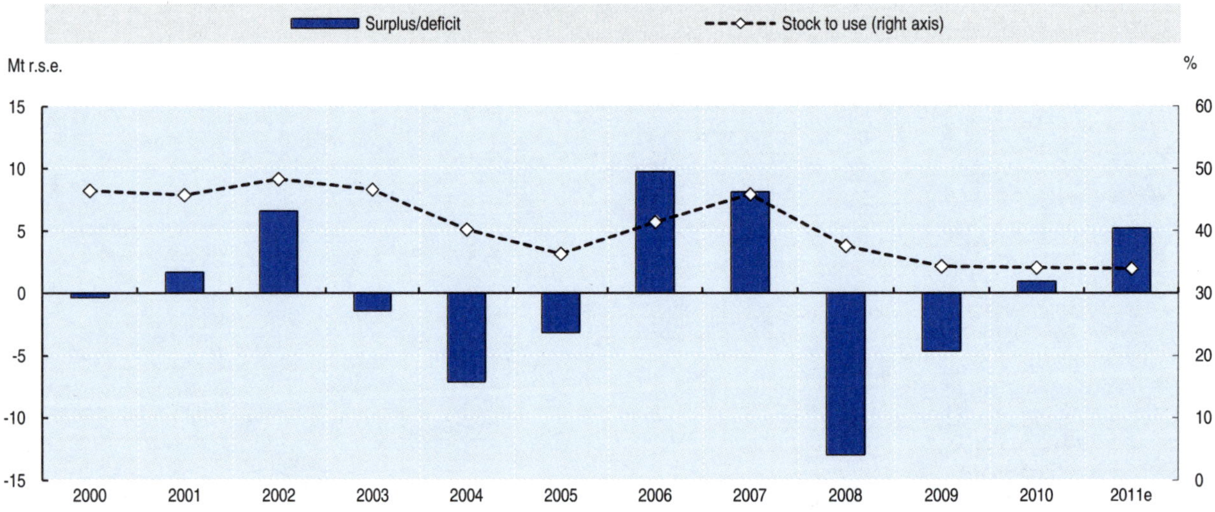

Source: ISO database, February 2012.

StatLink http://dx.doi.org/10.1787/888932639837

start of the outlook period, sugar prices are expected to remain on an elevated plateau and to average higher over the projection period in both nominal and real terms (when adjusted for inflation) than in the last decade. White sugar prices (Euronet, Liffe futures Contract No. 407, London) follow a similar pattern to raw sugar prices and are projected to be at USD 566/t (USD 26 cts/lb.) in 2021/22. A relatively large white sugar premium at the outset is expected to narrow over the course of the outlook period, averaging around USD 82/t, as additional white sugar supplies come on stream from new refineries (Figure 6.2).

- High price volatility has been a significant feature of world sugar markets. Looking forward, further bouts of price surges and volatility remain a clear possibility in response to unforeseen production shocks in major producing countries, and particularly while global stocks remain at historically low levels. Market volatility will also reflect the continuing production cycles in India and some neighbouring countries of Asia, existing government policies that intervene in sugar markets and fluctuations in oil and energy prices around a high and rising trend.
- Structurally higher and remunerative sugar prices are expected to encourage further growth in global sugar crops, mainly sugarcane. Global sugar production is projected to reach 208 Mt by 2021/22, up 43 Mt or 26% above the average for 2009-11. Larger sugarcane harvests are projected in Brazil to reach over 1.1 Bt by 2021/22. This will cement Brazil's position as the world's leading producer and trading nation, and, along with an increasing share of cane allocated to ethanol production, will be among the leading determinants of global sugar production and world prices over the coming decade. Increasing global production from Brazil and other countries should allow stock rebuilding to accelerate initially, but steady global consumption growth of 2.1% p.a., on average, is expected to eat into sugar supplies and lead to a decline in the global stocks-to-use ratio in the second half of the projection period, providing support for high market prices (Figure 6.3).

Figure 6.2. **World prices to remain on a higher plateau**

Evolution of world sugar prices in nominal (left figure) and real terms (right figure) to 2021[1]

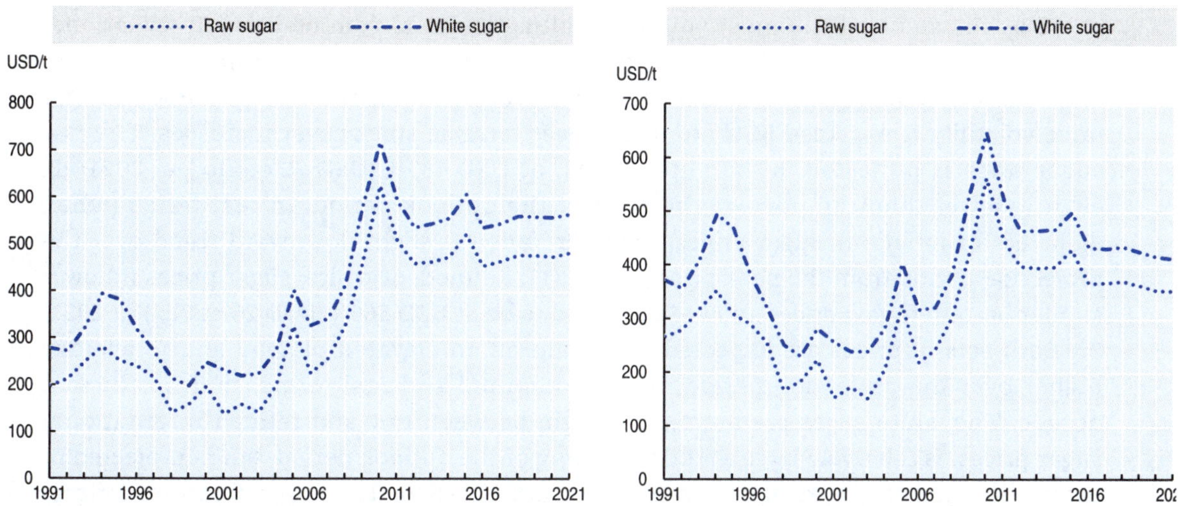

Notes: Raw sugar world price: Intercontinental Exchange contract. No. 11, nearby futures price. Refined sugar price: Euronet, Liffe futures Contract No. 407, London.

1. Real sugar prices are nominal world prices deflated by the US GDP deflator (2005 = 1).

Source: OECD and FAO Secretariats. StatLink http://dx.doi.org/10.1787/888932639856

- The outlook for high fructose corn syrup (HFCS), an alternative caloric sweetener to sugar, is for an increase of around 13% in both production and consumption, which are projected to reach 16.3 Mt and 15.9 Mt, respectively, by 2021/22. HFCS production and use is projected to expand in China and with consumption to increase in Mexico based mainly on imports from the United States, as part of two-way trade with sugar shipments to the United States, in an integrated sweetener market between these two countries under NAFTA.

Figure 6.3. **Global stocks-to-use to rise in the mid-term and then decline**

Evolution of world sugar production, consumption and stock-to-use ratio to 2021

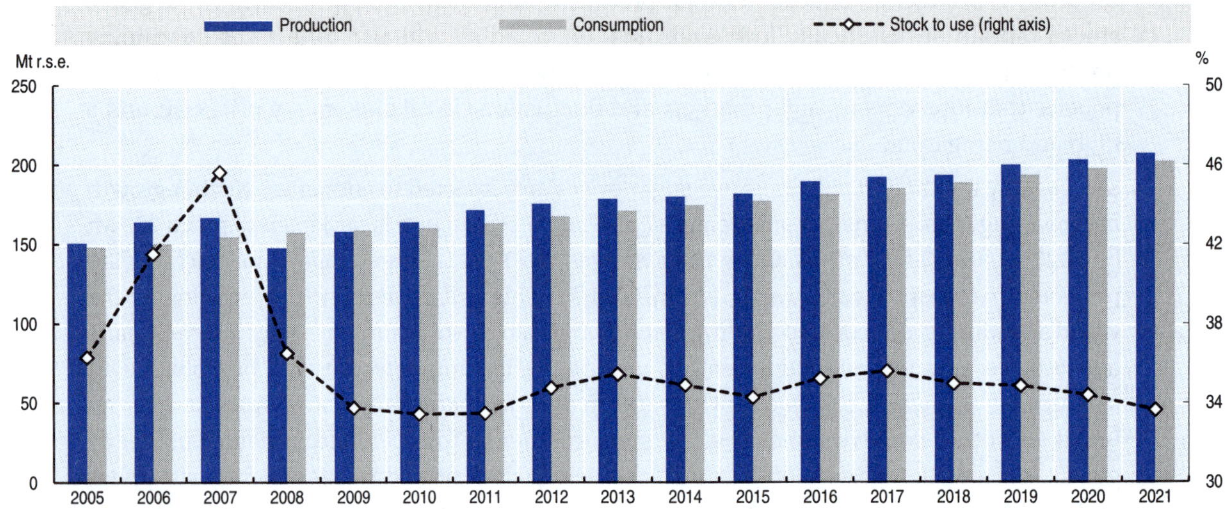

Source: OECD and FAO Secretariats.

StatLink http://dx.doi.org/10.1787/888932639875

Market trends and prospects

Prices are underpinned by low stocks and remain volatile

Sugar is amongst the most volatile of all agricultural commodities. International sugar prices are projected to remain relatively high over the projection period, underpinned by strong demand, the depreciation of the US dollar, and tight stocks. Sugar price volatility is expected to abate in the near term as stock cover increases. The raw sugar price is projected, in nominal terms, at USD 483/t (USD 22 cts/lb.) in 2021/22, slightly lower than the price at the start of the outlook period. In real terms (when adjusted for inflation), sugar prices, however, follow a slight downward trend over the projection period from the peak price of 2010/11. Refined, or white, sugar prices follow a similar pattern to raw prices and are projected to be at USD 566/t (USD 26 cts/lb.) in 2021/22. Both sets of prices are expected to remain on an elevated plateau and to average higher over the projection period, in both nominal and real terms, than in the last decade. The white sugar premium, or margin between raw and refined international sugar prices, is relatively large at the beginning of the outlook period, but is projected to narrow in following years and to average around USD 82/t. The lower premium reflects developments in the raw and white markets, and particularly prospects for additional white sugar supplies from new refineries that will come on stream in the coming years in the Middle East, Africa and other locations in Asia.

World sugar prices should continue to ease in the near term as the global market returns to a sizeable production surplus in 2011/12. Significantly higher production in a number of countries in response to earlier high prices is projected to more than offset a sharp contraction in output in Brazil and a more steady increase in global consumption. Prices are projected to recover in subsequent years and to oscillate around a generally rising trend in nominal terms over the rest of the outlook period. This price projection reflects market fundamentals of continuous demand growth in a context of more variable annual sugar production and a continuing tight global stock (and stock-to-use ratio) situation. Further bouts of high price volatility remain a clear possibility in a market context characterised by continuing low global stocks. Other factors will likely contribute to volatile sugar markets. The widespread reliance of national sugar industries on government support policies (such as border measures) that intervene in sugar markets and insulate domestic markets to varying degrees from world markets add to market volatility.

Sugar price movements will increasingly reflect developments taking place in world oil and energy markets. Higher oil and energy prices are projected over the outlook period and these will lead to rising input prices (chemical fertilisers, herbicides and machinery costs), adding to overall sugar production costs. In addition, there is an increasing link between oil, energy and sugar markets, particularly in Brazil, where more than half of the enormous sugarcane crop is devoted to ethanol production, a leading petrol substitute in transport fuel. Brazil is the world's largest sugar producer and in 2010/11 accounted for about 49% of world sugar trade as well as being the second largest ethanol producer. Although Brazil is no longer among the lowest cost sugar producers, following rising production and labour costs and currency appreciation, it remains unrivalled in its cane production and sugar supply potential. The size of the annual sugarcane harvest and its allocation between ethanol and sugar is, therefore, a key determinant of world sugar prices over the outlook period. Brazil's sugar supply costs effectively set a floor for world market prices. Sugar production with higher relative prices than ethanol was more attractive to Brazilian mills, which can produce both products, at the beginning of the outlook period. However, the price situation is expected to increasingly favour ethanol in future years. As a consequence, more of sugarcane production, from 51% in 2011/12 to 65% in 2021/22, is likely to be allocated to ethanol (made from sugarcane juice and molasses) to meet rising domestic and export demand. This will have direct implications for Brazil's sugar production, export availabilities, and world prices to 2021/22.

Another feature of sugar markets that contribute to international price volatility are production cycles in some Asian countries, particularly India, and to a lesser extent China and Pakistan. These cycles can cause large switches every few years between imports and exports of sugar. When such trade switches occur in a large sugar-producing country such as India, they can have a significant influence on global trade volumes and the prevailing movement of world sugar prices. Due to India's production cycle, world sugar prices are projected to decline to a trough in 2012/13 as production peaks in this country, and rises in others, and additional sugar exports are placed on the world market. If past practices are repeated, large mill payment arrears to growers are likely to develop in India as domestic sugar prices decline and become out of sync with high fixed sugarcane prices. The cycle would then enter a downward phase as sugarcane growers turn to other more remunerative crops, leading to a decline in production and possibly a shortfall in meeting annual consumption requirements. The shortage of sugar leads to domestic price increases resulting in India's return to the world market as a prominent importer to meet part of its consumption needs. This action, in turn, would help increase world prices. The upturn in the cycle would then recommence, with high prices contributing to larger domestic production, increasing export availability, and eventually lower world prices.

Production and use of sugar

Structurally high and remunerative prices, and an assumption of normal weather conditions with a mix of favourable and unfavourable climatic conditions that produce trend yields, are expected to lead to a further expansion in sugar crops in many parts of the world over the period to 2021. Perennial sugarcane production continues to dominate sugarbeets in the production of sugar. Although some increase in sugarbeet production is projected to take place at the beginning of the outlook period, the dominant share of sugarcane continues to increase over time to account for 89% of global sugar output by 2021/22. Global growth in sugar yields and increased area are projected to contribute to larger world sugarcane harvests to 2021; whereas yield growth alone is expected to account for essentially all the expected expansion of global sugarbeet production. World sugar production is projected to grow slightly faster in the coming ten years to reach 208 Mt in 2021/22, some 43 Mt or 26% above the average level for the base period. The bulk of this additional sugar production will come from the developing and emerging countries, with the share from the developed countries in the OECD region continuing to shrink. Sugarcane production in the developing and emerging countries will also account for most of the additional ethanol production from sugar crops over the outlook period (Figure 6.4). Relatively high sugar prices are expected to encourage increased production in a number of countries over the coming decade, including traditional export suppliers and emerging producers, but the main burden of growth is expected to continue to fall on Brazil. Sugar production in Brazil is expected to continue to account for less than 50% of its enormous sugarcane harvest which should grow by an average of 3% per year, slower than in the last decade, with sugar output to reach nearly 47 Mt by the close of the decade.

Figure 6.4. **Main increase in sugar crops will occur in developing and emerging economies**

Change in allocation of sugar crops between biofuel and sugar production to 2012

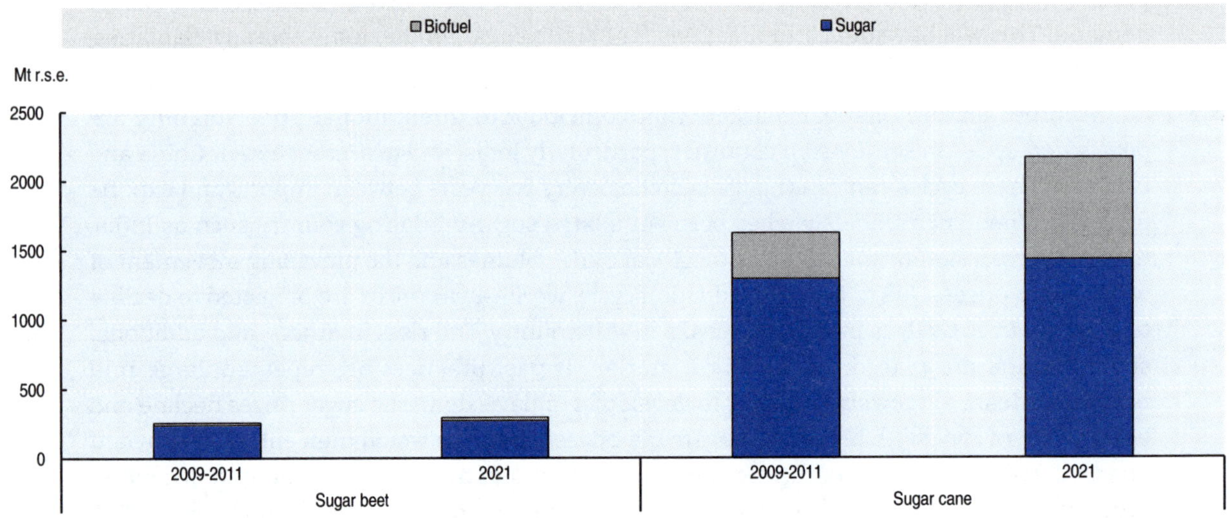

Source: OECD and FAO Secretariats.

StatLink http://dx.doi.org/10.1787/888932639894

Global consumption of sugar is projected to reach nearly 202 Mt in 2021/22, an increase of 42 Mt, or 26%, over that of the base period. Sugar consumption has continued to increase despite the economic difficulties faced by a number of developed countries, compounded by the period of high sugar prices and increased volatility. These conditions have slowed sugar off-take, in general, and in particular for industrial use of sugar in food manufacturing, food preparations and beverages that make up a majority of sugar use in developed countries, amounting to around 70% of sugar disappearance in the United States and the European Union. Industrial uses of sugar are more sensitive to economic activity and the state of the business cycle than is direct sugar consumption for household use. The current slowdown in consumption growth is expected to continue over the coming decade as world sugar prices average higher in real terms. In fact, global consumption is projected to grow at 2.1% p.a. to 2021/22, and down from the 2.5% p.a. that prevailed in the ten years before the 2009 crisis. Developing countries will continue to experience the strongest growth in sugar consumption, based more on direct household use and fuelled by rising incomes and growing populations, although with considerable variation between countries. The sugar deficit regions of Asia, the Far East, and Africa will be responsible for most of the expansion in use. In contrast, sugar consumption in many developed countries is expected to show little or no growth, consistent with their mature sugar markets. This reflects, among other things, slowing population growth and dietary shifts that are underway as a result of increasing health awareness and concerns with, for example, obesity. For these and other reasons, the share of industrial consumption of sugar in total use is expected to increase further, including in a number of higher income developing countries, by the close of the outlook period.

Brazil has rapidly expanded production over the past two decades, but low renewal rates for sugarcane plantings and the ageing of the national sugarcane crop, exacerbated by adverse weather conditions in the last two seasons, have reduced yields and will lower production prospects in the near term. In addition, a slowdown in investment in new mills following the financial crisis of 2008 is expected to cut the pace of sugar industry growth. Reduced investment, increasing production costs, a shortage of skilled labour, and emerging infrastructure bottlenecks are expected to slow the pace of expansion over the coming ten years. Despite this anticipated slowdown, Brazil will continue to be the largest producer with sugar production projected to increase by over 8 Mt to nearly 47 Mt by 2021/22; some 22% above the average for the base period. Consumption of sugar is projected to grow at an average of 1.9% per year to reach around 16.2 Mt in 2021/22. Argentina, another South American producer, is expected to undertake further investment in the sector with production projected to increase to 2.6 Mt in 2021/22, but with much of this devoted to satisfying faster domestic consumption growth.

India is the second largest global producer and the world's leading consumer. Relatively strong sugar consumption growth is projected for India at 2.7% p.a., on average, to reach nearly 32 Mt by 2021/22. However, sugar production is expected to continue to be subject to the longstanding production cycle leading to fluctuations in annual output and switching India from a surplus to deficit producer every few years. Variable annual sugar production is expected to average around 27.5 Mt per year over the coming decade, or some 11% higher than in base period; a time when production was in the rising phase of its cycle (Figure 6.5). Other countries of Asia, such as China and Pakistan, are also expected to continue to experience milder forms of production cycles, which contribute to fluctuations in production and their import volumes. Emerging resource supply constraints, including available land and water, are anticipated to lead to sugar production in China growing less rapidly than consumption in coming years. China's sugar production is projected to increase by an average of 1.9% per year to reach 15.2 Mt by 2021/22, or nearly 32% higher than the base period. This higher production will remain below the record crop of 16.1 Mt in the 2007/08 season. China is the world's second largest consumer, although *per capita* consumption levels remain low in comparison to other Asian countries and world standards. In addition, surging domestic prices have slowed consumption at the start of the outlook period. While sustained high GDP growth has increased sweetener consumption, this has not always been for sugar but also for artificial sweeteners and HFCS. However, with government controls being enforced on artificial sweeteners use, and further maize processing into HFCS limited by food security concerns, consumption of sugar is projected to expand further and to reach nearly 20 Mt by 2021/22.

Figure 6.5. **India's production cycle to influence world prices**

Evolution of India's sugar production, consumption and imports to 2021

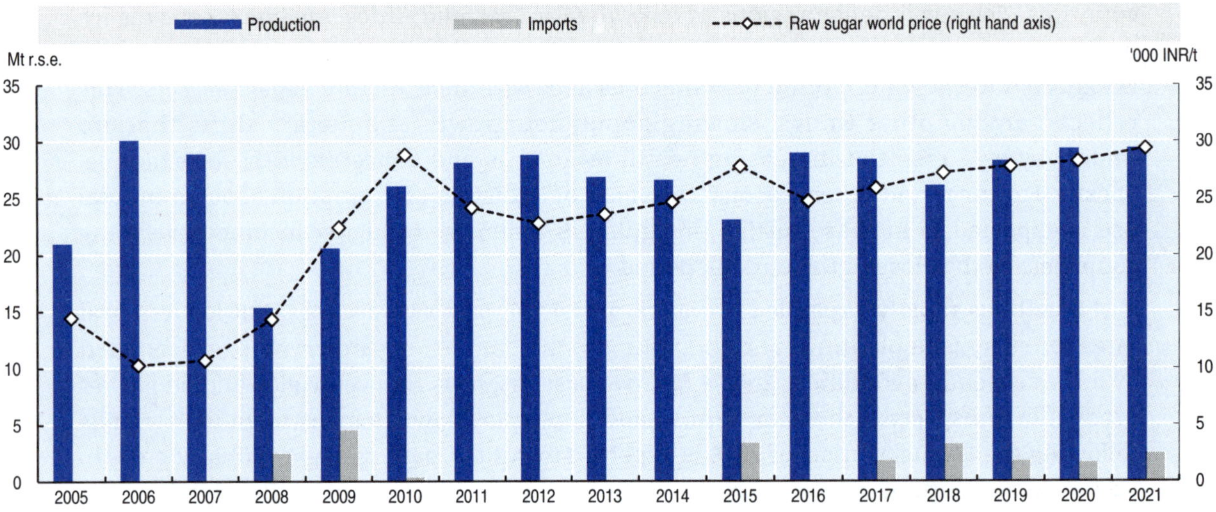

Source: OECD and FAO Secretariats.

StatLink http://dx.doi.org/10.1787/888932639913

Outside this group, an expansion drive is underway in Thailand in response to recent high prices, and poor prospects for competing crops, with production of sugarcane and sugar surging in recent seasons. Further growth in the sector is expected in coming years to 2021/22, provided sugarcane maintains its relative price advantage over cassava, its principal competitor (Box 6.1) and maintaining Thailand's position as the third largest global producer. Sugar production in the Russian Federation has benefited in recent years from increased government support to the industry in the form of production subsidies and tariff protection. With exceptional seasonal conditions, following a drought in the previous year, production has surged in 2011/12. Sugarbeet production is projected to grow by around 1% p.a., allowing sugar output to reach 5.4 Mt in 2021/22, and with rather static consumption this is expected to make the country largely self-sufficient in sugar. Indonesia is also striving for increased self-sufficiency through increased investment in the sector. As a result, sugar production is projected to increase by two-thirds to reach 4.4 Mt in 2021/22, when compared to the base period. Despite this rapid growth, it is not anticipated that production will be sufficient to meet the projected increase in consumption by 2021/22, with imports expected to increase to 4.2 Mt in 2021/22. Aggregate sugar production in Africa is forecast to increase by 3.7% per year to reach 13 Mt in 2021/22, a reflection mainly of production gains in South Africa, Egypt, Mozambique and Tanzania. The expansion is expected to be driven by rising domestic and regional demand as a result of growing population and increasing *per capita* incomes. South African production is projected to reach 2.3 Mt by 2021/22.

In contrast, to the strong expansion trends underway in much of the developing world, the traditional sugar industries in several developed countries are expected to witness less rapid or relatively low production and consumption growth over the coming decade. The share of the world sugar market held by developed countries in the OECD area will continue to shrink, to around 19% of global production by 2021/22 (down from 23% in the base period) and only 23% of consumption in the same period (down from 27% in the base period) (Figure 6.6). In the European Union (EU27), production quotas are assumed to expire in September 2015 (end of the 2014/15 crop year) in line with existing legislative proposals,* with beet production for sugar unrestricted thereafter. As a result, internal prices will likely decline and further adjustments will occur in national industries of member states within the European Union in response to changing price incentives, including the conversion of some former non quota beet production destined for ethanol to higher value sugar output. Unrestricted isoglucose production is also expected to increase. Production of sugar is projected at 16.1 Mt in 2021/22, and to remain well below annual consumption, projected at 18.5 Mt (rse) in the same period. Australian sugar production is expected to recover in the near term from both earlier adverse weather events and a rust outbreak as it continues to respond to remunerative sugar prices and improved profitability. Some area expansion and further farm consolidation along with yield growth is projected to result in sugar production increasing by an average of 0.8% p.a. to reach 5.2 Mt by 2021/22. Consumption is expected to increase by around 0.7% p.a. to reach 1.1 Mt, in the same period.

Production of sugar in the United States is projected to show little growth in part due to competition for land from other crops, and to remain well below the 85% minimum allotment level of the US 2008 FCE Act. Sugar production from sugarbeets and sugarcane is projected to increase by 10% to 7.9 Mt when compared to the average output in the base period. Beet sugar production is expected to account for around 61% of the total sugar output in 2021/22. Sugar consumption is also projected to show only a limited increase, to reach 11.4 Mt in 2021/22, with average growth in human use of around 1% p.a.

In a single, integrated sugar market between Mexico and the United States under NAFTA, sugar production in Mexico is projected to rise modestly by 10%, to 5.9 Mt in 2021/22 as a result of further investment in the industry and generally remunerative prices set in the US market. There is expected to be further substitution of domestic sugar use for lower cost HFCS, primarily as imports from the United States, for use in the manufacture of foodstuffs and beverages. As a result, domestic sugar consumption will continue to grow less rapidly, while HFCS off-take remains below the higher level of use in the United States.

* A discussion is taking place between certain EU member states concerning the possible temporary prolongation of quotas beyond 2015 rather than immediately terminating them as proposed by the Commission.

Figure 6.6. **Developing countries dominate sugar production and consumption**

Comparison of production and consumption between developing and developed countries

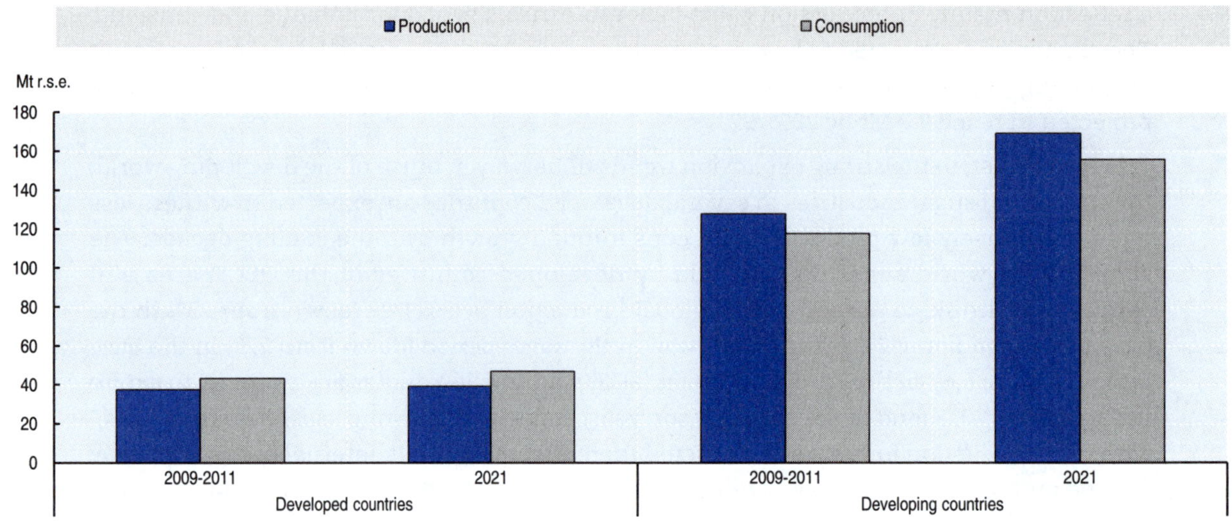

Source: OECD and FAO Secretariats.

StatLink http://dx.doi.org/10.1787/888932639932

Trade

Over the last decade or so there have been a number of structural changes that have affected the evolution of trade volumes and patterns and which will continue to influence international sugar transactions in the coming period. These include increased concentration in sugar export trade, with a smaller number of global exporters and the rising dominance of Brazil, as well as a decline in the volume of white sugar traded internationally. The reform of the sugar regime in the European Union led to an abrupt decline in white sugar exports, of 6.7 Mt in 2006/07, as production quotas were progressively reduced below consumption requirements and export subsidy limits applied. As a consequence, the European Union switched from being a large net exporter of white sugar to a large importer of mainly raw sugar for further refining and sale in the domestic market. In addition, the world has become increasingly dependent on Brazil, which accounted in 2010/11 for about 49% of world sugar exports to supply its sugar needs. The growth in exports from Brazil, made up almost exclusively of raw sugar, has been accompanied by a surge in investment in stand-alone destination and toll refining capacity in strategic locations around the world to process the raw sugar.

The white sugar trade is expected to recover over the coming years and lead to a narrowing of the white sugar premium, to USD 82/t compared to USD 100/t for the base period. This will occur as more refined sugar is exported by traditional exporters in response to the high white sugar premium at the start of the outlook period and as new toll and destination refineries in several Asian, African and the Middle East countries progressively come on stream and begin to export increasing quantities of white sugar to neighbouring countries and regional markets.

While Brazil is expected to witness a slowdown in the rate of expansion of its sugar industry, structurally higher world sugar prices are projected to lead to an acceleration in output growth in other major producing countries. Increased sugar production results in either higher exports, notably by Australia, Thailand and Mexico, or lower imports as in the case of the Russian Federation, to 2021/22. As a result, Brazil is expected to lose export

Figure 6.7. **Sugar exports remain highly concentrated and dominated by Brazil**

Comparison of export volumes of leading exporters between 2009-11 and 2021

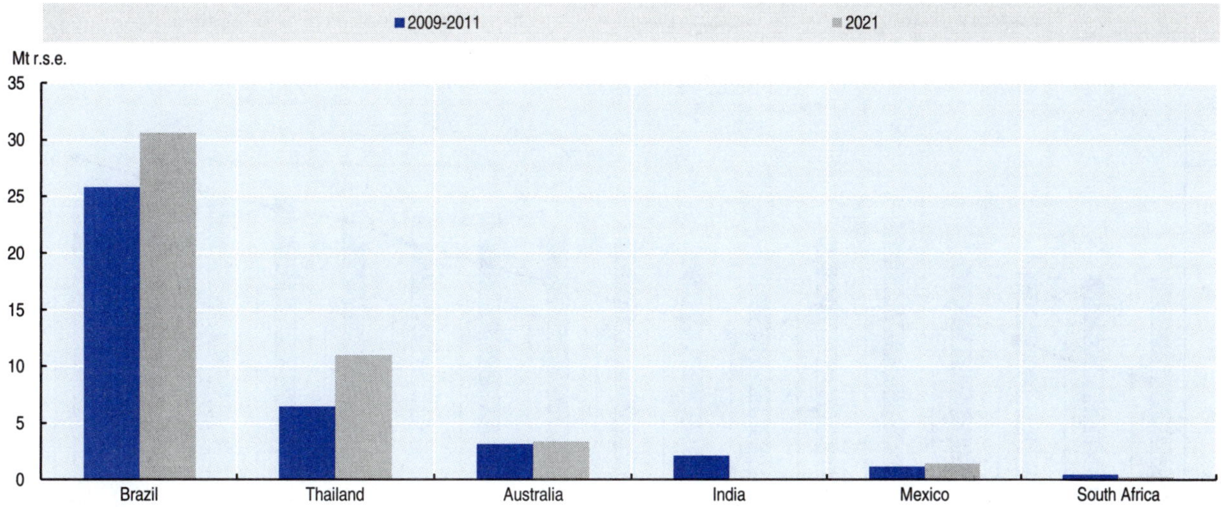

Source: OECD and FAO Secretariats.

StatLink http://dx.doi.org/10.1787/888932639951

share to other lower cost producers in the near term. However, provided sugar prices remain remunerative, and with an eventual increase of investment in new mills and cane production during the coming decade, Brazil is also expected to expand its sugar exports to reach 30.1 Mt by 2021/22 (Figure 6.7). The bulk of Brazil's exports will continue to comprise high quality raw sugar (VHP), which should increase to 20 Mt in 2021/22. However, the composition of Brazil's sugar exports is expected to include more white or refined sugar shipments as well. These are projected to increase by 63% relative to the base period and amount to over 10.6 Mt by the close of the outlook period. Whether Brazil is successful in regaining its absolute majority in sugar exports will depend on several factors, not least of which is the demand for ethanol for both domestic and exports, and the resulting share of any additional sugarcane production that is allocated to this end use (Figure 6.8.). In any event, a trend towards increasing market shares held by other exporters over the course of the outlook period should be a beneficial development for sugar importers as they become less dependent on production from a single country.

In terms of other leading exporters, Thailand will play a growing role in Asia as the only consistent producer of a large sugar surplus and with a natural trade advantage, along with Australia, to service the large and ballooning sugar deficit of that region. Exports from Thailand, ranked number two in the world, are projected to grow to around 11 Mt by 2021/22, an increase of over 69% on the base period. In the case of Australia, increased production over the projection period should support exports of around 4 Mt by 2021/22. Increasing use of HFCS in Mexico will substitute for higher cost domestic sugar in beverage manufactures, releasing surplus sugar for export to the usually higher priced US market. Mexican exports to the preferred US market are projected to average around 1.4 Mt per year over the ten years to 2021/22.

Sugar importers make up a more diverse and numerous group of countries (Figure 6.9). A significant feature of the *Outlook* is that China will become a larger consumer of sugar, mainly for industrial use in food manufacture and preparations rather than direct human consumption, as demand exceeds the growth in production from domestic sugar crops.

Figure 6.8. **Sugar production and exports to grow in Brazil as ethanol output expands**

Evolution of sugar production, exports and ethanol output from sugarcane in Brazil

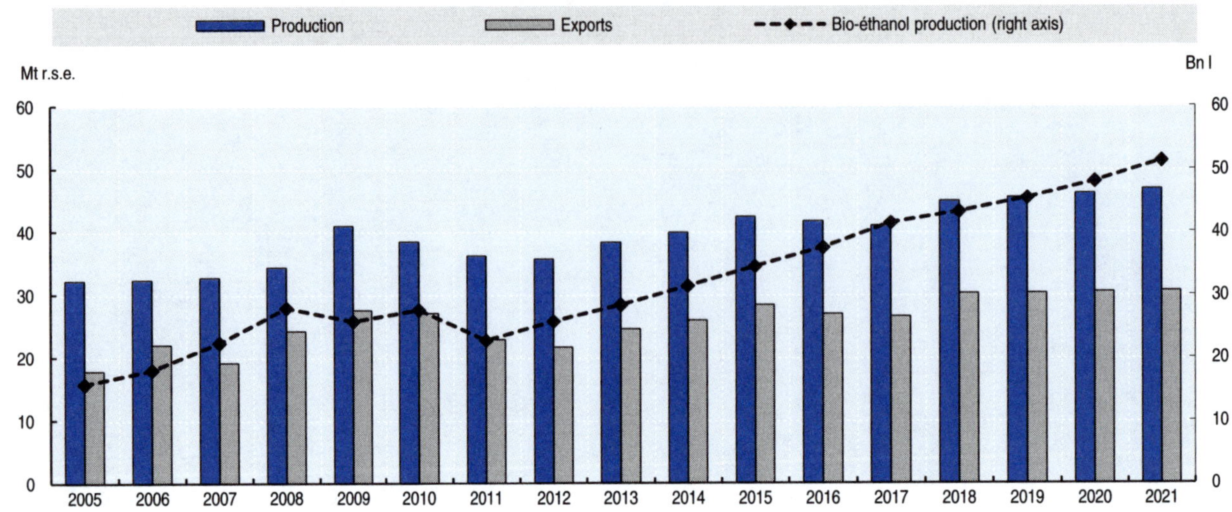

Source: OECD and FAO Secretariats.

StatLink http://dx.doi.org/10.1787/888932639970

Figure 6.9. **China to become the leading sugar importer**

Comparison of import volumes between 2009-11 and 2021

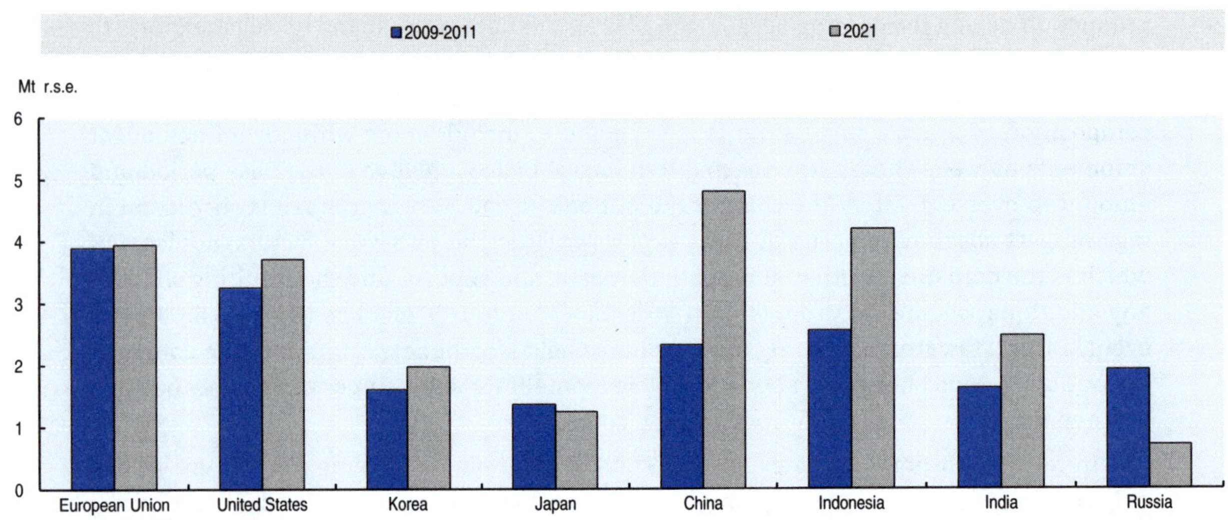

Source: OECD and FAO Secretariats.

StatLink http://dx.doi.org/10.1787/888932639989

The production of sugarcane and sugarbeets in China is expected to be increasingly constrained by domestic resource limitations, which will also lead to increased imports. Sugar imports in 2010/11 exceeded China's TRQ, established for sugar trade at the time of its entry in the WTO in 1998 of 1.95 Mt, for the first time and are projected to continue to do so in each year over the outlook period to reach 4.8 Mt in 2021/22 (Figure 6.10). This will make China the largest global sugar importer, exceeding imports projected for the European Union, the United States, Indonesia, and the Russian Federation. India, for its part, is expected to remain a periodic importer of mainly raw sugar in years of domestic shortfalls.

Figure 6.10. **China's imports to rise strongly**

Evolution of China's sugar production, consumption and imports to 2021

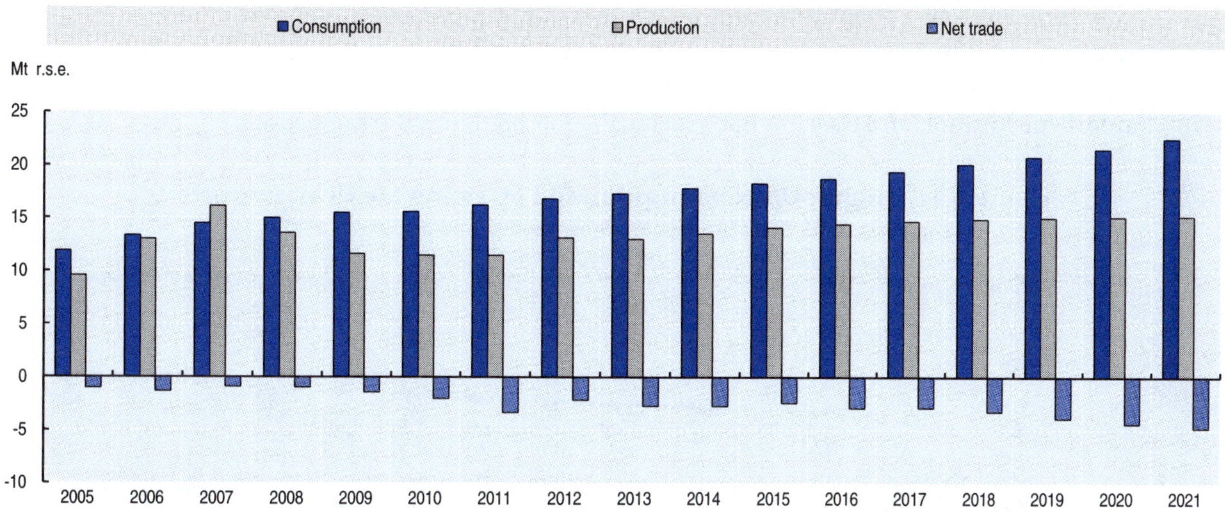

Source: OECD and FAO Secretariats.

StatLink http://dx.doi.org/10.1787/888932640008

High world sugar prices at the outset of the outlook period and declining internal support prices with sugar policy reform have made the European Union a less attractive destination for preferential exports from LDC countries under the EBA initiative and Economic Partnership Agreements. This outcome has left the European Union short of supplies with prices rising sharply and has necessitated some offsetting actions, such as conversion of non-quota sugar to quota sugar and the opening of reduced duty TRQs to attract sugar from the world market. As world prices decrease in the near term, the European Union, as an assured market, will likely once more become an attractive destination for many of these preferential sugar trading countries; although ongoing problems with infrastructure and technology adoption could constrain some from exploiting fully their export opportunities. The United States' sugar market remains heavily insulated from the world market with prohibitive tariffs and safeguard measures on imports in excess of minimum TRQ volumes. However, when world prices rise above US support levels, US prices also tend to increase to ensure that sufficient volumes of imports of sugar can be obtained from its preferential suppliers. Duty-free and unrestricted imports are projected to grow from Mexico under NAFTA over the coming decade, and from periodic increases above minimum levels, as required, in the US sugar tariff-rate quota. Total US imports are projected to reach 3.7 Mt in 2021/22, and with ending stocks not expected to exceed 14.5% of total use, internal prices remain above sugar support levels. As a consequence, imports are not expected to trigger the Feedstock Flexibility Program (FFP) under the FCE Act for converting excess sugar supplies to ethanol in order to avoid any stock accumulation or forfeitures by the Commodity Credit Corporation (Figure 6.11).

For its part, Mexico is expected to backfill periodically from the world market with tariff rate quotas (TRQs) to assure its domestic consumption requirements in periods of lower production. The Russian Federation has historically been a leading destination for white sugar imports, before switching in the early 1990s to raw sugar imports for domestic processing by beet factories at the end of the beet slicing season. With expanding domestic production and stable consumption, imports are projected to decline to around 0.7 Mt by 2021/22, leading to higher self-sufficiency. The Russian Federation joined the World Trade Organisation in December 2011. As part of its accession agreement, the Russian Federation will lower slightly its tariff on white sugar imports and has agreed to reform its sugar import arrangements in 2012 with a view to further market liberalisation. It is uncertain what effect, if any, these reform measures will have on the current pace of sugar import substitution and they are not specifically factored into the projections.

Figure 6.11. **Higher US consumption fed by rising Mexican imports**

Evolution of US sugar production, consumption and imports to 2021

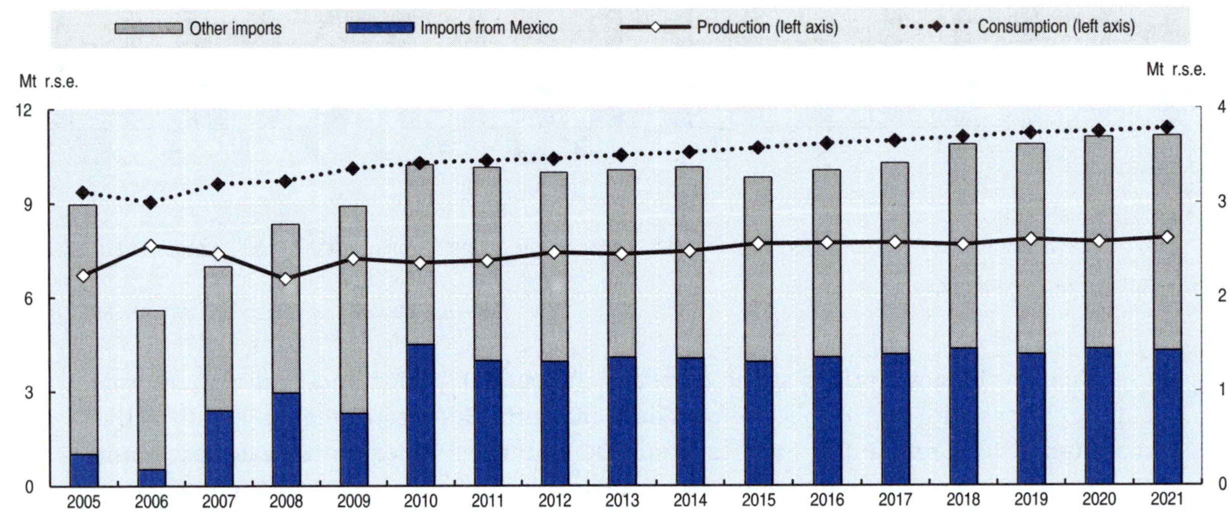

Source: OECD and FAO Secretariats.

StatLink http://dx.doi.org/10.1787/888932640027

Main issues and uncertainties

The medium term sugar projections discussed in this chapter are a conditional scenario of likely market developments based on macroeconomic, domestic sugar support arrangements, trade policies, and normal weather assumptions. Should any of these assumptions change, the resulting set of sugar projections would be different. A number of major uncertainties remain for the international sugar market. Despite some easing of the tight world market situation at the beginning of the outlook period and an expected acceleration of stock rebuilding as a substantial production surplus emerges, global sugar stocks are at historic lows. Until stocks are rebuilt, any further production disruptions in the main producing country of Brazil, or in other large sugar producers such as India or Thailand, could radically change the market outlook in the near term, igniting further price surges and bouts of high volatility, a leading characteristic of international sugar prices in recent years, and prolong the period of high world sugar prices.

The *Outlook* projection of structurally higher average sugar prices also remains dependent on world demand increasing more rapidly than production, as yield and area expansion are projected to slow in the next ten years when compared to the last decade. For the international sugar market an additional factor is the increasing demand for bio-ethanol, particularly in Brazil. This demand will consume an ever larger share of the domestic sugarcane crop in order to meet the fuel requirements from a rapidly increasing flex-fuel fleet of vehicles in Brazil and for prospective export destinations such as the United States. Higher oil prices and existing US biofuel mandates are expected to increase demand for Brazil's ethanol and help switch the current advantage enjoyed by sugar production in that country in favour of more ethanol output (which already uses over 49% of the sugarcane crop). This is likely to be beneficial to world sugar market prospects and producer returns, but less so for sugar consumers and users who will likely face higher prices. In the absence of this demand, the world sugar market would rapidly transform into the buyers' market of past years, characterised by high stocks and prolonged periods of low prices. Another uncertainty is the prospect for global economic growth, particularly in the near term. Economic prospects have weakened, particularly in Europe with the euro zone sovereign debt crisis and with rising oil and energy prices threatening growth prospects in other countries. A marked slowdown in economic activity would mean that aggregate demand for commodities, including sugar, would be negatively affected. Demand for sugar in the developed world is largely dominated by industrial use, comprising food manufacturing, food preparations, and beverages rather than direct consumption by households. These food and beverage sectors are most affected by changes in income and economic activity and the state of the business cycle. In addition, investment in new mills and the sugar sector generally is heavily dependent on access to credit which may become more difficult as banks deleverage to improve their balance sheets in a period of fiscal consolidation.

The world sugar market has undergone a number of reforms and structural changes over the past two decades. Nonetheless, it remains heavily distorted by government policy interventions that contribute to high world price volatility, particularly in periods of tight market fundamentals. Changes in domestic support policies and border measures, such as the imposition of export restrictions, can have a major bearing on trade volumes and international prices. Particular policy uncertainties for the world sugar economy during this outlook period are policy choices in the European Union on the future of production quotas after 2015 and the sugar provisions of the next US Farm Bill due in 2013. One issue with the elimination of production quotas in the European Union is the likely future production response in member states in the face of possibly lower internal sugar prices. The projections for the European Union in this *Outlook* assessment show production remaining stable with this policy reform. This reflects the outcome of some further internal adjustments in sugarbeet and sugar production, with some countries contracting production and others expected to expand output; for instance, as non-quota sugarbeet production for ethanol is attracted into higher value sugar production. This adjustment would be expected to occur in countries that are already significant producers of sugarbeet for non-sugar uses. As a consequence, following the removal of production quotas, and in the absence of any offsetting adjustments, there may be changes in the composition of feedstock use for ethanol production in the European Union.

Another uncertainty concerns the future demand for other caloric sweeteners. Associated with rising incomes, urbanisation and population growth in the developing world is a growing demand for soft drinks, beverages and processed foods which use sweeteners in there manufacture. Starch-based sweeteners such as high fructose corn syrup (HFCS) can be used as a direct substitute for sugar in liquid applications and this becomes attractive when sugar prices are maintained at high levels relative to grain prices, the key ingredients used in the production of HFCS. Changes underway in the Brazilian sugar industry, the leading sugar supplier to the world market, with the appreciation of the *real*, higher production costs and the need for the sugar to remain competitive with ethanol will help underpin continuing high world sugar prices in coming years. With a potentially larger margin between sugar prices and the raw material grain costs of HFS production, an increasing amount of the future growth in caloric sweeteners demand is likely to be meet by HFCS, particularly in the European Union, the Russian Federation, Mexico and China, at the expense of sugar.

Box 6.1. **Thailand: What are the prospects for the sugar industry?**[1]

Thailand took the world sugar economy by surprise in 2010/11 when it boosted its sugar production by 41% to over 10 Mt raw value and exports by 9% (2010/11 over 2009/10) in the space of a single season. The country is poised in 2011-12 for its second straight year of record production and exports that are well above the average of the past decade. Thailand is thus one of the few countries with the potential to expand in a situation where Brazil no longer has the lowest sugar production cost advantage over its competitors. Over the past decade, output has remained within a range of 5-8 Mt, raw value. The area under cane production has remained stable at around 1 Mha in recent years, but with cane yields and sugar extraction rates relatively low by international standards. However, good monsoon rains in the last two seasons have helped to improve the yield situation, along with some increase in planted area, and has lead to record production. This raises the question of whether this higher level of output represents the new standard from which there will be further growth and what it means for Thailand's export capacity.

Sugar output in Thailand tends to be constrained by the availability of sugarcane rather than crushing capacity which is in ample supply. As there is no zoning of cane areas in Thailand, farmers, the majority having small scale farms, can deliver to whichever mill offers the best price for cane. Sugar growers receive a high 70% of the revenue from sales of sugar and molasses (the residual 30% goes to sugar mills) and 13.9% of revenue from refined sugar export sales. Thailand is the second largest world sugar exporter, mainly of raw sugar. On the consumption side, wholesale and retail prices are controlled by a regulated structure for sugar values, including the allocation of sugar for the domestic and world markets by three quotas to ensure the domestic market is not oversupplied. These comprise Quota A – (domestic), Quota B – (long term contracts); and Quota C – (exportable surplus) covering the remainder. Domestic refined sugar prices are currently fixed at THB 23.50 (USD 0.73) per kg. Problems arise when world prices rise above this price, as has occurred in recent seasons, as it creates incentives for traders to buy sugar on the domestic market for sales in neighbouring countries, leaving the domestic market short of sugar. As a consequence, these revenue and pricing arrangements are currently under review by the government.

To a large extent the future level of sugar production in Thailand will be determined by how much cane farmers are willing to grow. This will be determined, amongst other things, by the long term competitiveness of sugarcane with cassava, the main competing crop. The outlook is for further growth in cane production and sugar output over the coming ten years, aided in the short term by lower yields for cassava due to insect infestation problems, and with an increase in new sugar mills, leading to a further expansion in processing capacity. At the same time, Thailand's sugar consumption, driven by more rapid growth in industrial use, could reach 3.5 Mt by the end of the ten year period; up some 32% on average for the 2009-11 period. This leaves export availabilities rising to 9.7 Mt by 2021, compared to a level of 7.3 Mt estimated for 2011/12.

The major destinations for Thai exports remain its ASEAN neighbours, as sugar import tariffs in the ASEAN Free Trade Area (and the common effective preferential treatment – AFTA-CEPT), enacted in 2010, will gradually be reduced from 28% in 2012 to 5% in 2015. This will benefit Thailand's access to markets within the region to the detriment of third country imports. The larger sugar crop will be accompanied by higher molasses production and a higher level of supplies available for ethanol production both domestically and for export. A summary of the results of the sugar sector in recent years and the projections for 2021 are presented in Figure 6.12.

Box 6.1. **Thailand: What are the prospects for the sugar industry?**[1] *(cont.)*

Figure 6.12. **Thailand's surge in production and exports to 2021**

Evolution of Thailand's sugar production, exports and price

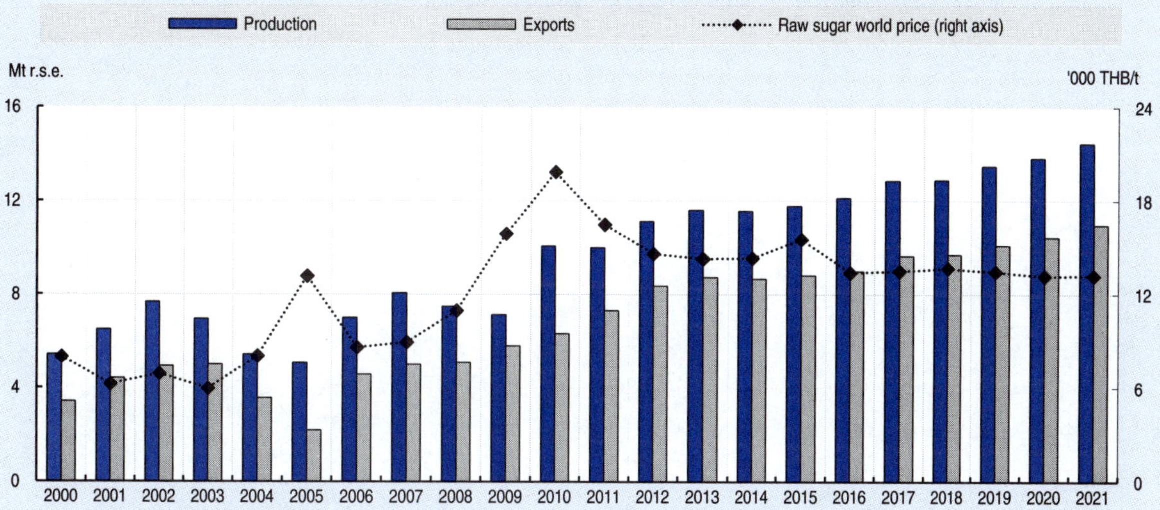

Source: OECD and FAO Secretariats.

StatLink http://dx.doi.org/10.1787/888932640046

1. *Thailand's Sugar and Fuel Ethanol Outlook*, Mecas (11)17, International Sugar Organisation, November 2011.

OECD-FAO Agricultural Outlook 2012

Chapter 7

Meat

Market situation

The market situation for the meat sector is characterised by high nominal output prices for all meats, underpinned on the demand side by rapidly growing developing economies and on the supply side by high input costs, notably for feed grain and energy related inputs such as transport and cold chain storage. As feed costs moderate somewhat, increased profitability should assure expansion. These factors tend to favour greater domestic supply responses in developing countries, particularly for cheaper meats and meat cuts (poultry), and also where low input systems, including pasture, predominate. On the policy front, the prospects of a further opening of international meat trade that may result from the accession to the WTO of the Russian Federation, which is among the world's largest meat importers, will render a favourable trade environment for the sector. While growth in production and trade is envisaged in the short term for poultry, pig and sheepmeats, beef will initially be constrained by herds which have depleted in recent years in major exporting regions.

Projection highlights

- The strong rise in feed grain prices in the past five years is now moving substantially through the market chain and, with the exception of poultry where adjustments have largely been made already, is being reflected in higher meat prices. Prices are projected to remain high throughout the next decade, and in real terms about 11%, 17%, and 4% above base period (2009-11) values for bovine, pigmeat, and sheepmeat respectively. Real prices for poultry are projected to remain close to current levels. For all meats, real prices are currently at their highest levels of the past 15 to 20 years, and little moderation is expected as long as feed and energy prices remain high.
- Higher prices for meat will induce supply response, albeit limited by higher input costs in addition to competition for land and water. The combined effect of these factors will slow global production growth for meat to 1.8% p.a. in the outlook period compared to 2.2% p.a in the previous decade. Bovine meat production is projected to increase 1.8% on average each year, while that for pigmeat and sheepmeat may grow 1.4% and 1.8% respectively. Poultry remains the fastest growing meat sector, with growth projected at 2.2% p.a. Developing countries will increase their share of global production in all meat categories, and by the end of the period will account for 58%, 64%, 63% and 78% of bovine, pig, poultry and sheepmeat production respectively. Increasing returns to scale will continue to concentrate production in fewer and larger farm units, not only in developed countries, but increasingly also in emerging countries. This structural change will continue to increase the reliance of meat production on feed grain inputs.
- World meat consumption continues to grow at one of the highest rates among major agricultural commodities. Growth in developing countries will capture 82% of the additional global consumption over the projection period. *Per capita* consumption will increase by 3.2 kg p.a., with poultry accounting for 70% of this increase. By 2021, consumers in developed countries will eat an extra 3.6 kg of meat *per capita* relative to the base period,

which will also be mostly poultry, except for Eastern Europe where consumption of red meats still has a substantial growth potential.

- Despite strong meat prices through the projection, meat imports by developing countries are expected to increase, driven by population and income growth and high income elasticity of demand. Equally so, strong prices will result in sustained export earnings, which will encourage large meat exporting countries to invest in international meat markets despite the high prevailing incidence of food-safety and sanitary import bans.

Market trends and prospects

Prices

Meat prices will remain on a high plateau during the outlook period under persistently high production costs due not only to high feed prices and energy related inputs including transport and cold chain supply costs, but also to increasingly more stringent food safety, environmental, and animal welfare regulations (traceability, housing, transportation, etc.). Nominal prices for beef and sheepmeat are projected to be USD 4 717/t c.w.e. and USD 4 812/t c.w.e. respectively in 2021, whereas already high pigmeat and poultry prices will increase to USD 2 380/t c.w.e. and USD 1 419/t r.t.c. respectively. Sheepmeat prices, which have recently seen a drop due to a higher supply and subdued demand in the European Union, are expected to remain at firm levels (Figure 7.1).

In real terms, meat prices in 2011 stood at 15-20 year highs. Feed costs have moderated in the past year, but they are anticipated to remain at high levels over the outlook period. Output price to feed price ratios for bovine and pigmeat, are anticipated to remain at low levels throughout the outlook period For poultry meat, which typically shows faster response, adjustment to higher feed costs has already taken place, and real prices over the next decade are anticipated to remain flat, compared to recent prices in markets (Figure 7.1).

Figure 7.1. **World prices in real terms expected to remain strong**

Nominal (left figure) *vs.* Real (right figure) meat prices[1]

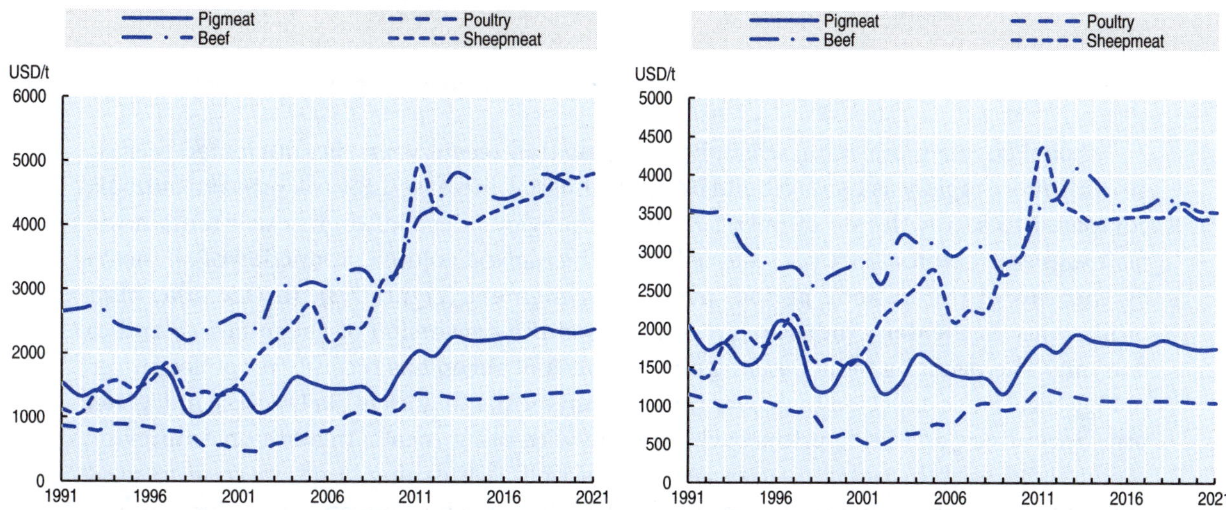

1. US Choice steers, 1100-1300 lb dressed weight, Nebraska. New Zealand lamb schedule price dressed weight, all grade average. US Barrows and gilts, No. 1-3, 230-250 lb dressed weight, Iowa/South Minnesota. Brazil average chicken producer price ready to cook.

Source: OECD and FAO Secretariats.

StatLink http://dx.doi.org/10.1787/888932640065

Production

Although meat prices will remain on a high, supply response may be limited in many countries and regions by constraints on natural resources, competition for land and water from alternative crops, and insufficient investments on infrastructure in key regions richly endowed with natural resources for livestock production (Brazil, the Russian Federation and Sub Saharan Africa). Annual world meat production growth during the outlook period is projected to slow from an average 2.2% the previous decade to 1.8% p.a., largely due to slower growth in Latin America, particularly Brazil and Argentina, which grew strongly in the previous decade. In the Russian Federation, growth in meat production is anticipated to slow as a result of accession to the WTO. Poultry and pigmeat production, which grew by a large 14% p.a. and 5% respectively in the last decade, are projected to grow in the 2% p.a. range over the outlook period. More generally, developing countries will capture 77% of the additional meat output growth over the outlook period (Figure 7.2).

Figure 7.2. **Meat production growth dominated by developing countries**

Production growth: By region and meat type, 2021 *vs.* base period (c.w.e. or r.t.c.)

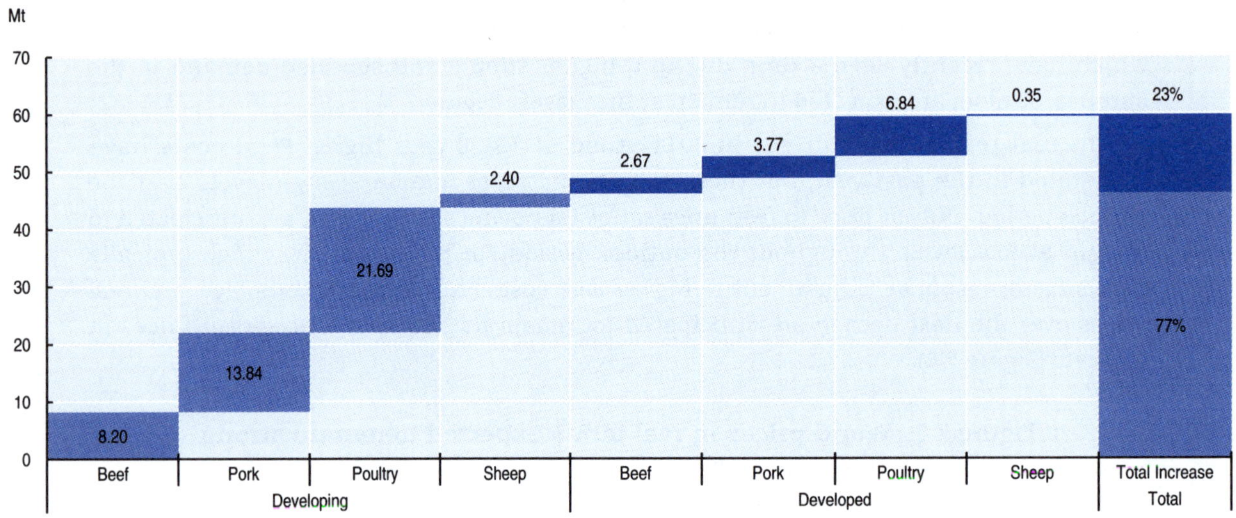

Source: OECD and FAO Secretariats.

StatLink http://dx.doi.org/10.1787/888932640084

Global bovine production, which has stagnated in recent years, is anticipated to start growing more rapidly as herds rebuild by 2014, and may increase 1.8% p.a. over the outlook, compared to a growth rate of only 1.2% in the previous decade. Poultry will remain the fastest growing sector (2.2% p.a.) and will have the highest volume production of all meats by the end of the outlook period, overtaking pigmeat (Figure 7.3). In Oceania, high profitability in the dairy industry has encouraged the conversion of sheep farms/land to dairying. The resulting reduction in global supplies of sheep has recently propelled prices to high levels. It is expected that, over the medium term, the flock is likely to expand, given price incentives, and bring prices back into line with other meats. In addition, improving productivity growth through better genetics as well as finishing lamb on grain to raise carcass weights, in certain countries, will contribute to raising sheepmeat production.

Productivity growth throughout the meat production chain has been significant in recent years. Despite rising costs, improved herds, breeding and herd management practices, especially improved feeding practices, have enabled growth in meat production

Figure 7.3. **Globally, poultry will overtake pigmeat as largest meat sector**

Production growth: By meat type, 1995-2021 (c.w.e. or r.t.c.)

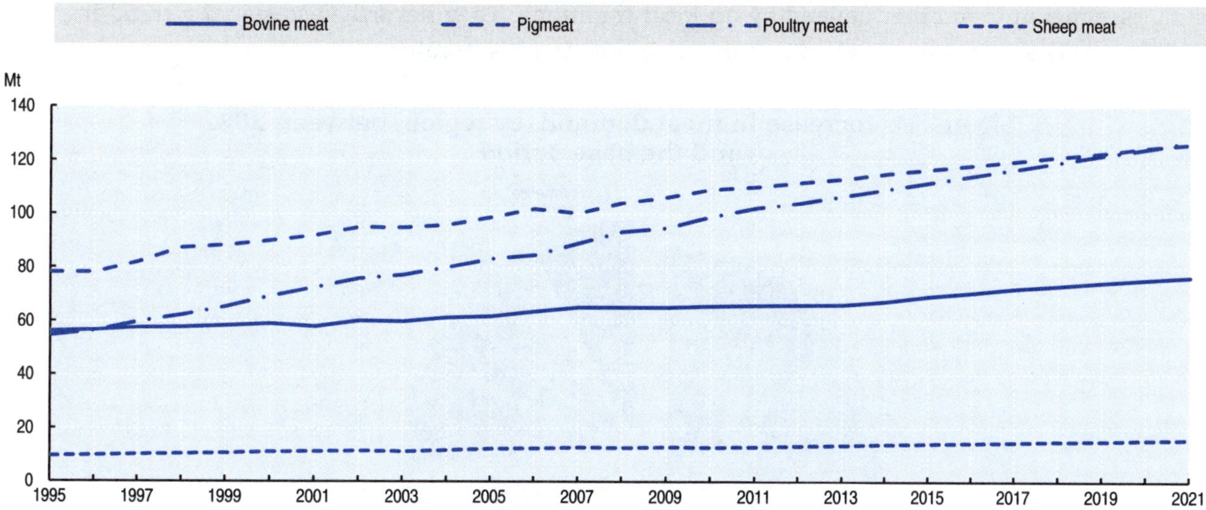

Source: OECD and FAO Secretariats.

StatLink http://dx.doi.org/10.1787/888932640103

and this trend is expected to continue through the outlook period (Box 7.1). Increasing productivity of livestock herds has been widespread, except in many African countries, where output per head has remained at very low levels for many years. Productivity in the meat sector is seen as critical in the long term, since it implies a lower level of inputs are required to produce a given output. For example, productivity gains that raise offtake ratios imply lower animal inventories that use extensive feed, land, water and other inputs, and help improve sustainability. The current projection foresees global increases in cattle numbers to 1.80 billion head, hog numbers to 0.97 billion, poultry to 24.3 billion, and sheep to 3.0 billion. Apart from increased farm productivity, improvements in supply chain management, in particular cold chain management, has and will continue to have a very important impact on the growth of this sector. This is especially true in many developing countries where storage and transportation of meat has been limited.

Consumption

Growth in demand will stem mostly from large economies in Asia, Latin America and oil exporting countries (Figure 7.4). Emerging economies will also increase their demand where income growth and urbanisation will strengthen the intake of animal proteins at the expense of foods of vegetal origin. Conversely in developed countries, where demand is largely saturated, a slowing down of income and population growth, ageing and the recurrence of food scares (E. coli and salmonella) will combine to curb demand for meats. The net result of these contrasting trends at the global level, points towards strong growth on a *per capita* basis, albeit lower than that experienced in previous decades.

Relative to the base period, by 2021 developed and developing country consumers put similar additional quantities of meat in their annual baskets: 3.6 kg and 3.2 kg r.w.t. respectively. However, the meats chosen by consumers are markedly different. Some 90% of the extra meat that consumers in developed countries put in their baskets is poultry, except for consumers in Eastern Europe where red meats have additional room for growth. Conversely, the extra meats which consumers in developing countries choose for their baskets is more heterogeneous, consisting of 62% poultry, 19% pigmeat, 13% beef, and 6% sheepmeat. These are averages, whereby *per capita* consumption is likely to change from one region to another depending on local traditions. They nevertheless signal a trend in markets as *per capita* meat consumption progressively saturates.

Figure 7.4. **Increase in meat demand, by region, between 2021 and the base period**

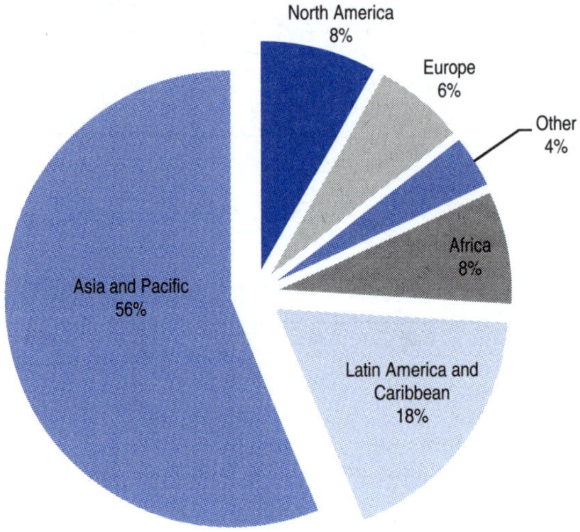

Source: OECD and FAO Secretariats.

StatLink http://dx.doi.org/10.1787/888932640122

Trade

Led mostly by an expansion of poultry and beef shipments, world meat exports will increase by 19% by 2021 relative to the 2009-11 base period (Figure 7.5). The bulk of the growth in meat exports is expected to originate largely from North and South America, which will account for nearly 70% of the total increase in all meat exported by 2021. These two regions will increase their combined share of total world trade in meat from 61% to 63%. US meat exports are projected to expand significantly in the outlook period, not only from the easing of BSE-related import restrictions, but also from the progressive lowering of import tariffs by Korea, following the coming into force of KORUS (bilateral free trade agreement) on 15 March 2012. Meat exports from the European Union are anticipated to stagnate over the outlook period due to a tight domestic supply of domestic produce following policy reforms. Japan is projected to remain the leading meat importing country by 2021, followed by China, Mexico and Saudi Arabia. The Russian Federation, expected to join the WTO shortly, remains one of the world's largest market players despite a significant fall in total meat imports.

Led by Brazil and the United States, beef trade during the outlook period will expand at 1.8% p.a. In the United States, larger exports stem from improved market access to FMD-free markets, including a lowering of tariffs by Korea and an expansion of the EU import quota for US beef free from growth-inducing hormones. A larger presence of US beef in the FMD-free "Pacific" market will affect the performance of Australia, Canada and New Zealand, whose exports are anticipated to stagnate. Exports of Brazilian beef to the Atlantic market, which in the past five years have been shrinking from the combined effect of strong domestic demand growth, sanitary export restrictions and falling output, will reach

Figure 7.5. **Evolution of world exports of beef, pigmeat, poultry and sheep**

Overall meat export to reach nearly 32 Mt by 2021, a 19% increase from the base period (c.w.e. or r.t.c.)

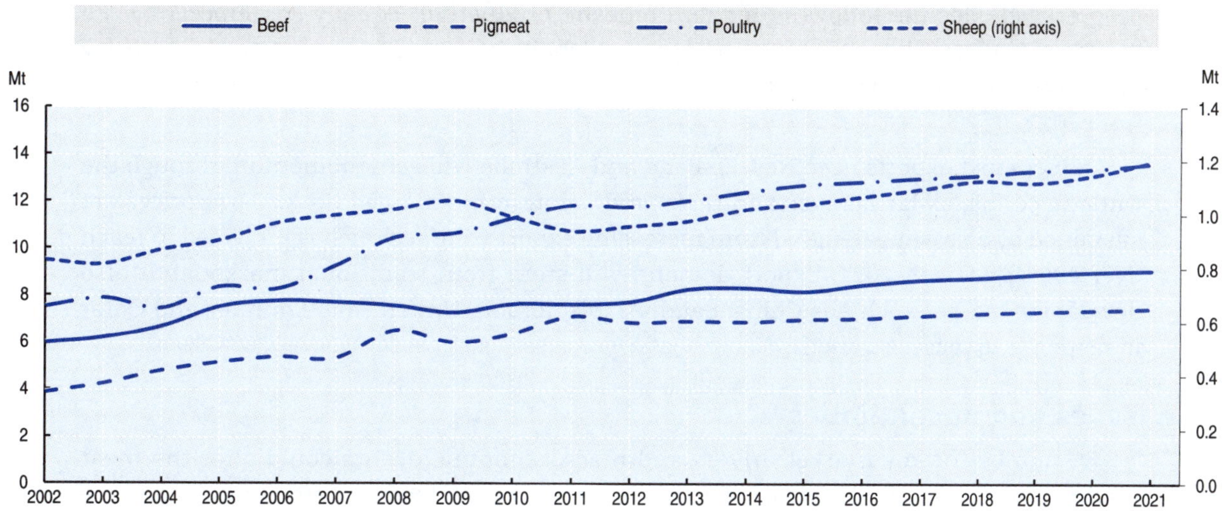

Source: OECD and FAO Secretariats.

StatLink http://dx.doi.org/10.1787/888932640141

an inflexion point early in the outlook period. The steady recovery of Brazilian beef exports will be due to an expansion of domestic output, better compliance to sanitary import regulations, and sustained import demand from the Middle East. India will benefit from an increasing consumer interest in buffalo meat and will rank as the fourth largest exporter of beef by 2021.

Aggregate growth on pigmeat trade will be relatively modest during the outlook period, but some changes in the composition of trade are to be expected. North American pork shipments will increase faster than those from the rest of the world, and the region will become the dominant player in world markets. Exports from Brazil, which until recently were expanding rapidly, will slow down as a result of an increase in domestic consumption and the recurrence of sanitary import restrictions. Net trade in China, where half of the world's pigmeat is produced and consumed, is not expected to change during the outlook period. In China, government policies will continue to support the pork industry through the scaling up of production and the modernisation of markets. These include buying into intervention stocks, setting up futures markets, supporting R&D and the scaling up of production facilities. Export growth from the European Union will stagnate as producers adapt their farms to new animal housing regulation to be enforced from 2013.

In poultry, a significant slowing down of annual trade growth is anticipated, from an annual rate of 5.5% in last decade to less than 2% during the outlook period. During the first part of the projection, exports will stagnate due to high poultry meat prices. Nevertheless, the adaptation of production to higher feed and energy costs is expected to induce structural and technological changes in the industry, boosting productivity, production and export growth during the second half of the projection period, when annual growth rates are expected to accelerate once again. The largest contributors to growth are the United States and Brazil, both of which will strengthen their dominance of world trade. By 2021, the United States and Brazil will generate nearly 80% of the expansion of world poultry trade.

World poultry import growth will be led by countries in the Middle East, Southeast Asia and Latin America. Purchases by Russia, once the world's largest importer, will progressively decline following higher domestic production. Equally so, imports by the European Union will stagnate from a larger availability of domestic produce as the industry quickly adapts to the EU Directive which banned the use of conventional battery cages from 2012.

Sheepmeat exports from New Zealand and Australia will gain momentum through the Outlook period, as farmers expand their production from enlarged flocks. Firm prices are sustained by a growing demand from increasing export volume which is projected to reach over 499 Kt c.w.e by 2021. Export demand will stem from traditional markets but also developing market such has China, Saudi Arabia, Jordan, United Arab Emirates and Qatar in response to income growth.

Main issues and uncertainties

A number of key market drivers and macroeconomic events could alter the meat market projections of this *Outlook*. The key issue will be the feed market situation, and the factors that will affect its evolution over the outlook period (Chapter 4). Given the sensitivity of the meat sector to macroeconomic conditions, any disturbance over the outlook period, particularly but not exclusively in emerging developing countries, could have a large impact.

The Russian Federation has traditionally been a top meat importer, but in recent years the pigmeat and poultry sectors have experienced sustained growth. This *Outlook* assumes that this trend will continue, albeit reduced, during the outlook period, with the Russian Federation achieving a higher degree of self-sufficiency and some exportable surpluses. However, if accession to the WTO does not occur as assumed, trade will be reduced even more than projected. China's net trade position *vis-à-vis* pigmeat is also a key uncertainty for world markets. Due to its extraordinary volumes both in terms of production and consumption, unforeseen events in China could easily induce import surges of pigmeat from the world market, with the potential to severely impact international markets. In North Africa and the Middle East, large importers of sheepmeat, poultry and beef, changes in oil prices, or as recently experienced the fallout from civil unrest, have the potential to impact world meat trade.

A certain number of animal diseases have the potential to affect domestic and regional meat production. Although the eradication of some has proved technically possible, it has also proved to be expensive. Countries and regions are therefore protective of their disease-free status and make strenuous efforts to sustain this situation. Foot and Mouth disease (FMD) is a case in point.

In addition to affecting domestic production, some animal disease outbreaks have also caused radical and lasting effects on trade. The world market for beef, for example, has for decades been divided into FMD-free trade routes (the "Pacific" market) and the rest of the world (the "Atlantic" market). The relatively recent episode of BSE (year 2004) is an example of the severity of the impact on world meat trade when the country affected is a large exporter. Disease outbreaks of zoonotic scope, such as H1N1, also loom as potential factors that could impact significantly not only across meat markets, but also on consumer behaviour.

The world meat market is also highly fragmented by country-specific legislation on food safety, and import restrictions pose a significant risk to the validity of the projections. For example, in May 2011, the Russian Federation imposed sanitary restrictions on meat imports from a number of Brazilian states. The ban resulted in a substantial contraction of bilateral trade on beef and pigmeat, and the end of two years of (almost) uninterrupted monthly increases of world meat prices. These projections assume that no import bans with significant and long-lasting effects on trade will occur during the outlook period.

Finally, environmental costs are rising for the production of virtually all meats, and novel legislation that conditions production to environmental protection may affect the growth of the sector. The livestock sector is considered by analysts and policy makers as a key contributor to anthropogenic greenhouse gas (GHG) emissions. As world population and income growth expand the demand for livestock products, these emissions are expected to increase. It remains uncertain the extent to which livestock production may be subject to carbon mitigation constraints over the next decade.

Box 7.1. **Productivity change in the meat sector**

Meat production has grown about 300% in the past 50 years, and as noted in this *Outlook*, is anticipated to be one of the fastest growing commodities, due largely to growing incomes and the westernisation of diets in many emerging economies. At the same time, livestock capital – the number of bovine, porcine, poultry and ovine animals has grown, at 57%, 137%, 400% and 49%, respectively. The change in "off-take", or the quantity of meat produced per animal, in inventory has therefore increased substantially over time. This means that fewer animals are required to achieve a given level of meat production. This partial productivity measure captures a number of changing characteristics in the meat sector including, the number of offspring per breeding animal, length of feeding period, and of course, the yield of meat for each animal slaughtered. Ultimately, higher off-take ratios imply a lower inventory of animal number or capital which is required to produce meat, and has considerable resource implications. Table 7.1 provides selected country examples of off-take ratios for different meats, recent trends, and projected future growth rates over the next decade.

Off-take ratios by country and by animal type may vary for a number of reasons. Meat production characteristics vary by animal and by country depending on pasture and arable land availability, social norms and the state of development. Large differences in off-take ratios can be observed, particularly noting that intensive operations normally indicated higher off-take ratios than less intensive ones. Grain-fed operations typically show higher off-take ratios, as animals may be slaughtered at a younger age and at higher weights. In the main, off-take ratios appear much lower in developing countries, particularly for bovine meat. Ratios appear very low for African countries, where growth rates are also much lower. Often, animals are kept for other reasons than simply meat production, such as for providing a source of wealth or, in the case of sheep, fibre such as wool. Historical growth in off-take ratios has been high for a number of emerging countries, particularly Brazil and China (and India and the Russian Federation for pigmeat). As these countries increase their meat production, higher off-take ratios will be important in controlling the size of their animal inventories and associated problems.

Estimated trend projections provided in Table 7.1 generally indicate that the rate of partial productivity growth is slowing in most countries. It should be noted that this lower growth is from a higher base. In general, except for many African countries, the gaps in off-take ratios have been converging to some degree, although not rapidly. There would appear to be substantial scope for increasing this measure of productivity in many countries, offering the potential to limit the growth of animal numbers over the long term, and minimise resource and environmental costs associated with larger numbers.

Box 7.1. Productivity change in the meat sector *(cont.)*

Table 7.1. Trends in meat off-take ratios in selected countries

	Bovine meat			Pigmeat			Poultry meat			Sheepmeat		
	Offtake ratio	Growth	Projected	Offtake ratio	Growth	Projected	Offtake ratio	Growth	Projected	Offtake ratio	Growth	Projected
	2005-09	1985-2011	2012-21	2005-09	1985-2011	2012-21	2005-9	1985-2011	2012-21	2005-09	1985-2011	2012-21
	kg/hd	%/yr	%/yr	kg/hd	%/yr	%/yr	kg/hd	%/yr	%/yr	kg/hd	%/yr	%/yr
Algeria	75	1.4	0.8	22	0.5	0.3	2	–0.6	–0.3	8	1.5	0.9
Argentina	60	0.5	0.3	122	4.2	2.3	14	4.6	2.5	3	0.7	0.4
Australia	158	1.3	0.7	144	1.2	0.7	9	1.4	0.8	1	3.2	1.8
Brazil	45	2.5	1.5	83	5.3	3.2	9	4.4	2.7	3	–1.5	–0.9
Canada	126	0.1	0.0	174	2.0	1.1	7	0.7	0.4	19	0.9	0.5
China	55	8.5	4.9	108	3.0	1.7	3	3.2	1.8	13	5.3	3.0
E27	91	0.1	0.0	144	1.0	0.7	8	0.5	0.3	10	–0.7	–0.4
Egypt	77	1.5	0.8	46	–6.5	–3.4	6	1.6	0.9	7	–4.1	–2.2
India	8	0.4	0.2	39	0.3	0.2	4	7.5	4.2	4	0.1	0.0
Indonesia	24	0.6	0.3	93	1.4	0.8	1	0.8	0.5	5	–0.3	–0.2
Japan	133	–0.3	–0.2	130	–0.2	–0.1	13	1.8	1.1	6	0.8	–
Malaysia	15	0.0	–0.1	101	0.5	0.3	5	0.1	0.1	1	1.3	0.6
Nigeria	15	–0.4	–0.3	31	–0.6	–0.4	1	0.3	0.2	8	3.0	1.7
Russia	80	–2.7	–1.7	122	3.2	2.0	5	8.4	5.2	9	–1.6	–1.0
South Africa	57	1.1	0.5	139	2.9	1.5	7.7	0.1	0.0	4.5	0.2	0.1
Tanzania	14	0.0	0.0	33	0.7	0.4	2	1.8	1.0	3	–0.2	–0.1
Thailand	37	1.2	0.6	108	1.4	0.7	5	–1.0	–0.6	4	–1.2	–0.6
United States	120	0.8	0.5	153	0.9	0.5	8	1.4	0.8	10	–1.5	–0.8
World	42	0.3	0.3	111	1.5	1.2	4.6	1.2	0.8	5.0	2.3	2.3

Note: Off-take ratios are computed as gross indigenous meat production divided by all animal inventories.
Trend growth rates are computed from trend regression over the period indicated, but shorter in where data are limited. Growth estimates for E-27 and world are limited to the period from 1996, the Russian Federation from 1992.
Source: OECD and FAO Secretariats.

StatLink http://dx.doi.org/10.1787/888932640578

Chapter 8

Fish and Seafood[1, 2]

Market situation

After the recovery in 2010, the seafood sector further expanded in 2011 and early 2012, as evidenced by higher fish production, demand, trade and prices. Preliminary data for 2011 indicate that total fishery production should reach about 154 Mt, with capture fisheries recovering by about 2% and aquaculture further growing by 6% compared to 2010. Total exports of fish and fishery products (including fishmeal and fish oil) set a new record in 2011, reaching over USD 126 billion, 16% more than in 2010.

Due to growing demand, for the first time the average world apparent *per capita* fish consumption reached 18.8 kg (live weight equivalent). Fish accounted for about 16% of the global intake of animal protein and 6% of all protein consumed.

Fish prices sharply rose during the first part of 2011, decreasing slightly towards the end of the year and into early 2012, but still remaining higher than year earlier levels. The FAO Fish Price Index indicates that current fish prices, on average, are at record highs, with an absolute peak reached in August 2011 (14% more than in August 2010), after which aggregate price levels slightly declined.

Over the past decades, global markets for fish and fish products have changed considerably. This is an ongoing process in which operators along the fisheries value chain (fishers, fish farmers, traders, processors and retailers) seek new opportunities, reduce production costs and profitable investments in an increasingly internationalised business environment. New products and production methods, fragmentation and outsourcing of production processes, and changing value chains are characteristics of the ever evolving nature of global fish markets. The particularity about fisheries is that it is part of a global commons that demands regional and international co-operation to ensure that fish stocks are exploited in a sustainable and responsible way.

Projection highlights

- World fisheries and aquaculture production is projected to reach about 172 Mt in 2021, a growth of 15% above the average level for 2009-11. The increase should be mainly driven by aquaculture, which will rise by 33% over the *Outlook* period compared to the 3% growth of capture fisheries. However, a slowing down of aquaculture growth is anticipated, from an average annual rate of 5.8% in the last decade to 2.4% during the period under review.
- The fish sector is expected to enter into a decade of higher prices, but also higher production costs. Due to the growing prices of fishmeal, fish oil and other feeds, the average price of farmed species should increase slightly more than that for wild fish during the *Outlook* period.
- Fisheries supply chains will continue to be globalised, with a significant share (34%) of total fishery production being exported.

- Apparent *per capita* fish consumption is expected to reach 19.6 kg in 2021, 16% more than the average level for 2009-11. Due to high fish prices, fish consumption growth is projected to slow to 0.3% per year over the projection period compared to 1.7% per year over the previous decade.

Market trends and prospects

Prices

World prices for capture, aquaculture and traded fish products are projected to continue increasing in the outlook period (Figure 8.1). The main drivers will be the underlying positive trend in demand, income and population growth, increasing meat prices, a generally weak US dollar and limited growth of capture fisheries production, as well as rising costs for some of the most important input factors such as energy, including crude oil and feed. In particular, as a consequence of slightly declining capture fisheries for reduction and preference for fishmeal and oil in certain animal production, prices for fishmeal and fish oil are expected to grow by about 59% and 55%, respectively, in nominal terms during the projection period. Against the backdrop of stagnant supplies increasing demand is expected to lead to an increase in the price ratio of fish to oilseed meal and oil, especially in assumed years of *El Niño*. This is a naturally occurring climatic event resulting in warmer sea surface temperatures in the Pacific Ocean and off the coast of South America that generally reduces fish catches, in particular of anchoveta (*Engraulis ringens*), a species mainly used for fishmeal and fish oil processing.

Figure 8.1. **Higher fish prices due to higher feed costs and strong demand**

Nominal terms

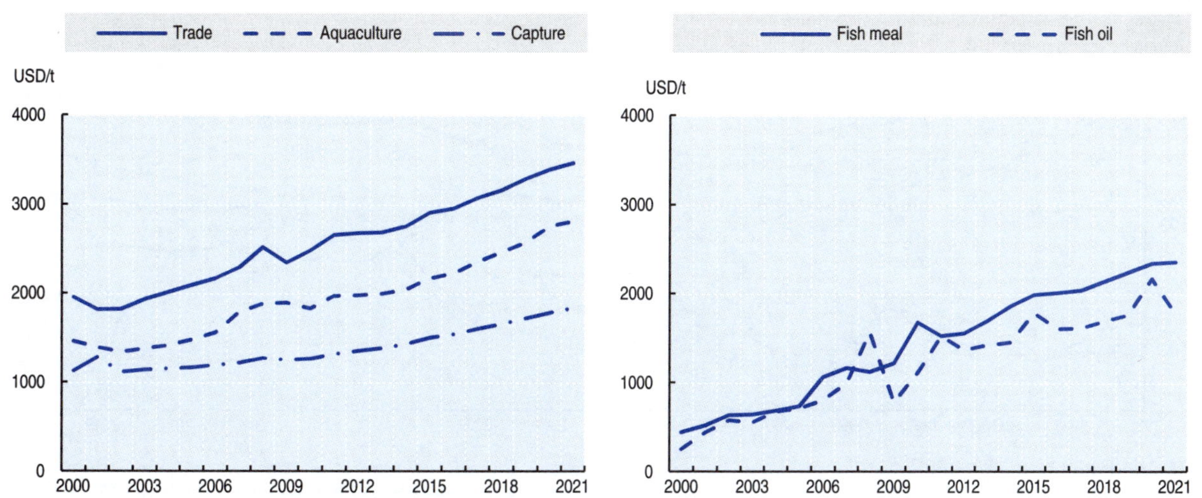

Note: Trade: world unit value of trade (sum of exports and imports). Aquaculture: world unit value of aquaculture fisheries production (live weight basis). Capture: FAO estimated value of world ex-vessel value of capture fisheries production excluding for reduction. Fishmeal: 64-65% protein, Hamburg, Germany. Fish oil: any origin, N.W. Europe.

Source: OECD and FAO Secretariats.

StatLink http://dx.doi.org/10.1787/888932640160

The impact of the coarse grain price on the price of aquaculture products will continue to be relatively modest, although it is expected to increase somewhat over the outlook period, as in the case of meat. The price ratio of aquaculture compared to fishmeal will gradually stabilise over the period under review. In order to compensate for the higher costs foreseen for fishmeal, fish oil and other feeds (crucial ingredients for carnivorous fish species), the average price for aquaculture production should increase slightly more than that for capture fisheries (excluding fish for reduction), by 48% compared to 43%) during the next decade. The difference in growth rates is also influenced by the increasing share of lower value fish in overall catch.

Higher prices for substitutes, meat in particular, will stimulate demand for fish and fishery products for human consumption. This in turn, will increase fish prices which will encourage more aquaculture production, in particular in developing countries, for exports as well as for local and regional consumption. During the last few decades, increased aquaculture production has shifted the demand for, and consumption of, species primarily wild-caught to primarily aquaculture-produced, with a decrease in their prices and a strong increase in their commercialisation. This was particularly evident in the 1990s and early 2000 (Figure 8.2), but subsequently due to higher costs, prices have started to rise again. Over the next decade, with aquaculture becoming a much larger share of total fish supply, the cyclical nature of certain aquaculture production and disease outbreaks could have a significant impact on price formation in the sector overall. This could lead to more volatility in the future. Furthermore, high feed prices could have an impact on the species composition of aquaculture output, towards those requiring less expensive feed or any feed for their production.

Figure 8.2. **Falling fish prices reversed by higher costs**

Real terms (2005)

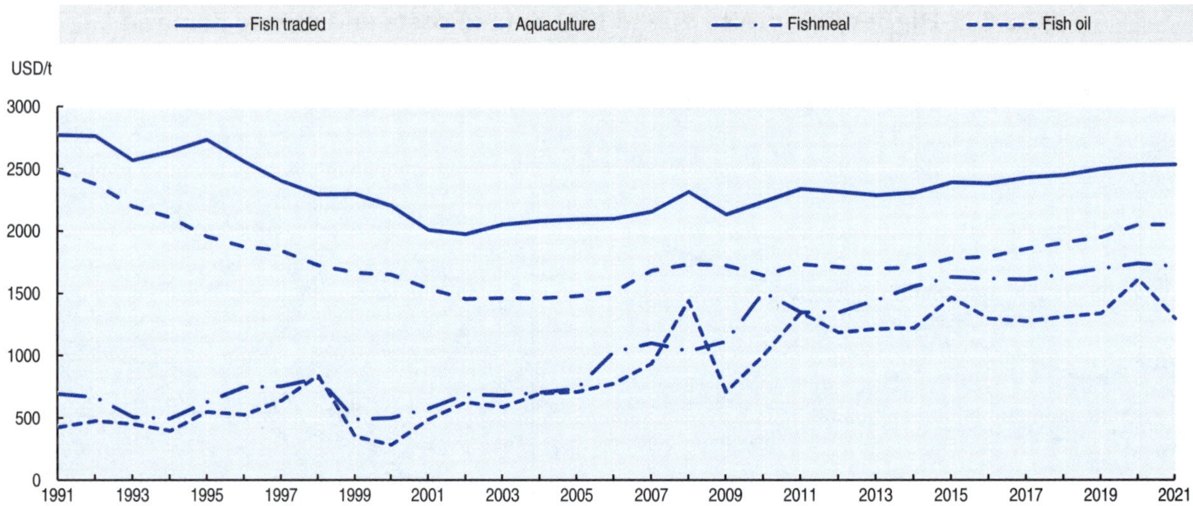

Source: OECD and FAO Secretariats.

StatLink http://dx.doi.org/10.1787/888932640179

As for many other commodities, the price of traded fish products will continue its upward trend over the outlook period. As international fish trade is expected to remain relatively liberal with low or minimal import tariffs in the main importing countries, global price changes should continue to be readily transmitted from one market to another. Price changes in international markets will over time have spillover effects on non-traded species as well. For individual fisheries commodities, price volatility could be more pronounced as they can be influenced by supply swings caused by changes in catch quotas, from disease outbreaks in the aquaculture sector as well as volatile feed prices.

Production

Under the set of assumptions used in this *Outlook* and stimulated by higher demand for fish, world fisheries production will continue to expand over the course of the projection period, reaching 172 Mt in 2021. This represents a growth of 15% over the average 2009-11 level. World capture fisheries should slightly increase from around 90 Mt to about 93 Mt over the next decade. This limited rise should derive from improved catches in some fishing areas where stocks of certain species are recovering as well as from reduced onboard waste and discards, stimulated by higher prices in the market. A 2% decline of capture fisheries is assumed in years affected by the *El Niño* phenomenon.[3]

Surging demand for fish and fishery products will mainly be met by growth in supply from aquaculture production, which is expected to more than offset the effects of rather static capture fisheries production. Aquaculture production is projected to reach about 79 Mt in 2021, with a 33% growth over the average level for 2009-11. However, overall the annual growth rate for the projection period is estimated to slow down from 5.8% in the previous decade to 2.4%. This decline will be mainly caused by water constraints, limited availability of optimal production locations and the rising costs of fishmeal, fish oil and other feeds. Notwithstanding the slower growth rate, aquaculture will still continue to be one of the fastest growing animal food-producing sectors, with total production volume (capture and aquaculture) exceeding that of beef, pork or poultry (Figure 8.3). Products derived from aquaculture will contribute to an increasing share of global fishery production, growing from 40% on average in 2009-11 to 46% in 2021. By 2018, farmed fish is expected to exceed captured fish for human consumption[4] for the first time, and its share is projected at 52% in 2021 (Figure 8.4).

Figure 8.3. **Aquaculture keeps total fishery production volume above beef, pork or poultry**

Meat and fishery production in dressed weight or eviscerated basis

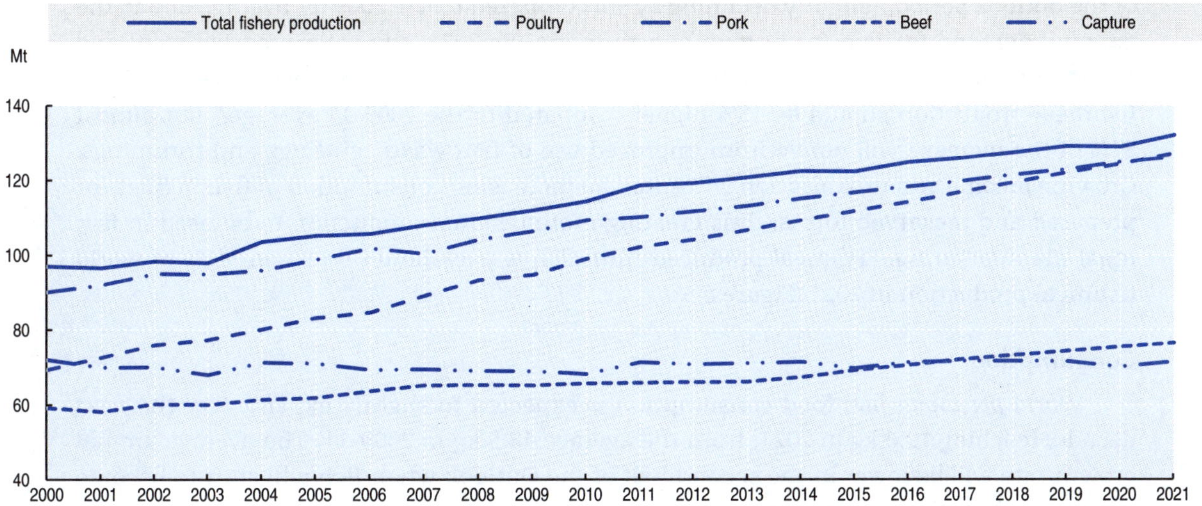

Note: Total fishery production: capture + aquaculture. Beef and pork on a dressed weight basis; poultry and fish on an eviscerated basis.

Source: OECD and FAO Secretariats.

StatLink http://dx.doi.org/10.1787/888932640198

Aquaculture production is expected to continue to expand on all continents, with variations across countries and regions in the product range of species and product forms. Asian countries will continue to dominate world aquaculture production, with a share of

Figure 8.4. **Aquaculture overtakes capture fisheries for human consumption**

Fishery production in live weight equivalent

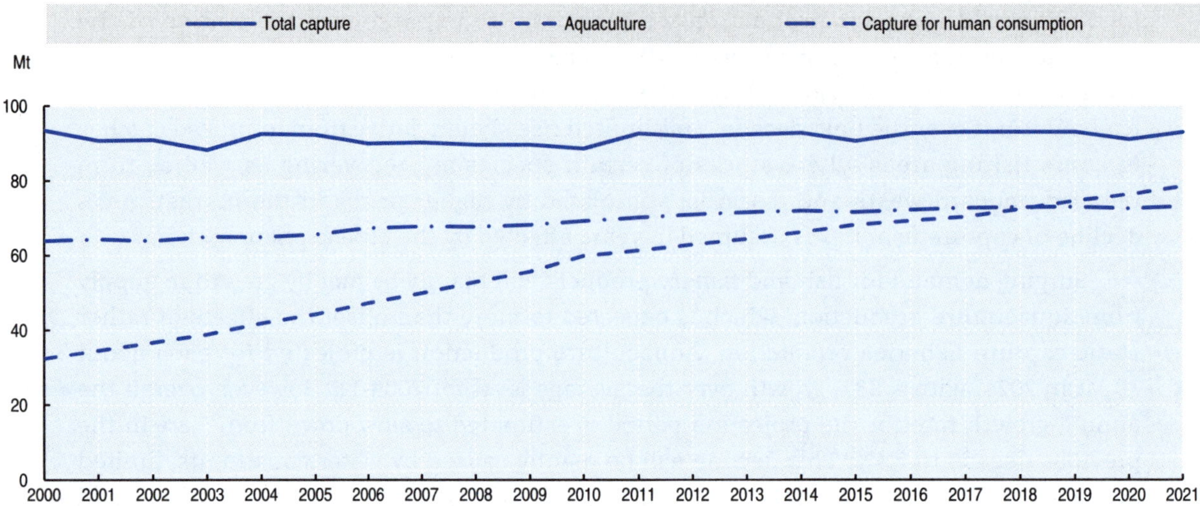

Source: OECD and FAO Secretariats.

StatLink http://dx.doi.org/10.1787/888932640217

89% in 2021, with China alone representing 61% of total production. Other major increases are expected in Latin America, in particular in Brazil (+68%) due to significant economic investments in the sector. African production should also expand over the projected period by 39% (reaching 1.7 Mt) due to the private sector capacity put in place in the 2000s, but also in response to economic growth, rising local demand and local policies promoting aquaculture.

The portion of capture fisheries used to produce fishmeal will be about 17% by the end of the outlook period,[5] slightly declining by 6% compared to the 2009-11 average due to the growing demand for fish for human consumption. By 2021, the estimated fishmeal and fish oil production, in product weight, should reach 6.8 Mt and 1.1 Mt, respectively. In 2021, fishmeal production should be 15% higher compared to the 2009-11 average,[6] but almost 87% of the increase will derive from improved use of fish waste, cuttings and trimmings. Growing income and urbanisation will entail an increasing consumption of fish in fillets or prepared and preserved forms, thus creating more residual production to be used in fish meal manufacturing. Fishmeal produced from fish waste should represent 43% of world fishmeal production in 2021 (Figure 8.5).

Consumption

World *per capita* fish food consumption is expected to slightly expand over the next decade, reaching 19.6 kg in 2021, from the average 18.5 kg in 2009-11. The average annual growth rate will be lower in the second half of the Outlook when fish will start to become more expensive than red meats. *Per capita* fish consumption will increase in all continents, except in Africa due to population growing faster than supply, with Oceania showing the highest growth rate (Figure 8.6). Notwithstanding *per capita* fish consumption will generally continue to be higher in more developed economies than in developing countries (25.0 kg as against 18.4 kg in 2021), it will grow more rapidly in developing countries during the period under review (+7.1% *versus* +5.8%).

Figure 8.5. **Increasing share of fishmeal from fish residues**

Fishmeal production in product weight

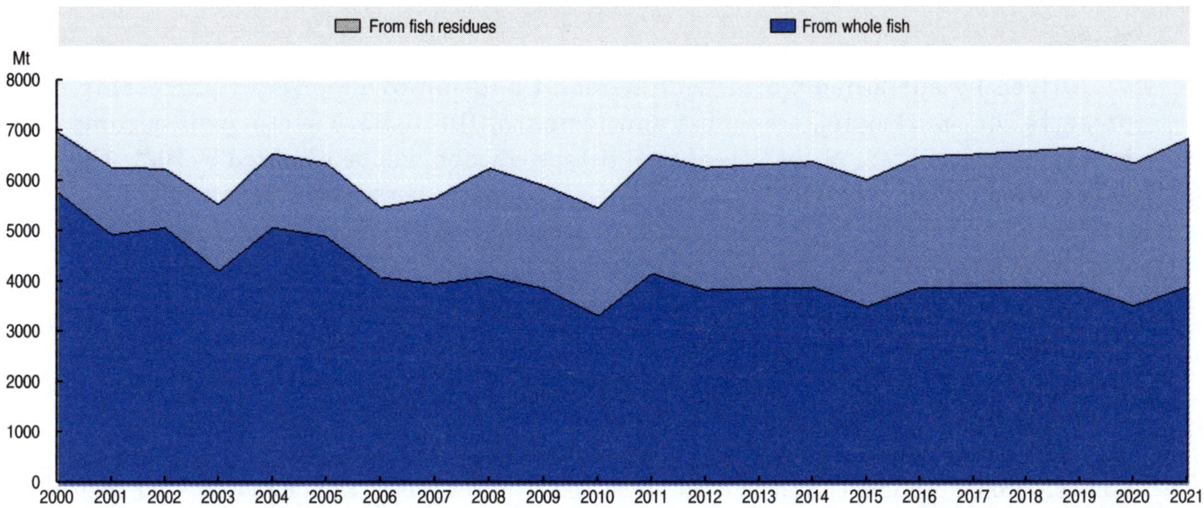

Source: OECD and FAO Secretariats.

StatLink http://dx.doi.org/10.1787/888932640236

Figure 8.6. **Growth of *per capita* fish consumption, except Africa**

Kg/capita in live weight equivalent

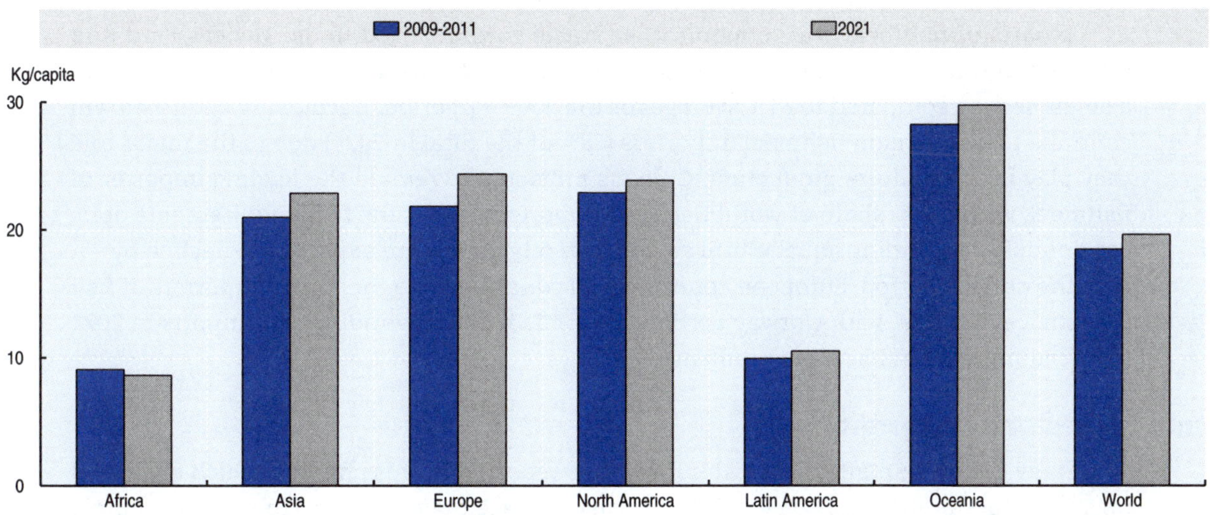

Source: OECD and FAO Secretariats.

StatLink http://dx.doi.org/10.1787/888932640255

The total amount of fish consumed will differ among regions and countries, reflecting the different levels of availability of fish and other alternative products, including the accessibility of aquatic resources in adjacent waters, diverse food traditions, tastes, prices and seasons. Growth in consumption will also be the result of complex interactions between several factors, including rising living standards, population growth, increased trade and transformations in food distribution and in the retail sectors, dietary changes linked to rapid urbanisation with increase in demand for animal proteins and growing emphasis on fish as a healthy and nutritious food. For example, annual *per capita* apparent fish consumption varies from less than 1 kg in Ethiopia to more than 70 kg in Norway and the Republic of Korea.

Trade

Driven by sustained world-wide demand and improvements in processing, preservation, packaging, transport and logistics, the fishery sector will become increasingly globalised. About 34% of total fish production will be exported by 2021. The fishery supply chain will remain rather complex, primarily because of the outsourcing of processing activities to several developing countries, including China, India, Thailand and Viet Nam.

In quantity terms, world trade of fish for human consumption is expected to expand moderately over the outlook period (+25%). However, the annual growth rate of exports will decline from 3.6% for the last decade to 1.9% per year over the next ten years. The share of world fish imports for human consumption going to developed countries will gradually fall from 59% to 56%. This is mainly due to the growing imports by developing countries for domestic consumption as well as of unprocessed fish to be used as raw material for their processing industries. Developing countries will continue to account for about 67% of world exports. Exports will be driven by Asian countries, which remain very competitive and are expected to benefit from growing investment in the aquaculture sector. In 2021, 55% of world fish exports for human consumption will originate from Asia, with China as the leading world exporter.

Exports of fishmeal will remain rather stable in the period under review. Peru and Chile will continue to be the leading exporters, with a 58% share of total fishmeal exports, a slight decline compared to 61% average for the 2009-11 period. Developing countries will continue to be the main fishmeal importers (68% of the total in 2021) due to the major role they play in aquaculture production. China is projected to remain the leading importer of fishmeal, with a 40% share of world fishmeal imports, to meet the feed requirements of its continuously expanding aquaculture sector. Fish oil exports are expected to decline by 4% over the outlook period. European countries will continue to be the major importers of fish oil with a 47% share, with Norway accounting for 21% of the world fish oil imports in 2021 to be mainly used in its salmon industry.

Main issues and uncertainties

Many factors can affect the fish projections reported in this chapter, which are based on specific assumptions. The next decade is likely to see major changes in macroeconomic environment, international trade rules and tariffs, market characteristics, resources and social conduct, whose effects can influence fish markets in the medium-term. Climate change impacts may also bring increasing uncertainty in many food sectors and might represent a compounding threat to the sustainability of capture fisheries and aquaculture development. These possible events take place in the context of other global social and economic pressures on natural resources and ecosystems, including environmental degradation and increasing land and water scarcity. New climate adaptation approaches will likely have to be integrated into the processes of improving fisheries governance. Action may also be required to secure conservation of aquatic ecosystems and safeguard stocks and productivity through technological innovation, investment in R&D and a more closely controlled approach to fisheries management. Furthermore, increased risks of species invasions and the spread of diseases raise additional concerns. Fish diseases could have major impacts on supply, demand and trade at domestic and international markets, since resulting trade restrictions might alter markets for extended periods of time.

Considerable benefits can accrue from rebuilding fisheries, an urgent task which is high on the international policy agenda. The OECD Fisheries Committee decided to contribute to efforts by member countries to rebuild their fisheries, where needed, by providing an analysis of the main policy issues. The focus was on rebuilding fisheries, which is a broader approach than rebuilding fish stocks, and took into consideration the social, economic and environmental dimensions. The outcome of this project, the study "The Economics of Rebuilding Fisheries", is a set of principles and guidelines that help policymakers in their rebuilding efforts, taking into account the economic and institutional aspects. Those practical and evidence-based principles and guidelines aim to ensure that rebuilding plans are examples of good governance which implies inclusiveness, empowerment, transparency, flexibility and predictable sets of rules and processes. Rebuilding of fisheries may imply a change in fisheries management settings and reform towards the use of market based instruments. The Principles and Guidelines have been adopted as an OECD Council Recommendation.

As production from capture fisheries in essence has remained virtually constant, further aquaculture growth will be needed to meet rising global demand for seafood. However, many constraints might affect the production prospects of this sector, including growing scarcity of water and limited opportunities for sites for new operations given multiple users of coastal and riparian areas, carrying capacity of the environment for nutrient and pollution loading and a less permissive regulatory environment. If not guided and monitored adequately, aquaculture expansion may contribute to environmental problems including degradation of land and marine habitats, chemical pollution, endangering biodiversity through escapees and reduction of fish resistance to diseases. Inadequate biosecurity measures and disease outbreaks can also cause large economic losses to the sector. Meeting the future demand for food from aquaculture will also depend on the availability of inputs, including fish seeds[7] and feeds in the requisite quality and quantities. Continued progress in developing terrestrially-sourced substitutes for fishmeal and oils will help support continued growth in aquaculture.

Consumer concerns related to such issues as animal welfare, food quality, production and processing methods may further cause uncertainties in the fish sector. Especially in the more affluent markets, consumers are increasingly requiring high standards of quality assurance and demanding guarantees that the fish they purchase are produced sustainably (Box 8.1). The stringent quality and safety-related import standards, together with requirements for products meeting international animal health and environmental standards and social responsibility requirements, might act as barriers for small-scale fish producers and operators to penetrate international markets and distribution channels. Future prices might be influenced not only by higher feed prices, but also by the growing introduction of more rigorous regulations related to environment, food safety, environmental traceability and animal welfare regulations.

Box 8.1. **Certification and sustainability of fisheries**

Concerns about sustainability and the effectiveness of fisheries management have resulted in demand from NGOs, retailers and consumers for assurances that the food they purchase has been sustainably produced. This has led to the development and implementation of a number of primarily private initiatives that have established eco-labels and certification schemes that claim to provide credible information to the consumer. These schemes establish a sustainability standard that fisheries may be certified against. Their label on fish products tells consumers that the fish they purchase meet the standards set by the labelling organisation, helping consumers make responsible choices when they purchase fish. The Marine Stewardship Council (MSC), established by the World Wildlife Fund, is the most popular, but several competing initiatives exist. The existence of different eco-labels in the marketplace raises concerns regarding consumer confusion, which weakens the effectiveness of the labelling effort. Retailers play an important role in bringing certified fish products to the marketplace, and often do so as part of their overall corporate social responsibility commitments. Retailers must choose which labelling organisation to work with in sourcing products, which risks segmenting markets according to where different labels are in use. This can in some cases limit the options available to the consumer-termed "choice editing". This is problematic when the retailer makes choices that are most profitable for them but may not serve the public interest and policy objectives of governments. The challenge is for private labelling initiatives to provide consistent and trusted information to consumers that helps foster sustainability throughout the fishery production chain.

Figure 8.7. **Volume of MSC certified fisheries**

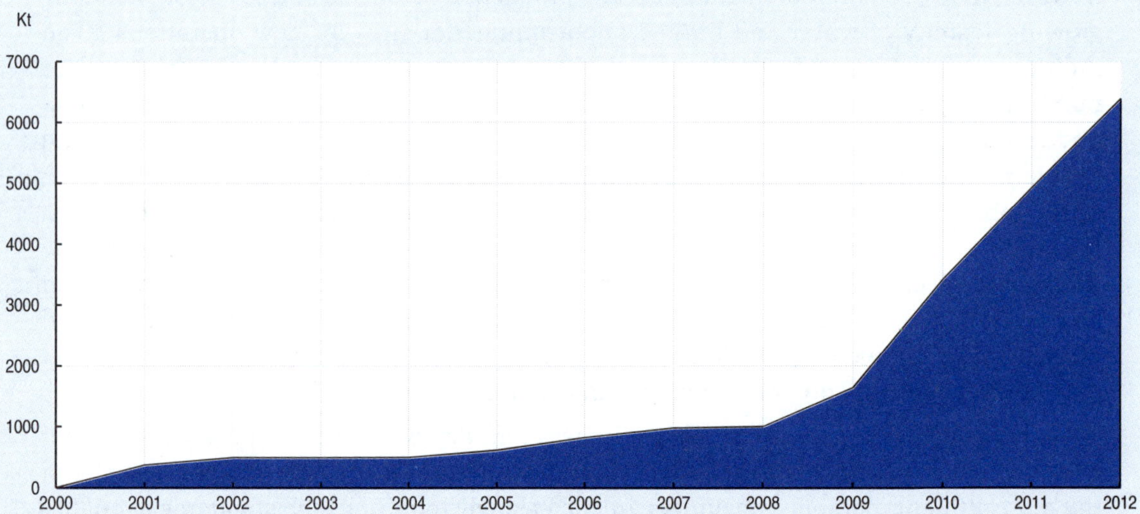

Source: Marine Stewardship Council.

StatLink http://dx.doi.org/10.1787/888932640274

Notes

1. The terms "fish and seafood" or simply "fish" indicate fish, crustaceans, molluscs and other aquatic invertebrates, but excludes aquatic mammals and aquatic plants.
2. For the second time, this *Outlook* publication includes a chapter illustrating the main results of the dynamic policy specific partial equilibrium model on fish. At present, it is a standalone model using the same macroeconomic assumptions, the same feed and food prices employed or generated by the agricultural market model Aglink-Cosimo. The baseline is deterministic and assumes normal weather and production conditions, with the exception of the impact of the *El Niño* phenomenon set for selected Latin American countries in 2015 and 2020.
3. In the model the impact of the *El Niño* is set in the years 2015 and 2020.

4. In last year's *Outlook* aquaculture was expected to exceed capture by 2015, however, this projection has been revised due entirely to a downwards revision in the amount of non-food estimates for China's apparent consumption starting from 2000 data, to reflect improved national information on the sector.
5. That share will be lower in years of *El Niño* (set in 2015 and 2020) due to reduced catches of anchoveta.
6. The reference point is low because of *El Niño* in 2010.
7. Fish seeds indicate eggs, spawn, offspring, progeny or brood of the aquatic organism (including aquatic plants) being cultured. At this infantile stage, seed may also be referred to or known as fry, larvae, postlarvae, spat, and fingerlings. They may originate from two principal sources: from captive breeding programmes or caught from the wild.

OECD-FAO Agricultural Outlook 2012

Chapter 9

Dairy

Market situation

After the dramatic fall in 2009 of international dairy prices, these rebounded in 2010 and significantly strengthened through to the first half of 2011. Sustained imports of milk powder by South East Asia, Mexico and North Africa (mainly Algeria) have been predominantly behind this price firmness. China, in particular, has continued to underpin the global dairy markets as the demand fuelled by a rapidly growing middle class with increased disposable income continued to outstrip the domestic supply, which continues to be confronted with milk safety issues. Global supply response, stimulated by high returns and excellent pasture conditions in Oceania and parts of Latin America, eventually caused prices to decline in the second half of 2011. The production gains were translated into increased exports, confirming the remarkable growth in dairy trade recorded since 2009. Export volumes increased mainly from New Zealand, Argentina, the United States, and the European Union. The decline in international dairy prices are expected to be nowhere near that experienced in 2009 as global markets continue to enjoy strong demand growth in developing countries.

Projection highlights

- The majority (70%) of global milk production gains over the outlook period is anticipated to come from developing countries, particularly India and China. The projections indicate that the total milk production of developing countries will overtake that of developed countries from 2013 onwards. The average growth rate of global milk production for the projection period is estimated at 2%, only slightly below the 2.1% level witnessed in the last decade. The dairy sector remains among the fastest growing sectors covered in the *Outlook*.
- After a downward correction from peak 2011 levels, prices in nominal terms are projected to increase by about 2% annually from 2014 onwards, reflecting increasing production costs and growing demand driven by rising population and incomes (Figures 9.1 and 9.2). In real terms, world market prices are expected to average 6% (cheese) to 30% (butter) higher over the projection period compared with the previous decade. Price volatility is likely to remain an issue for the outlook as dairy markets remain thin in volumes of milk traded and the small number of players that dominate export trade.
- The recent growth in trade is expected to continue, particularly for milk powder; the trade of SMP and WMP is expected to increase from the 2009-11 base period by 34% and 30%, while global cheese and butter trade is projected to increase by 27% and 20%, respectively. Butter markets will remain thin with volumes bypassing only 1 million tonnes by the end of the projection period.
- Dairy product consumption in developed countries may increase only modestly (with the exception of cheese), while in developing regions the consumption of all products is expected to increase vigorously at around 30% from the base period. This increase is driven by increasing population, income levels, and the growing influence of retail chains and multinational companies which facilitate improved consumer access to dairy products.

Figure 9.1. **World dairy prices in nominal terms**

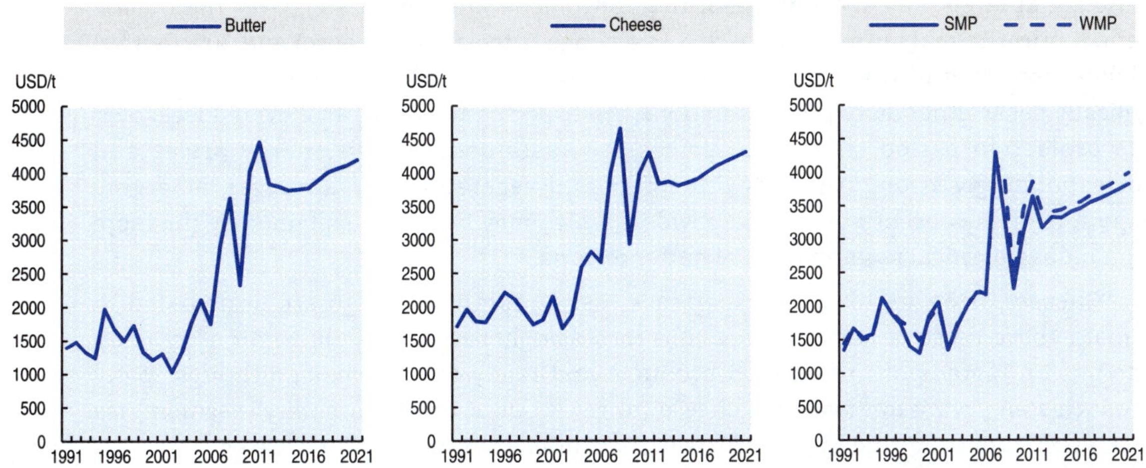

Note: Butter: f.o.b. export price, 82% butterfat, Oceania. Cheese: f.o.b. export price, cheddar cheese, 39% moisture, Oceania. SMP: f.o.b. export price, non fat dry milk, 1.25% butterfat, Oceania. WMP: f.o.b. export price, 26% butterfat, Oceania.

Source: OECD and FAO Secretariats.

StatLink http://dx.doi.org/10.1787/888932640293

Figure 9.2. **World dairy prices in real terms (2005 USD)**

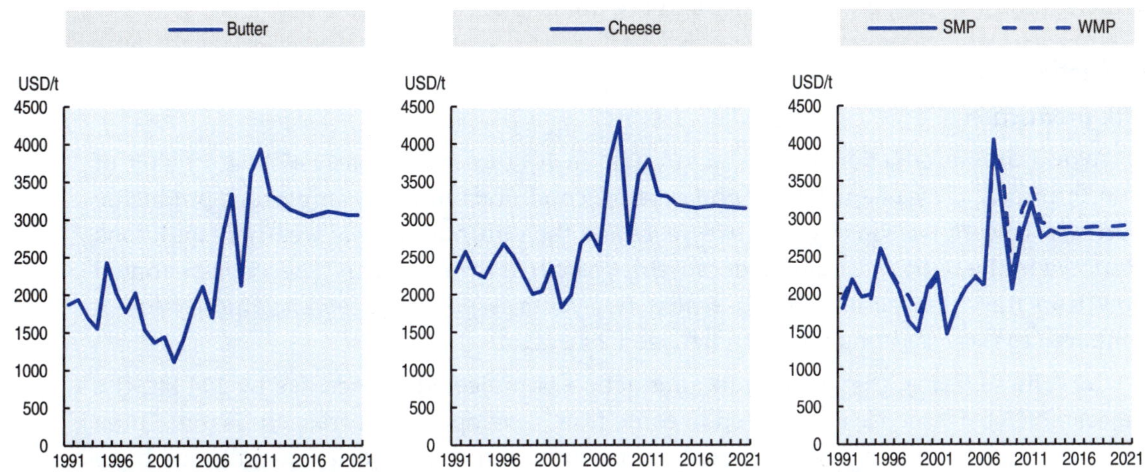

Note: Butter: f.o.b. export price, 82% butterfat, Oceania. Cheese: f.o.b. export price, cheddar cheese, 39% moisture, Oceania. SMP: f.o.b. export price, non fat dry milk, 1.25% butterfat, Oceania. WMP: f.o.b. export price, 26% butterfat, Oceania.

Source: OECD and FAO Secretariats.

StatLink http://dx.doi.org/10.1787/888932640312

Market trends and prospects

Prices

International dairy prices are expected to decrease from the elevated levels of 2011. The projected decline in international prices should be moderate as strong demand in developing countries, particularly in South East Asia, continues to underpin markets, despite a large supply response. As demand and supply adjust and markets rebalance at lower price levels, the international dairy quotations are projected to start increasing by about 2% annually from 2014 onwards, reflecting increasing production costs and growing demand driven by rising population and incomes (Figure 9.1)

Prices in real terms are anticipated to stay relatively flat over the next decade, although at levels well above those of the previous decade (Figure 9.2). Over the *Outlook* period, prices in real terms are expected to average between 6% (cheese) and 30% (butter) higher compared to the average levels of the last decade. Butter prices, for decades typically below other dairy price quotations, are expected to stay at an elevated level over the projection period, as milkfat prices are sustained by high energy prices and correspondingly strong prices of other fats and vegetable oils. Moreover, emerging exporters often concentrate on milk powders which may be partly explained by the more demanding logistical requirements of butter exports.

Cheese prices are likely to experience less strength, although are anticipated to remain above those of butter. Cheese has increasingly become a bulk commodity often used as an ingredient in fast food products and ready-to-eat-meals, although many producers are trying to re-valorise cheese (to get it out of the "commodity" segment) *via* increased innovation, and improved traditional varieties and speciality cheeses.

The *Outlook* price projections reflect the usual assumptions of stability in weather and in economic and policy conditions. It follows that actual price outcomes are likely to exhibit significant annual variations about these trend projections. Nevertheless, following dramatic market swings over the period 2007-10, a strategy to mitigate the volatility and manage risk could be anticipated on the part of dairy companies, traders and dairy farmers, thus lowering the probability and the impacts of future instability.

Production

Milk production

World milk production increased strongly in 2011 (estimated at 2.4%) as a result of good returns and excellent fodder and pasture conditions in many important producing countries. Lower prices in the next two years and the return to normal weather conditions (assumed in the *Outlook*) would moderate this growth in the short run. The average annual growth in milk production over the next ten years is projected at 2%, slightly lower compared to the 2.1% annual growth between 2002-11.

In volume terms, the global milk production is projected to increase by 154 Mt, the majority (70%) of which is anticipated to come from developing countries. India and China alone account for nearly 40% of global gains. Although the majority of milk has been traditionally produced in the developed world, the projections indicate that developing countries will bypass these levels in 2013 (Figure 9.3).

Regional differences in the growth of milk production are expected to persist. Growth levels are driven by market and policy context, the milk-feed price ratio, competition for feed and land, as well as water and other environmental constraints. In the context of higher energy and feed prices, pasture-based milk-producing systems, such as those in Oceania and Latin America, are expected to strengthen their comparative advantage (Figure 9.4). The growth in milking animal inventories is expected to slow down in developing countries, while the negative trend in developed countries of the last decade is expected to abate, mainly as herd declines in Europe moderate and stabilise in Australia.

In New Zealand, milk production increased in 2011 (year ending 30 May) by 6.5% after a poor 2010, with excellent autumn pasture conditions balancing the impact of a dry spring. Pasture conditions in 2012 are likely to be the best in a decade, driving a further increase in production. The projection assumes a return to normal weather conditions hence the future

Figure 9.3. **Evolution of milk production in developing and developed countries**

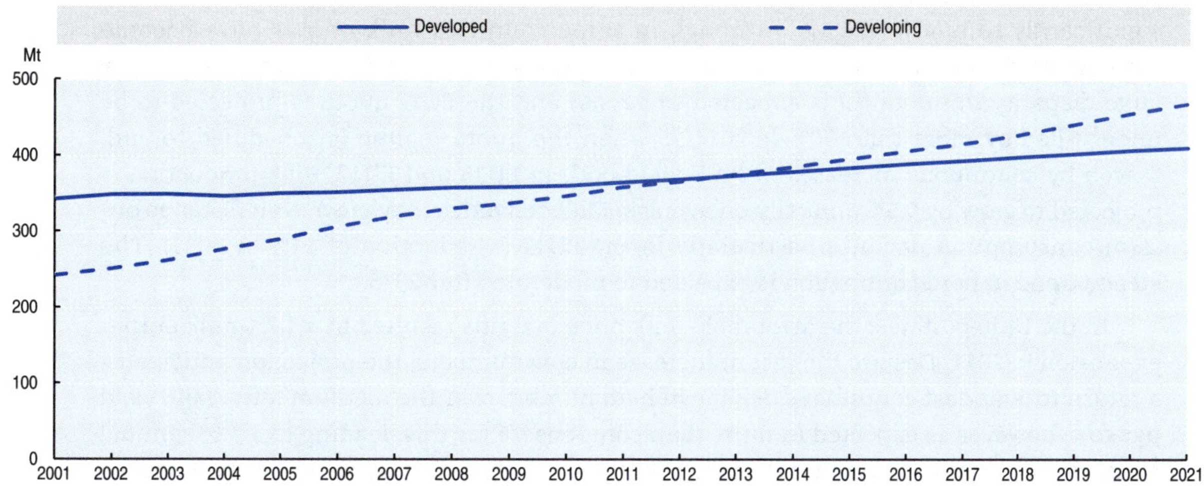

Source: OECD and FAO Secretariats.

StatLink http://dx.doi.org/10.1787/888932640331

Figure 9.4. **Outlook for milk production growth**

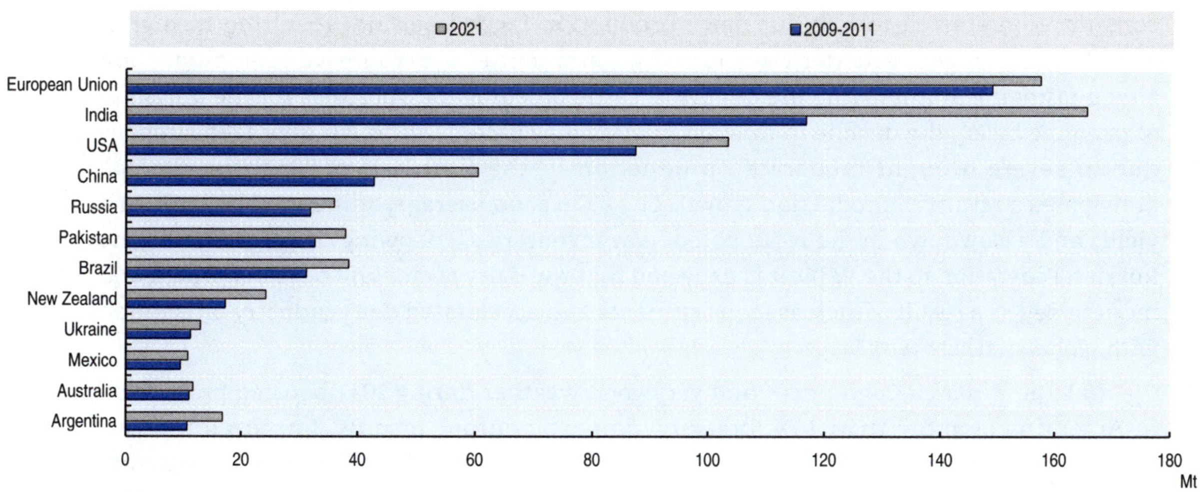

Source: OECD and FAO Secretariats.

StatLink http://dx.doi.org/10.1787/888932640350

growth would return to lower levels and average 2.4% p.a. which is slightly below the percentage growth recorded in the last decade. Installations of new dairy farms (often *via* conversions from beef or sheep farms), continued cow inventory growth, and per cow and per hectare productivity growth trends are behind the steady milk production increases.

After several years of drought conditions, milk producers in Australia have enjoyed much improved pasture growth, particularly in the south-eastern dairy regions. The last decade trend of declining cow inventories was reversed last year and numbers increased by 0.5%. A higher increase in cow inventories in the next few years, reflecting a good hydrologic situation, will be hindered by a shortage of replacement heifers and competition for land. As a result, only a modest growth in herd numbers is projected. Nevertheless, with productivity gains assumed on trend, milk production is expected to increase by 1.2% p.a. over the projection period.

Despite the drought in many parts of Europe in early 2011, stimulated by favourable returns, milk output in the European Union recovered and, after years of decline, increased significantly to nearly 151 Mt. Although in some countries milk quotas have become binding, production levels in others remain well below the available delivery quotas. The large disparity in quota fill is expected to persist and the EU12 quota is projected to be under-filled by more than 20% in 2014/15 when the quota regime is scheduled to end. Driven by continuous increases in milk yield both in EU15 and EU12, milk production is projected to grow by 0.5% annually on average. Milk deliveries may grow even faster as on-farm consumption declines particularly in the EU12 (a reduction of 11% by 2021). The steady trend in herd contraction is expected to moderate after 2015.

In the United States, the favourable milk price margins resulted in a 1.7% milk output expansion in 2011. Despite the recent increase in cow numbers, the projection anticipates a return to a modest continuous decline in herd number over the medium term. Milk yield per cow, however, is expected to more than compensate for this, leading to a 1.6% annual production increase, a growth rate only slightly lower than that achieved in the previous decade. Milk production gains will be largely driven by new investment, increased economies of scale, and improved management.

In the 1990s, milk and dairy product production and consumption in the Russian Federation contracted by more than 60%. Driven by increased incomes over the last decade, consumption started to rise but dairy production lagged behind, resulting in higher imports. Encouraged by better returns and government efforts to revive milk production aimed primarily at improving the quality of cattle breeding (*i.e.* subsidies for the purchase of pedigree bulls), the decline in milk production was halted. Although poor feed supplies due to severe drought moderated production in the first half of 2011, the *Outlook* anticipates a return to production growth of 1.6% p.a. on average, stimulated by improved yields and a slowdown in the reduction of cow inventories. Following the accession of the Russian Federation to the WTO, it is expected that the dairy sector will be transformed and modernised as a result of increased investments and accelerated dairy industry integration with global markets (Box 9.1).

In Latin America, high prices and very good weather during 2011 boosted production in Argentina by more than 10%, breaking domestic output records. Improved returns, investment, increased economies of scale, and management efficiency are anticipated to drive milk production gains in the future. Milk production is expected to grow by more than 3.4% p.a. over the *Outlook* period. Brazil's milk production is projected to grow by 1.7% annually, stimulated by firm prices and domestic demand, but also by development programmes aimed at increasing productivity through animal breeding and pasture improvements. In Mexico, farmers had to cope with reduced forage supplies due to dry weather which led to stagnating milk output in 2011. The dairy sector is expected to be modernised with investments in infrastructure and genetics, supported by government, and it is projected that milk production will increase annually by 0.5% on average.

The milk farming base in China has been very fragmented and the rapid growth of the dairy industry allowed profit-seeking intermediaries (often with no background in dairy) to enter the supply chain. The 2009 melamine milk adulteration cases brought dramatic changes with the government stepping up efforts to restructure the domestic dairy industry, and to improve milk quality and confidence in domestically-produced products. Milk production is expected to grow by less than 3% annually on average. This is much

Box 9.1. Russian Federation WTO membership implies lower dairy tariffs, but the impacts on global dairy markets are likely to be modest

After many years of negotiations, the terms of the WTO Russian Federation membership were agreed and a protocol for the country's accession signed on 16 December 2011. The country has until 15 June 2012 to ratify its accession package and will become a full member 30 days after notification to the WTO of the ratification. As a WTO member, a position assumed in the baseline, the Russian Federation would undertake a series of commitments to further open its trade regime.

The Russian Federation is an important importer of butter and cheese, and the baseline applied tariffs assumptions for those two products are already at 15%, which is the level of agreed bound WTO tariffs to be operational after the accession implementation period. These tariffs were already lowered from 20% under the Custom Union Agreement between Russia, Kazakhstan and Belarus which came into effect on 1 January 2010 – all customs borders were removed on 1 July 2011. Nevertheless, Custom Union tariffs for milk powders are higher than those of the WTO agreement. The baseline assumes that milk powder tariffs will be reduced from 20% to 15% following the agreed WTO schedule from 2012 until 2015. The baseline also assumes that, as a result of WTO accession, subsidy payments on milk production will be halved after 2017.

The results of a counterfactual scenario, in which milk powder tariffs and subsidy levels are not lowered, illustrates that the impact of WTO accession on dairy markets can be expected to be relatively modest. In this scenario, milk production in the Russian Federation will be 1% higher by 2021 as compared to the baseline, while butter and cheese imports will be reduced by 2-4%. SMP and WMP imports would decrease more dramatically, by 7-14%, but given the relatively small importance of these powder imports on the global markets, the impact on world dairy prices would be minor and amount to a reduction of less than 1%. It should be noted that the terms of accession related to trade do not only include tariffs but other policies that affect trade. For example, Sanitary and Phytosanitary (SPS) Measures and the Agreement on Technical Barriers to Trade (TBT), which are not explicitly treated in this scenario, can be expected to have an impact on trade.

WTO membership is expected to accelerate integration in global dairy markets and stimulate the flow of investments, and hence benefit the Russian Federation dairy industry in the long-term. For example, under WTO membership, foreign-owned dairy companies would be allowed to operate in the wholesale, retail and franchise sectors of the Russian Federation.

slower growth than seen in the last decade as it is assumed that restructuring and the withdrawal of backyard operations will moderate domestic output expansion.

Growth in milk production in other developing countries continues to be strong, particularly in India, Pakistan and other central Asian countries where dairying has a strong traditional base. India will further consolidate its position as the world's largest producing country, growing at 3.4% p.a., while Pakistan will grow more slowly at 1.9%. Milk production will also continue to grow strongly in other developing areas in South East and East Asia, where, for example, milk production in Thailand, Indonesia, Malaysia, and the Philippines will respond to recent high returns, albeit from a lower base. In North Africa and the Middle East, local milk production systems have responded to high domestic demand growth. The milk sectors in Egypt, Algeria and Saudi Arabia are projected to grow in the 1.5-2.5% range over the *Outlook* period. This region will nevertheless increase its reliance on dairy product imports given strong domestic demand in these countries.

Milk production plays an important role in the rural fabric of Sub-Saharan Africa. More than in other regions, milk production is derived from milk cows, buffalos, sheep and goats. Productivity levels are very low and unchanging, with high animal numbers. Milk output may grow in the 3.3% range, more than in step with population growth in the region. As commodity prices have risen, so have average incomes, creating higher demand for milk and milk products in urban environments. In some countries, higher prices in the last ten years have encouraged greater participation in the formal milk sector, thereby increasing milk pooling and milk quality needed for commercial sector growth.

Dairy products production

Global whole milk powder (WMP) production continued to grow strongly in 2011 as China recovered from the melamine-related problems and New Zealand continued its recent strong growth trend of nearly 20% p.a. Over the projection period, WMP is expected to be the fastest growing product followed by Fresh Dairy Products (FDP), which is by far the largest user of raw milk. The market for FDP remains dynamic, sustained by expansion in the production of fermented products. As compared to the base period, 2009-11, WMP output is expected to grow by 32%, while FDP, butter and SMP gain 26%, 24% and 23% respectively by 2021. Cheese output is expected to grow by 19% over the outlook period.

China and New Zealand are projected to cover nearly two-thirds of all WMP production expansion. Brazil and Argentina may account for 17% of the expansion. New Zealand WMP production is expected to continue to expand as a result of good WMP returns and strong demand for New Zealand WMP in South East Asia. New Zealand is expected to outpace China as the leading world WMP producer.

India and Pakistan are expected to provide 70% of all butter production gains, the majority of which is in the form of ghee. The United States is expected to step up its butter production, which will account for 8% of global output, while an additional 5% would come from New Zealand. The SMP production was recently discouraged by large overhanging stocks from 2009, but growth recovered in 2011 in response to steady global demand. Most of the additional global production of SMP is expected from the United States (39%) with New Zealand contributing 21%. Boosting its SMP production, the United States is projected to surpass the European Union by 2018 to become the world's largest producer.

Cheese production is dominated by the European Union (44%) and the United States (23%), which supply two-thirds of the global output. Together they will account for 59% of the additional global cheese output produced over the projection period. Brazil and Argentina, the largest cheese producers after the European Union and the United States, together contribute less than 7% of global cheese production. Their production is expected to record a steady 1.8% annual growth over the projection period.

Consumption

Dairy product consumption in developed countries may increase only modestly with the exception of cheese, which is expected to be 15% higher compared to the base period. In developing regions, the consumption of all products may increase markedly at around 30% from the base period. Dairy products remain among the agricultural commodities for which production and consumption exhibit the highest growth rates.

However, the rate of growth and *per capita* consumption of milk and milk products significantly differs among regions (Figure 9.5). Europe and North America *per capita* consumption levels are twice those of other countries, but are projected to grow by less than 5% over the projection period. Asia and Pacific and Latin America are expected to

Figure 9.5. **Despite strong growth in *per capita* consumption (in milk equivalent), an important gap among regions remains**

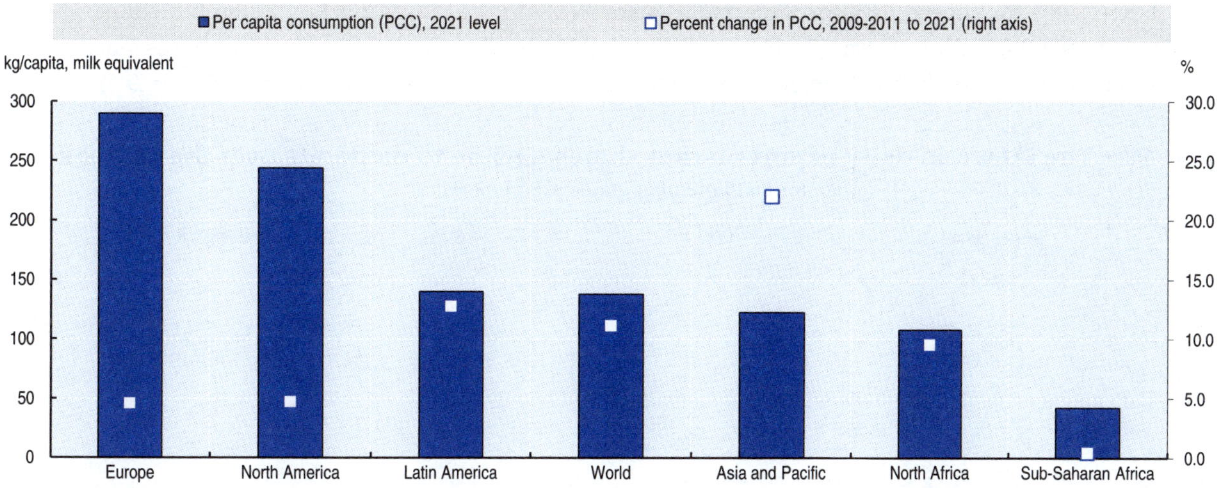

Source: OECD and FAO Secretariats.

StatLink http://dx.doi.org/10.1787/888932640369

narrow the gap and increase the *per capita* consumption by 22% and 13%, respectively. The *per capita* consumption of milk and dairy products is also expected to remain strong in North Africa and the Middle East. Least Developed Countries and countries in the Sub-Saharan region remain a notable exception with very low *per capita* consumption levels. The consumption growth only marginally exceeds the population growth, which results in small increases in *per capita* consumption over the outlook period.

The large differences are expected to remain not only in total consumption (in milk equivalent) but also across individual products. This gap is driven by income, product attributes and differences in diets. The developed countries continue to dominate cheese consumption, maintaining a 75% share of the world total. The *per capita* consumption of cheese in the European Union or the United States is above 15 kg per person, while in developing countries it is often negligible and reaches only 0.9 kg per person on average in 2021. However, developing countries are expected to consume more than 85% of global WMP consumption and account for nearly all additional WMP consumption over the outlook period.

Increasing population and income, together with the growing popularity of dairy products, particularly among consumers in developing countries, is a key factor behind strong demand in the medium term. Demand continues to be encouraged by the growing influence of retail chains and multinational companies in these countries, which is facilitating improved consumer access to dairy products. In many countries consumption is enhanced by government programmes (*i.e.* school milk). Finally, among the important factors affecting the consumption of dairy products is the increasing trend towards a greater variety in the choice of food and increased health, nutrition and diet concerns, as well as higher awareness of animal welfare and environmental issues.

Trade

The landscape of international trade has changed tremendously in the last decade, with the reduction in intervention buying and withdrawal of export measures. Trade has become more influenced by regional developments and strategic decisions of large international dairy companies. Marked changes in the EU market shares illustrate the important shift on the international markets. The EU dairy export shares amounted to 11% for butter, 28% for cheese, 18% for SMP and 21% for WMP in 2011 (data for EU15 member states prior to 2004), falling from 50-60% levels in the 1980s. The decline in the EU shares has moderated from early 2000 and is projected to slow further over the outlook period (Figure 9.6).

Figure 9.6. **The EU world dairy export market shares decline to moderate over the Outlook**

Data for EU15 member states prior to 2004

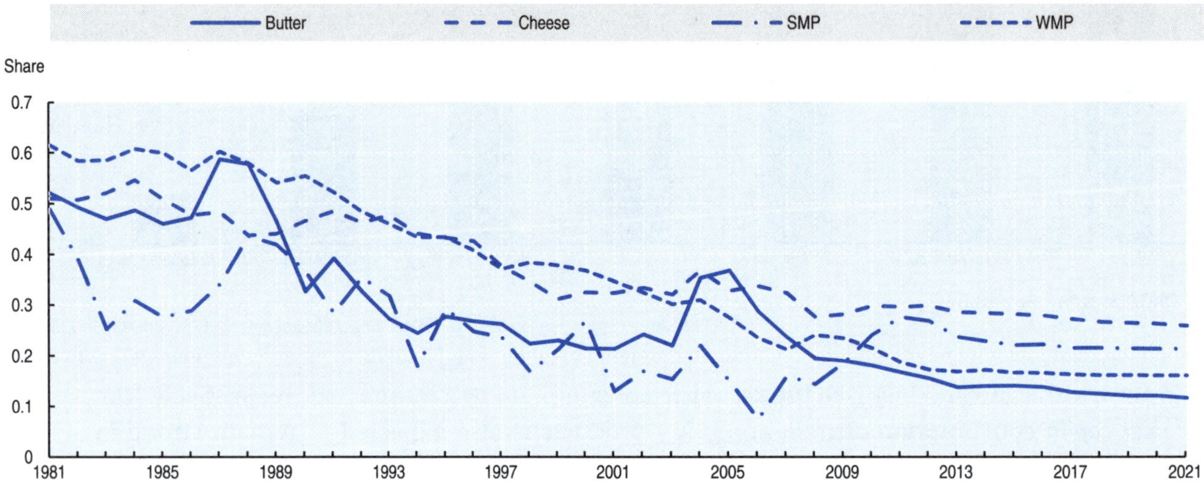

Source: OECD and FAO Secretariats.

StatLink http://dx.doi.org/10.1787/888932640388

Global trade consistently outpaced milk production growth in the last decade and recorded remarkable gains in the last two years. All major global players increased exports. The European Union strengthened its position as the leading cheese exporter. The United States and the European Union boosted SMP exports while New Zealand concentrated mainly on WMP, thus limiting export expansion of other dairy products.

The strong growth in exports is expected to continue, particularly for powders (Figure 9.7). The trade of SMP is expected to increase by 34% while that of WMP by 30% from the 2009-11 base period. The United States is expected to increase significantly volumes on the SMP export market, up 70% from the base period. New Zealand is expected to consolidate its dominant WMP exporter position with volumes growing by more than 50% from the base period. By the end of the outlook period, New Zealand is estimated to account for more than half of the global WMP trade.

Global cheese trade is projected to increase by 27% over the projection period. A strong growth in cheese exports is expected from the United States which is to more than double its export volumes. The majority of the additional US exports is destined for Mexico. Exports from Australia and New Zealand, other important traders, are anticipated to grow over the outlook period by 20% and 28%, respectively. The recent growth in exports is expected to continue for emerging exporters, most notably Argentina, Uruguay, Ukraine and other Eastern European countries (*i.e.* Belarus).

Butter exports have stagnated over much of the last decade but increased recently, and the volume growth is expected to continue. Global butter exports are projected to increase by nearly 20% from the base period. The butter markets will remain among the thinnest

Figure 9.7. **Major dairy product exporters**

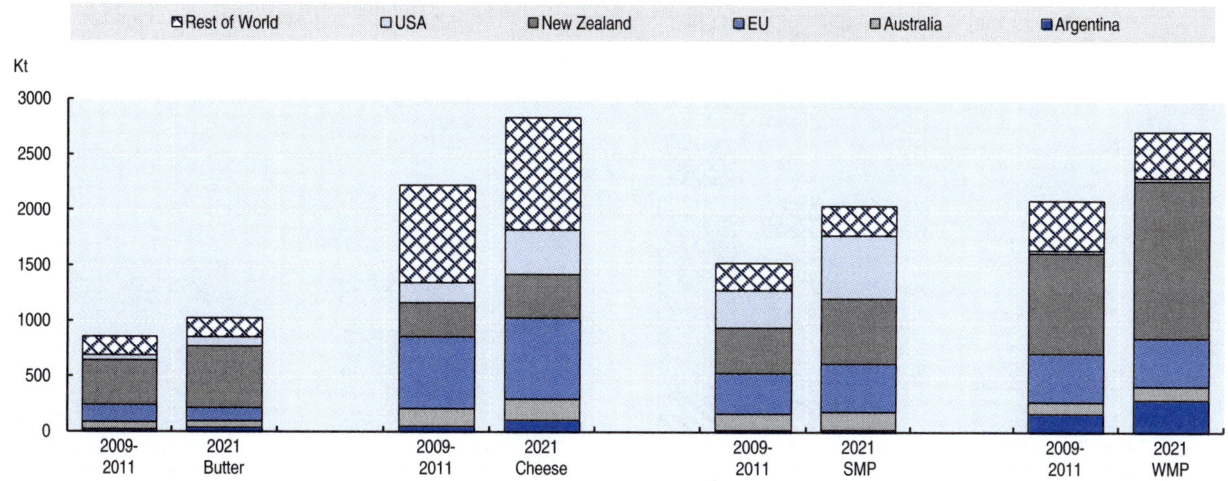

Source: OECD and FAO Secretariats.

StatLink http://dx.doi.org/10.1787/888932640407

dairy markets, with the fewest global exporters and volumes bypassing 1 million tonnes only at the end of the projection period. New Zealand remains by far the dominant country servicing more than 50% of the market. The *Outlook* anticipates the United States to reverse its net butter importer position – often attained in the last decade – and become an increasingly more important butter exporter.

The import markets are fragmented compared to the more consolidated export situation, where the largest importers of dairy products continue to account for less than 50% of the global trade (Figure 9.8). The Russian Federation is to remain the key importer of cheese. Japan, Saudi Arabia and US cheese import demand is also anticipated to remain strong over the outlook period (despite being a significant cheese exporter, the United States imports large volumes of cheese mainly from the European Union). Other countries, such as Mexico, Korea and parts of North Africa are slowly increasing their presence on cheese import markets. Cheese imports are aided by relatively lower cheese prices and "commoditising" of cheese which is increasingly consumed as an ingredient in fast food type products.

Although the Russian Federation was traditionally the most important butter importer, such imports have declined and are projected to decline further over the projection period. Middle East countries are increasingly more important in butter import markets, and given thin butter markets, any political instability in this region may result in strong uncertainties for the butter market over the outlook period.

The recent sharp increase in imports of WMP by China has slowed but annual growth of 7% in 2011, implied volumes reaching 350 000 tonnes, making China the most important importer, with a 16% global import share, ahead of Algeria which has about a 10% share. China has also recently boosted its SMP imports that are destined for domestic infant formulas. Increasing incomes, a growing appetite for dairy products, concerns for food safety and quality but also urbanisation and higher participation of women at work, are expected to keep milk powder imports at elevated levels over the projection period. Whey powder imports by China are used as a cheaper alternative to powders and as an important

Figure 9.8. **Major dairy product importers**

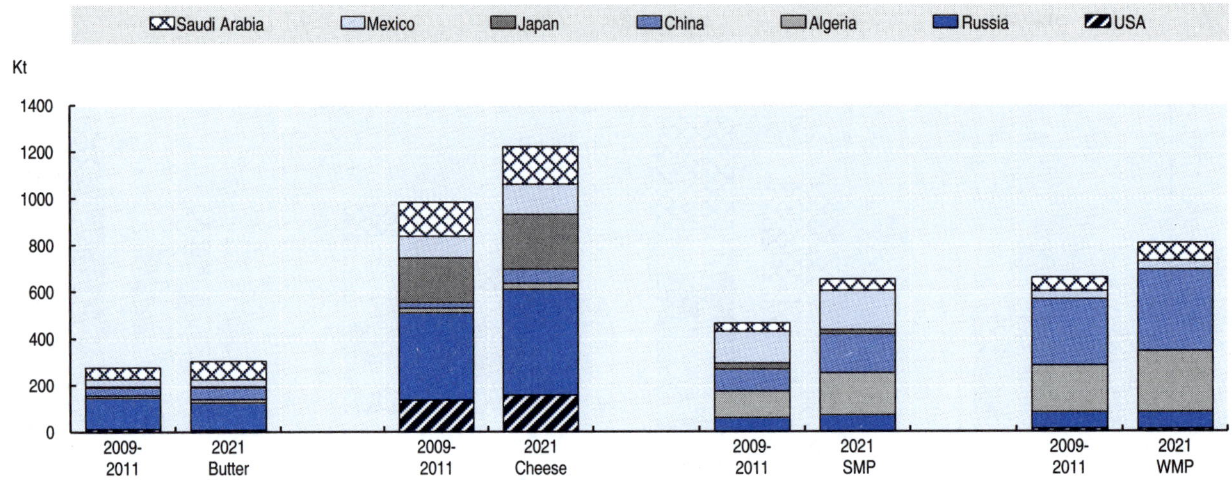

Source: OECD and FAO Secretariats.

StatLink http://dx.doi.org/10.1787/888932640426

source of protein. They form the backbone of whey trade and are expected to increase over the projection period on average by nearly 6% annually.

Market issues and uncertainties

The dairy outlook, as in previous years, is conditioned by weather, the economic situation, and evolution in policies. For example, a severe drought in an important dairy producing region could have an important impact on the projections in any one year, strengthening prices. Slower economic growth compared to that assumed in this *Outlook* would moderate international prices. Developments in energy prices (such as crude oil prices, gas, etc.) can be expected to influence trade patterns and the comparative advantage of market players. The oil price affects livestock markets through both higher cost of energy use, especially for energy and capital intensive production systems, and through the impact on feed ingredient costs. Higher energy and feed costs affect the competitiveness of pasture *versus* intensive feed based systems, and may encourage the growth in market shares of countries with relatively abundant pasture and water. However, in the future, higher costs of feedstuffs may, to a certain extent, be mitigated by the increased availability of distilled dry grains (DDGs), a by-product of bioethanol production.

In the context of prevailing border measures on dairy markets, a key uncertainty for future dairy trade is the potential outcome of the Doha Development Agenda (DDA) negotiations. However, with little progress on the Doha Round negotiations, many countries have opted for regional (or bilateral) arrangements. Although traditional trade barriers such as tariffs and TRQs remain important determinants of dairy trade flows, attention has increasingly shifted towards non-tariff measures and to regulatory mechanisms often linked through the Agreement on the Application of Sanitary and Phytosanitary (SPS) Measures and the Agreement on Technical Barriers to Trade (TBT). Moreover, the growth of private standards, adopted by the agri-food chain can be also expected to exert an impact on dairy markets. Although private voluntary standards schemes can contribute to improving food system efficiency, so as to deliver and ensure specific product and process attributes at a reasonable cost to consumers, the increasing use of private standards also raises the issue of effective market access both for developed and developing country suppliers. Compliance with private voluntary standards schemes may exclude those producers who, due to lack of potential scale economies or otherwise, cannot easily meet the standards' requirements and remain economically viable.

Over the outlook period, the shape of dairy markets will continue to be influenced by the spread of multinational dairy companies, retail chains and international investment which would lead to further narrowing of production and consumption differences across regions. Research and technological development can be expected to have a key role, although progress remains uncertain. New initiatives and innovations are likely to emerge in the fight for new and existing consumers. For example, increased use of dairy product fortification with minerals and vitamins to add "health" values to cheese and butter *via* probiotic cultures, extension of the shelf-life of dairy products, improvements of texture and flavour, and an increase in the absorption and bio-availability of nutrients are some potential avenues. Technological development and the spread of existing production technologies will also play an important role in narrowing the productivity gap (Box 9.2).

Box 9.2. **Productivity change in the dairy sector**

It is estimated that world milk production increased by about 100% over the last 50 years. With global population growth at 123%, average milk output *per capita* has fallen over this same period. While this fall can be somewhat attributed to a large decline in milk output in the transition countries of the former USSR, it also reflects population growth in regions where milk production has not been significant in *per capita* terms, particularly in Asia. In the last decade or so, however, milk production *per capita* has started to increase significantly, and this trend is anticipated to continue over the outlook period at a rate of about 1% p.a. to 2021. Milk and dairy products are anticipated to be one of the fastest growing sectors, placing larger demands on agricultural resources. Changing productivity, particularly as measured by milk yields, has considerable implication for resource use by the sector. Milking animals represent an important value of total farm capital and producer wealth.

In 2011, global milk production was produced by some 626 million milking animals, about 83% of which were milk cows, 13% buffalo cows, 2% goats, 1% sheep, and the remainder camels. The distribution of animal types varies significantly by country and region with developed regions using almost exclusively cow milk, while developing regions in South Asia rely more on buffalo animals and many African countries milk both sheep and goats. For example, whereas virtually 100% of milk production in the United States is derived from milk cows (and about 97% in the European Union), milk cows account for only 82% in Ethiopia, 78% in Algeria, and 40% in India (FAOSTAT, 2010). From a productivity perspective, milk yields have not grown significantly for goat and sheep milk, compared to buffalo and cow milk (Table 9.1).

Based on productivity estimates for milk cows, it would appear that growth in milk cow yields for many developed countries have slowed in recent years. However, in transition countries and many developing countries, growth in cow yields has accelerated considerably, albeit from a low base. As shown in Figure 9.9, there is substantial scope to increase cow yields given the large gaps. It is important to note, however, that technologies for milk production vary substantially and high yield does not necessarily mean low cost. Intensive grain-fed milk operations may be most efficient in some areas, particularly those where land is scarce and population density is higher, whereas pasture fed operations may be most efficient where land is in ample supply and there are no cropping alternatives. For example, from a unit cost perspective, New Zealand is considered one of the most efficient competitive suppliers to world markets, but its average milk yield is well below those sectors based on grain feeding regimes. Given the prospects for growth in demand for milk and milk products, future growth in milk yields, no matter what the animal type or feeding regime, may have considerable implications for resource use, particularly for land, water and labour inputs.

Box 9.2. **Productivity change in the dairy sector** *(cont.)*

Table 9.1. **Growth in milk yields, selected countries and animal types**

	Mean yield	Growth rate	Growth rate	Growth rate
	2005-09	1961-79	1980-99	2000-09
	Milk/yr	%/yr	%/yr	%/yr
Algeria	1.8	2.8	1.3	5.6
Argentina	4.6	0.3	5.5	1.7
Australia	5.4	1.9	2.8	1.2
Brazil	1.3	0.1	2.4	1.4
China	2.7	1.5	–1.3	4.8
E27	6.1	1.6	1.9	1.5
Egypt	1.6	0	2.1	5.7
India (milk cow)	1.1	1.2	2.9	2.2
India (buffalo cow)	1.6	0.3	1.7	1
Indonesia	1.6	–1.8	2.9	2.1
Japan	7.3	0.8	2.2	1.2
Mexico	4.8	4.7	0.9	1.1
New Zealand	3.6	0.8	0.5	–0.7
Nigeria	0.2	0	0.2	–0.3
The Russian Federation	3.5		1.1	4.4
South Africa	3.5	0.2	1.3	2.8
USA	9.1	2.5	2.1	1.6
Ukraine	3.7		0.2	5.6
World – cows	2.3	0.5	0.8	0.5
World – buffalo	1.5		1.6	0.9
World – sheep	0.04		0.1	0.4
World – goats	0.08		0.1	0.1

Note: For Russian Federation, Ukraine and world numbers, yield growth estimates in the 1980-99 column are based on years 1992-2000 only.
Source: OECD and FAO Secretariats.

StatLink http://dx.doi.org/10.1787//888932642516

Figure 9.9. **Cow milk yields, selected countries**

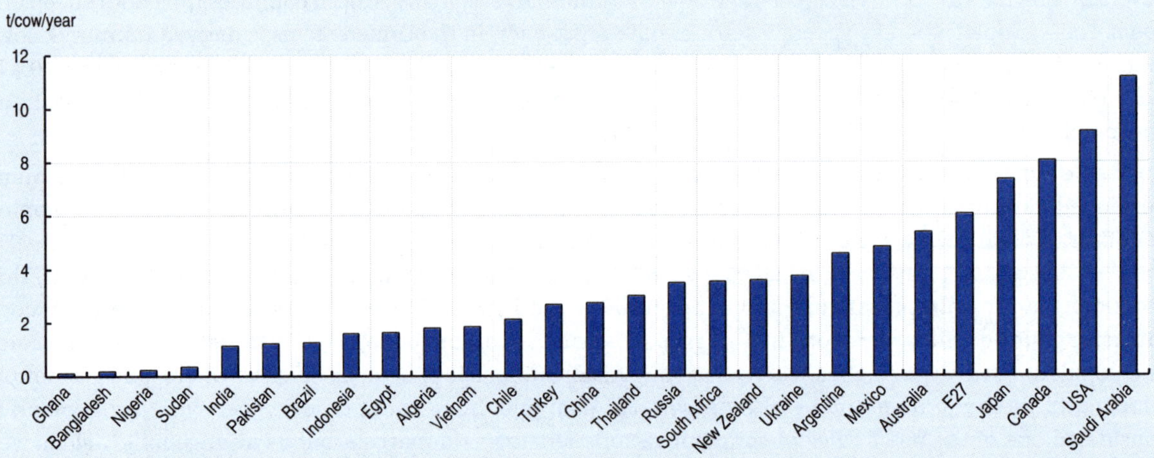

Source: OECD and FAO Secretariats.

StatLink http://dx.doi.org/10.1787/888932640445

Glossary of Terms

A-H1N1

This is an influenza virus that had never been identified as a cause of infections in people before the current H1N1 pandemic. Genetic analyses of this virus have shown that it originated from animal influenza viruses and is unrelated to the human seasonal H1N1 viruses that have been in general circulation among people since 1977.

Average Crop Revenue Election (ACRE) program

A new programme introduced with the 2008 US FCE Act allowing farmers to choose revenue-based protection against yield and market fluctuations.

AMAD

Agricultural Market Access Database. A co-operative effort between Agriculture and Agri-food Canada, EU Commission-Agriculture Directorate-General, FAO, OECD, The World Bank, UNCTAD and the United States Department of Agriculture, Economic Research Service. Data in the database is obtained from countries' schedules and notifications submitted to the WTO.

Aquaculture

The farming of aquatic organisms including fish, molluscs, crustaceans and aquatic plants, etc. Farming implies some form of intervention in the rearing process to enhance production, such as regular stocking, feeding and protection from predators. Farming also implies individual or corporate ownership of the stock being cultivated. For statistical purposes, aquatic organisms that are harvested by an individual or corporate body that has owned them throughout their rearing period contribute to aquaculture, while aquatic organisms that are exploitable by the public as a common property resource, with or without appropriate licenses, are the harvest of capture fisheries.

ASEAN

The Association of Southeast Asian Nations, or ASEAN, was established on 8 August 1967 in Bangkok, Thailand, with the signing of the ASEAN Declaration (Bangkok Declaration) by the Founding Fathers of ASEAN, namely Indonesia, Malaysia, Philippines, Singapore and Thailand. Brunei Darussalam then joined on 8 January 1984, Viet Nam on 28 July 1995, Lao PDR and Myanmar on 23 July 1997, and Cambodia on 30 April 1999, making up what is today the ten Member States of ASEAN.

Australia-US Free Trade Agreement (AUSFTA)

A Bilateral Agreement negotiated between the United States and Australia that came into force on 1 January 2005. AUSFTA covers goods, services, investment, financial services, government procurement, standards and technical regulations, telecommunications, competition-related matters, electronic commerce, intellectual property rights, labour and the environment.

Avian influenza

Avian influenza is an infectious disease of birds caused by type A strains of the influenza virus. The disease, which was first identified in Italy more than 100 years ago, occurs worldwide. The quarantining of infected farms, destruction of infected or potentially exposed flocks, and recently inoculation are standard control measures.

Atlantic beef/pigmeat market

Beef/pigmeat trade between countries in the Atlantic Rim.

Baseline

The set of market projections used for the outlook analysis in this report and as a benchmark for the analysis of the impact of different economic and policy scenarios. A detailed description of the generation of the baseline is provided in the chapter on Methodology in this report.

Biofuels

In the wider sense defined as all solid, fluid or gaseous fuels produced from biomass. More narrowly, the term biofuels comprises those that replace petroleum-based road-transport fuels, *i.e.* bioethanol produced from sugar crops, cereals and other starchy crops that can be used as an additive to, in a blend with or as a replacement of gasoline, and biodiesel produced mostly from vegetable oils, but also from waste oils and animal fats, that can be used in blends with or as a replacement of petroleum-based diesel.

Biomass

Biomass is defined as any plant matter used directly as fuel or converted into other forms before combustion. Included are wood, vegetal waste (including wood waste and crops used for energy production), animal materials/wastes and industrial and urban wastes, used as feedstocks for producing bioproducts.

Bovine Spongiform Encephalopathy (BSE)

A fatal disease of the central nervous system of cattle, first identified in the United Kingdom in 1986. On 20 March 1996 the UK Spongiform Encephalopathy Advisory Committee (SEAC) announced the discovery of a new variant of Creutzfeldt-Jacob Disease (vCJD), a fatal disease of the central nervous system in humans, which might be linked to consumption of beef affected by exposure to BSE.

BRIICs

Refers to the emerging economies of Brazil, the Russian Federation, India, Indonesia and China.

Capture fisheries

Capture fisheries refer to the hunting, collecting and gathering activities directed at removing or collecting live wild aquatic organisms (predominantly fish, molluscs and crustaceans) including plants from the oceanic, coastal or inland waters for human consumption and other purposes by hand or more usually by various types of fishing gear such as nets, lines and stationary traps. The production of capture fisheries is measured by nominal catches (in live weight basis) of fish, crustaceans, molluscs and other aquatic animals and plants, killed, caught, trapped or collected for all commercial, industrial, recreational and subsistence purposes.

Cereals

Defined as wheat, coarse grains and rice.

CAFTA

CAFTA is a comprehensive trade agreement between Costa Rica, the Dominican Republic, El Salvador, Guatemala, Honduras, Nicaragua, and the United States

Common Agricultural Policy (CAP)

The European Union's agricultural policy, first defined in Article 39 of the Treaty of Rome signed in 1957.

Coarse grains

Defined as barley, maize, oats, sorghum and other coarse grains in all countries except Australia, where it includes triticale and in the European Union where it includes rye and other mixed grains.

Country of Origin Labelling (COOL)

A provision of the 2008 US Farm Act that requires retailers to inform consumers of country of origin of different commodities, among them meats.

Conservation Reserve Program (CRP)

A major provision of the United States' Food Security Act of 1985 and extended under the Food and Agriculture Conservation and Trade Act of 1990, the Food and Agriculture Improvement and Reform Act of 1996, and the Farm Security and Rural Investment Act of 2002 is designed to reduce erosion on 40 to 45 million acres (16 to 18 million hectares) of farm land. Under the programme, producers who sign contracts agree to convert erodable crop land to approved conservation uses for ten years. Participating producers receive annual rental payments and cash or payment in kind to share up to 50% of the cost of establishing permanent vegetative cover. The CRP is part of the *Environmental Conservation Acreage Reserve Program*. The 1996 FAIR Act authorised a 36.4 million acre (14.7 million hectares) maximum under CRP, its 1995 level. The maximum area enrolled in the CRP was increased to 39.2 million acres in the 2002 FSRI Act.

Commonwealth of Independent States (CIS)

The heads of twelve sovereign states (except the Baltic states) have signed the Treaty on establishment of the Economic Union, in which they stressed that the Azerbaijan

Republic, Republic of Armenia, Republic of Belarus, Republic of Georgia, Republic of Kazakhstan, Kyrgyz Republic, Republic of Moldova, Russian Federation, Republic of Tajikistan, Turkmenistan, Republic of Uzbekistan and Ukraine on equality basis established the Commonwealth of Independent States.

Common Market Organisation (CMO) for sugar

The common organisation of the sugar market (CMO) in the European Union was established in 1968 to ensure a fair income to community sugar producers and self-supply of the Community market. At present the CMO is governed by Council Regulation (EC) No. 318/2006 (the basic regulation) which establishes a restructuring fund financed by sugar producers to assist the restructuring process needed to render the industry more competitive.

Crop year, coarse grains

Refers to the crop marketing year beginning 1 April for Japan, 1 July for the European Union and New Zealand, 1 August for Canada and 1 October for Australia. The US crop year begins 1 June for barley and oats and 1 September for maize and sorghum.

Crop year, oilseeds

Refers to the crop marketing year beginning 1 April for Japan, 1 July for the European Union and New Zealand, 1 August for Canada and 1 October for Australia. The US crop year begins 1 June for rapeseed, 1 September for soyabeans and for sunflower seed.

Crop year, rice

Refers to the crop marketing year beginning 1 April for Japan, Australia, 1 August for the United States, 1 September for the European Union, 1 November for Korea and 1 January for other countries.

Crop year, sugar

A common crop marketing year beginning 1 October and extending to 31 September, used by ISO (International Sugar Organisation).

Crop year, wheat

Refers to the crop marketing year beginning 1 April for Japan, 1 June for the United States, 1 July for the European Union and New Zealand, 1 August for Canada and 1 October for Australia.

Decoupled payments

Budgetary payments paid to eligible recipients who are not linked to current production of specific commodities or livestock numbers or the use of specific factors of production.

Developing countries

Refers to all countries that are not developed countries; therefore, includes BRIICs.

Direct payments

Payments made directly by governments to producers.

Doha Development Agenda

The current round of multilateral trade negotiations in the World Trade Organisation that were initiated in November 2001, in Doha, Qatar

Domestic support

Refers to the annual level of support, expressed in monetary terms, provided to agricultural production. It is one of the three pillars of the *Uruguay Round Agreement on Agriculture* targeted for reduction.

Eastern Europe

Refers to Russia, Ukraine and Kazakhstan.

Economic Partnership Agreements (EPAs)

Free trade agreements currently being negotiated between the EU and the African, Caribbean Pacific (ACP) group of developing countries to replace the Cotonou Agreement which expired in 2007.

El Niño

In this publication, El Niño is used to indicate a broader term of quasi-periodic ocean climate conditions including La Niña, Southern Oscillation, or ENSO, which are characterised by anomalies in the temperature of the surface of eastern coast of Latin America (centred on Peru) – warming or cooling known as *El Niño* and *La Niña* respectively – and air surface pressure in the tropical western Pacific (the Southern Oscillation), often around Christmas time. The abnormal warm ocean climate conditions are accompanied by dramatic changes in species abundance and distribution, higher local rainfall and flooding, massive deaths of fish and their predators (including birds).

Energy Independence and Security Act (EISA) 2007

US legislation passed in December 2007 that is designed to increase US energy security by lessening dependence on imported oil, to improve energy conservation and efficiency, expand the production of renewable fuels, and to make America's air cleaner for future generations.

Ethanol

A biofuel that can be used as a fuel substitute (hydrous ethanol) or a fuel extender (anhydrous ethanol) in mixes with petroleum, and which is produced from agricultural feed-stocks such as sugarcane and maize.

Everything-But-Arms (EBA)

The Everything-But-Arms (EBA) Initiative eliminates EU import tariffs for numerous goods, including agricultural products, from the least developed countries. The tariff elimination is scheduled in four steps from 2006/07 to 2009/10.

Export credits (with official support)

Government financial support, direct financing, guarantees, insurance or interest rate support provided to foreign buyers to assist in the financing of the purchase of goods from national exporters.

Export restitutions (refunds)

EU export subsidies provided to cover the difference between internal prices and world market prices for particular commodities.

Export subsidies

Subsidies given to traders to cover the difference between internal market prices and world market prices, such as for example the EU *export restitutions*. Export subsidies are now subject to value and volume restrictions under the *Uruguay Round Agreement on Agriculture*.

FCE Act, 2008

Officially known as the Food, Conservation and Energy Act of 2008. This US farm legislation replaces the FSRI Act of 2002 and covers the period 2008-13.

FSRI Act, 2002

Officially known as the Farm Security and Rural Investment Act of 2002. This US farm legislation replaces the FAIR Act of 1996, covering a wide range of commodity programs and policies for US agriculture for the period 2002-07.

Gur, jaggery, khandasari

Semi-processed sugars (plantation whites) extracted from sugarcane in India.

Health Check Reform of the Common Agricultural Policy

On 20 November 2008 the EU agriculture ministers reached a political agreement on the Health Check of the Common Agricultural Policy. Among a range of measures, the agreement abolishes arable set-aside, increases milk quotas gradually leading up to their abolition in 2015, and converts market intervention into a genuine safety net. Ministers also agreed to increase modulation, whereby direct payments to farmers are reduced and the money transferred to the Rural Development Fund.

High Fructose Corn Syrup (HFCS)

Isoglucose sweetener extracted from maize.

Historical Price Volatility

Historical price volatility is calculated following the method used by the Chicago Board of Trade (CBOT) from the following formula:

$$\sigma = \sqrt{\sum_{i=1}^{n} [r_t - \mu]^2 / n-1}$$

Where r_t are the logarithmic returns on prices P_t: $r_t = \ln(P_t) - \ln(P_{t-1})$ and μ is the average return, and n is the number of sample observations. In *annualized* terms, multiplied by the inverse of the square root of time, $1/\sqrt{T}$, where T represents the frequency of the observation (*e.g.* daily, monthly, etc).

Implied volatility

The concept of implied volatility is based on the Black-Scholes option pricing formula. Given the exercise price, current price, risk free rate and maturity of an option, there is some value for volatility that makes the price determined by the Black Scholes formula equal to the current price. This is called implied volatility. For further reference, refer to Mayhew, S. (1995), "Implied volatility", *Financial Analysts Journal* **51** (4): 8–20.

Industrial oilseeds

A category of oilseed production in the European Union for industrial use (*i.e.* biofuels).

Intervention purchases

Purchases by the EC Commission of certain commodities to support internal market prices.

Intervention purchase price

Price at which the European Commission will purchase produce to support internal market prices. It usually is below 100% of the intervention price, which is an annually decided policy price.

Intervention stocks

Stocks held by national intervention agencies in the European Union as a result of *intervention* buying of commodities subject to market price support. Intervention stocks may be released onto the internal markets if internal prices exceed intervention prices; otherwise, they may be sold on the world market with the aid of *export restitutions*.

Inulin

Inulin syrups are extracted from chicory through a process commercially developed in the 1980s. They usually contain 83 per cent fructose. Inulin syrup production in the European Union is covered by the sugar regime and subject to a production quota.

Isoglucose

Isoglucose is a starch-based fructose sweetener, produced by the action of glucose isomerase enzyme on dextrose. This isomerisation process can be used to produce glucose/fructose blends containing up to 42% fructose. Application of a further process can raise the fructose content to 55%. Where the fructose content is 42%, isoglucose is equivalent in sweetness to sugar. Isoglucose production in the European Union is covered by the sugar regime and subject to a production quota.

La Niña

Climatic condition associated with the temperature of major sea currents.

Least squares growth rate

The least-squares growth rate, r, is estimated by fitting a linear regression trend line to the logarithmic annual values of the variable in the relevant period, as follows: $Ln(xt) = a + r * t$.

Live weight

The weight of finfish and shellfish at the time of their capture or harvest. Calculated on the basis of conversion factors from landed to nominal weight and on rates prevailing among national industries for each type of processing.

Loan rate

The commodity price at which the *Commodity Credit Corporation* (CCC) offers *non-recourse loans* to participating farmers. The crops covered by the programme are used as collateral for these loans. The loan rate serves as a floor price, with the effective level lying somewhat above the announced rate, for participating farmers in the sense that they can default on their loan and forfeit their crop to the CCC rather than sell it in the open market at a lower price.

Market access

Governed by provisions of the *Uruguay Round Agreement* on *Agriculture* which refer to concessions contained in the country schedules with respect to bindings and reductions of tariffs and to other minimum import commitments.

Marketing allotments (US sugar program)

Marketing allotments designate how much sugar can be sold by sugar millers and processors on the US internal market and were established by the 2002 FSRI Act as a way to guarantee the US sugar loan program operates at no cost to the Federal Government.

Marketing year, oilseed meals

Refers to the marketing year beginning 1 October.

Marketing year, vegetable oils

Refers to the marketing year beginning 1 October.

Market Price Support (MPS) Payment

Indicator of the annual monetary value of gross transfers from consumers and taxpayers to agricultural producers arising from policy measures creating a gap between domestic market prices and *border prices* of a specific agricultural commodity, measured at the farm gate level. Conditional on the production of a specific commodity, MPS includes the transfer to producers associated with both production for domestic use and exports, and is measured by the price gap applied to current production. The MPS is net of financial contributions from individual producers through producer levies on sales of the specific commodity or penalties for not respecting regulations such as production quotas (*Price levies*), and in the case of livestock production is net of the market price support on domestically produced coarse grains and oilseeds used as animal feed (*Excess feed cost*).

Methyl Tertiary Butyl Ether (MTBE)

A chemical gasoline additive that can be used to boost the octane number and oxygen content of the fuel, but can render contaminated water undrinkable.

Milk quota scheme

A supply control measure to limit the volume of milk produced or supplied. Quantities up to a specified quota amount benefit from full *market price support*. Over-quota volumes may be penalised by a levy (as in the European Union, where the "super levy" is 115% of the target price) or may receive a lower price. Allocations are usually fixed at individual producer level. Other features, including arrangements for quota reallocation, differ according to scheme.

North American Free Trade Agreement (NAFTA)

A trilateral agreement on trade, including agricultural trade, between Canada, Mexico and the United States, phasing out tariffs and revising other trade rules between the three countries over a 15-year period. The agreement was signed in December 1992 and came into effect on 1 January 1994.

Oilseed meals

Defined as rapeseed meal (canola), soyabean meal, and sunflower meal in all countries, except in Japan where it excludes sunflower meal.

Oilseeds

Defined as rapeseed (canola), soyabeans, sunflower seed, peanuts and cotton seeds in all countries, except in Japan where it excludes sunflower seed.

Pacific beef/pigmeat market

Beef/pigmeat trade between countries in the Pacific Rim where foot and mouth disease is not endemic.

Payment-In-Kind (PIK)

A programme used in the US to help dispose of public stocks of commodities. Under PIK, government payments in the form of Commodity Credit Corporation (CCC)-owned commodities are given to farmers in return for additional reductions in harvested acreage.

PROCAMPO

A programme of direct support to farmers in Mexico. It provides for direct payments per hectare on a historical basis.

Producer Support Estimate (PSE)

Indicator of the annual monetary value of gross transfers from consumers and taxpayers to agricultural producers, measured at farm gate level, arising from policy measure, regardless of their nature, objectives or impacts on farm production or income. The PSE measure support arising from policies targeted to agriculture relative to a situation without such policies, *i.e.* when producers are subject only to general policies (including economic, social, environmental and tax policies) of the country. The PSE is a gross notion implying that any costs associated with those policies and incurred by individual producers are not deducted. It is also a nominal assistance notion meaning that increased costs associated with import duties on inputs are not deducted. But it is an indicator net of producer contributions to help finance the policy measure (*e.g.* producer levies) providing a given transfer to producers. The PSE includes implicit and explicit payments. The percentage PSE is the ration of the PSE to the value of total gross farm receipts, measured by the value of total production (at farm gate prices), plus budgetary support. The nomenclature and definitions of this indicator replaced the former Producer Subsidy Equivalent in 1999.

Purchasing Power Parity (PPP)

Purchasing power parities (PPPs) are the rates of currency conversion that eliminate the differences in price levels between countries. The PPPs are given in national currency units per US dollar.

Non-Recourse loan programme

Programme to be implemented under the US FAIR Act of 1996 for butter, non-fat dry milk and cheese after 1999 in which loans must be repaid with interest to processors to assist them in the management of dairy product inventories.

Renewable Energy Directive (RED)

EU directive legislating binding mandates of 20% for the share of renewable energy in all Member States' energy mix by the year 2020, with a specific mandate of 10% for the renewable energy share in transport fuels.

Renewable Fuel Standard (RFS and RFS2)

A standard in the United States for the use of renewable fuel use in the transport sector in the Energy Act (EISA). RFS2 is a revision of the RFS program for 2010 and beyond.

Saccharin

A low calorie, artificial sweetener used as a substitute for sugar mainly in beverage preparations.

Scenario

A model-generated set of market projections based on alternative assumptions than those used in the baseline. Used to provide quantitative information on the impact of changes in assumptions on the outlook.

Set-aside programme

European Union programme for cereal, oilseed and protein crops that both requires and allows producers to set-aside a portion of their historical base acreage from current production. Mandatory set-aside rates for commercial producers are set at 10% until 2006.

Single Farm Payment

With the 2003 CAP reform, the EU introduced a farm-based payment largely independent of current production decisions and market developments, but based on the level of former payments received by farmers. To facilitate land transfers, entitlements are calculated by dividing the reference amount of payment by the number of eligible hectares (incl. forage area) in the reference year. Farmers receiving the new SFP are obliged to keep their land in good agricultural and environmental condition and have the flexibility to produce any commodity on their land except fruits, vegetables and table potatoes.

SPS Agreement

WTO Agreement on Sanitary and Phyto-sanitary measures, including standards used to protect human, animal or plant life and health.

Stock-to-use ratio

The stock-to-use ratio for cereals is defined as the ratio of cereal stocks to its domestic utilisation

Stock-to-disappearance ratio

The stock-to-disappearance ratio is defined as the ratio of stocks held by the main exporters to their disappearance (*i.e.* domestic utilisation plus exports). For wheat the eight major exporters are considered, namely the United States, Argentina, the European Union, Canada, Australia, Russian Federation, Ukraine and Kazakhstan. In the case of coarse grains, United States, Argentina, the European Union, Canada, Australia, the Russian Federation, Ukraine and Brazil are considered. For rice Vietnam, Thailand, India, Pakistan and the United States enter this ratio calculation.

Support price

Prices fixed by government policy makers in order to determine, directly or indirectly, domestic market or producer prices. All administered price schemes set a minimum guaranteed support price or a target price for the commodity, which is maintained by associated policy measures, such as quantitative restrictions on production and imports; taxes, levies and tariffs on imports; export subsidies; and public stockholding.

Total Factor Productivity (TFP)

Is defined as total outputs over total inputs. It can be decomposed in two components: technical change (*i.e.* productivity technology frontier) and efficiency change (*i.e.* productivity level of a certain country/region with respect to its underlying technology frontier).

Tariff-rate quota (TRQ)

Resulted from the *Uruguay Round Agreement on Agriculture*. Certain countries agreed to provide minimum import opportunities for products previously protected by non-tariff barriers. This import system established a quota and a two-tier tariff regime for affected commodities. Imports within the quota enter at a lower (in-quota) tariff rate while a higher (out-of-quota) tariff rate is used for imports above the concessionary access level.

Uruguay Round Agreement on Agriculture (URAA)

The terms of the URAA are contained in the section entitled the "Agreement on Agriculture" of the Final Act Embodying the Results of the Uruguay Round of Multilateral Trade Negotiations. This text contains commitments in the areas of *market access*, domestic support, and *export subsidies*, and general provisions concerning monitoring and continuation. In addition, each country's schedule is an integral part of its contractual commitment under the URAA. There is a separate agreement entitled the Agreement on the Application of Sanitary and Phyto-sanitary Measures. This agreement seeks establishing a multilateral framework of rules and disciplines to guide the adoption, development and the enforcement of sanitary and phyto-sanitary measures in order to minimise their negative effects on trade. See also *Phyto-sanitary regulations* and *Sanitary regulations*.

Vegetable oils

Defined as rapeseed oil (canola), soyabean oil, sunflower seed oil, coconut oil, cotton oil, palm kernel oil, peanuts oil and palm oil, except in Japan where it excludes sunflower seed oil.

Voluntary Quota Restructuring Scheme

Established as part of the reform of the European Union's Common Market Organisation (CMO) for sugar in February 2006 to apply for four years from 1 July 2006. Under the scheme, sugar producers receive a degressive payment for permanently surrendering sugar production quota, in part or in entirety, over the period 2006-07 to 2009-10.

WTO

World Trade Organisation created by the Uruguay Round agreement.

Yield gap

Deviations from potential yield in a country or region due to different conditioning factors such as different soil, moisture and temperature environments, different farm sizes and management capacities, access to markets, and legislative/institutional factors.

OECD-FAO Agricultural Outlook 2012

Methodology

This section provides information on the methodological aspects of the generation of the present *Agricultural Outlook*. It discusses the main aspects in the following order: First, a general description of the agricultural baseline projections and the *Outlook* report is given. Second, the compilation of a consistent set of the assumptions on macroeconomic projections is discussed in more detail. A third part presents how production costs are taken into account in the model's supply equations. Then the fourth part presents the methodology developed for the stochastic analysis conducted with the AGLINK-COSIMO model.

The generation of the OECD-FAO *Agricultural Outlook*

The projections presented and analysed in this document are the result of a process that brings together information from a large number of sources. The use of a model jointly developed by the OECD and FAO Secretariats, based on the OECD's Aglink model and extended by FAO's Cosimo model, facilitates consistency in this process. A large amount of expert judgement, however, is applied at various stages of the *Outlook* process. The *Agricultural Outlook* presents a single, unified assessment, judged by the OECD and FAO Secretariats to be plausible given the underlying assumptions, the procedure of information exchange outlined below and the information to which they had access.

The starting point of the outlook process is the response by OECD countries (and some non-member countries) to an annual questionnaire circulated at mid-year. Through these questionnaires, the OECD Secretariat obtains information from these countries on future commodity market developments and on the evolution of their agricultural policies. The starting projections for the country modules handled by the FAO Secretariat are developed through model based projections and consultations with FAO commodity specialists. External sources, such as the World Bank and the UN, are also used to complete the view of the main economic forces determining market developments. This part of the process is aimed at creating a first insight into possible market developments and at establishing the key assumptions which condition the outlook. The main economic and policy assumptions are summarised in the Overview chapter and in specific commodity tables of the present report. The sources and assumptions for those assumptions are discussed in more detail further below.

As a next step, the modelling framework jointly developed by the OECD and FAO Secretariats is used to facilitate a consistent integration of this information and to derive an initial set of global market projections (baseline). In addition to quantities produced, consumed and traded, the baseline also includes projections for nominal prices (in local currency units) for the commodities concerned. Unless otherwise stated, prices referred to in the text are also in nominal terms. The data series for the projections is drawn from

OECD and FAO databases. For the most part information in these databases has been taken from national statistical sources. For further details on particular series, enquiries should be directed to the OECD and FAO Secretariats.

The model provides a comprehensive dynamic economic and policy specific representation of major world producing and trading countries for the main temperate-zone commodities as well as rice and vegetable oils. The Aglink and Cosimo country and regional modules are all developed by the OECD and FAO Secretariats in conjunction with country experts and, in some cases, with assistance from other national administrations. The initial baseline results for the countries under the OECD Secretariat's responsibility are compared with those obtained from the questionnaire replies and issues arising are discussed in bilateral exchanges with country experts. The initial projections for individual country and regional modules developed by the FAO Secretariat are reviewed by a wider circle of in-house and international experts. In this stage, the global projection picture emerges and refinements are made according to a consensus view of both Secretariats and external advisors. On the basis of these discussions and of updated information, a second baseline is produced. The information generated is used to prepare market assessments for biofuels, cereals, oilseeds, meats, dairy products and sugar over the course of the outlook period, which is discussed at the annual meetings of the Group on Commodity Markets of the OECD *Committee for Agriculture*. Following the receipt of comments and final data revisions, a last revision is made to the baseline projections. The revised projections form the basis of a draft of the present *Agricultural Outlook* publication, which is discussed by the *Senior Management Committee* of FAO's Department of Economic and Social Development and the OECD's *Working Party on Agricultural Policies and Markets of the Committee for Agriculture*, in May 2012, prior to publication. In addition, the *Outlook* will be used as a basis for analysis presented to the FAO's *Committee on Commodity Problems* and its various *Intergovernmental Commodity Groups*.

The Outlook process implies that the baseline projections presented in this report are a combination of projections developed by collaborators for countries under the OECD Secretariat's responsibility and original projections for the 42 countries and regions under the FAO Secretariat's responsibility. The use of a formal modelling framework reconciles inconsistencies between individual country projections and forms a global equilibrium for all commodity markets. The review process ensures that judgement of country experts is brought to bear on the projections and related analyses. However, the final responsibility for the projections and their interpretation rests with the OECD and FAO Secretariats.

Sources and assumptions for the macroeconomic projections

Population estimates from the 2010 Revision of the *United Nations Population Prospects Database* provide the population data used for all countries and regional aggregates in the *Outlook*. For the projection period, the medium variant set of estimates was selected for use from the four alternative projection variants (low, medium, high and constant fertility). The *UN Population Prospects Database* was chosen because it represents a comprehensive source of reliable estimates which includes data for non-OECD developing countries. For consistency reasons, the same source is used for both the historical population estimates and the projection data.

The other macroeconomic series used in the AGLINK-COSIMO model are real GDP, the GDP deflator, the private consumption expenditure (PCE) deflator, the Brent crude oil price (in US dollars per barrel) and exchange rates expressed as the local currency value of

USD 1. Historical data for these series in OECD countries (except Turkey, Chile and Israel) as well as Brazil, Argentina, China and the Russian Federation are consistent with those published in the *OECD Economic Outlook* No. 90, December 2011 and No. 89. For other economies, historical macroeconomic data were obtained from the IMF, *World Economic Outlook*, September 2011. Assumptions for 2012-2021 are based on the recent medium term macroeconomic projections of the OECD Economics Department, projections of the *OECD Economic Outlook No. 89* and projections of the IMF.

The model uses indices for real GDP, consumer prices (PCE deflator) and producer prices (GDP deflator) which are constructed with the base year 2005 value being equal to 1. The assumption of constant real exchange rates implies that a country with higher (lower)inflation relative to the United States (as measured by the US GDP deflator) will have a depreciating (appreciating) currency and therefore an increasing (decreasing) exchange rate over the projection period, since the exchange rate is measured as the local currency value of 1 USD. The calculation of the nominal exchange rate uses the percentage growth of the ratio "country-GDP deflator/US GDP deflator".

The oil price used to generate the *Outlook* is based on information from the *OECD Economic Outlook No. 90* until 2012 (short term update) and the growth rate of the *OECD Economic Outlook No. 89* for future paths.

The representation of production costs in AGLINK-COSIMO

Changes in production costs are an important variable for farmers' decisions on crop and livestock production quantities, in addition to output returns and, if applicable, policy measures.

While supply in AGLINK-COSIMO is largely determined by gross returns, production costs are represented in the model in the form of a cost index used to deflate gross production revenues. In other words, supply equations in the model in most cases depend on gross returns per unit of activity (such as returns per hectare or the meat price) relative to the overall production cost level as expressed by the index. Consequently, equations for harvested areas in crop production and for livestock production quantities take the following general forms:

$$AH = f\left(\frac{RH}{CPCI}\right);\ QP = f\left(\frac{PP}{CPCI}\right)$$

with:

AH	area harvested (crop production)
RH	returns per hectare (crop production)
CPCI	commodity production cost index
QP	production quantity (livestock production)
PP	producer price (livestock production)

Among others, energy prices, increased by rising crude oil prices, have fostered attention to agricultural production costs in agricultural commodity models. Energy prices can significantly impact on international markets for agricultural products as production costs for both crops and livestock products are highly dependent on energy costs. Fuels for tractors and other machinery, as well as heating and other forms of energy are directly used in the production process. In addition, other inputs such as fertilisers and pesticides have high energy content, and costs for these inputs are driven to a significant extent by energy prices. It is therefore important to explicitly consider energy prices in the representation of production costs.

The production cost indices employed in AGLINK-COSIMO for *livestock products* is constructed from three sub-indices representing non-tradable inputs, energy inputs, and other tradable inputs, respectively. While the non-tradable sub-index is approximated by the domestic GDP deflator, the energy sub-index is affected by changes in the world crude oil price and the country's exchange rate. Finally, the tradable sub-index is linked to global inflation (approximated by the US GDP deflator) and the country's exchange rate. This relationship is shown in the following equation:

$$\begin{aligned} CPCI_{r,t} = {} & CPCS_{r,t}^{NT} * GDPD_{r,t} / GDPD_{r,bas} \\ & + CPCS_{r,t}^{EN} * \left(XP_{t}^{OIL} * XR_{r,t}\right) / \left(XP_{bas}^{OIL} * XR_{r,bas}\right) \\ & + \left(1 - CPCS_{r,t}^{NT,I} - CPCS_{r,t}^{EN,I}\right) * (XR_{r,t} * GDPD_{USA,t}) / (XR_{r,bas} * GDPD_{USA,bas}) \end{aligned}$$

with:

CPCI	commodity production cost index for livestock
$CPCS^{NT}$	share of non-tradable input in total base commodity production costs
$CPCS^{EN}$	share of energy in total base commodity production costs
GDPD	deflator for the gross domestic product
XP^{OIL}	world crude oil price
XR	nominal exchange rate with respect to the US Dollar
r,t	region and time index, respectively
bas	base year (2000 or 2005 or 2008) value

The production cost index is different for each *crop products* and is constructed from five sub-indices representing seeds inputs, fertiliser inputs, energy inputs, other tradable inputs and non-tradable inputs, respectively.

$$\begin{aligned} CPCI_{r,t}^{c} = {} & CPCS_{r,t}^{NT} * GDPD_{r,t} / GDPD_{r,bas} \\ & + CPCS_{r,t}^{EN} * \left(XP_{t}^{OIL} * XR_{r,t}\right) / \left(XP_{bas}^{OIL} * XR_{r,bas}\right) \\ & + CPCS_{r,t}^{FT} * \left(XP_{t}^{FT} * XR_{r,t}\right) / \left(XP_{bas}^{FT} * XR_{r,bas}\right) \\ & + CPCS_{r,t}^{TR} * (XR_{r,t} * GDPD_{USA,t}) / (XR_{r,bas} * GDPD_{USA,bas}) \\ & + CPCS_{r,t}^{SD} * PP_{r,t}^{c}(-1) / PP_{r,bas}^{c} \end{aligned}$$

with:

$CPCI^{C}$	commodity production cost index for crop product c
$CPCS^{NT}$	share of non-tradable input in total base commodity production costs
$CPCS^{EN}$	share of energy in total base commodity production costs
$CPCS^{FT}$	share of fertiliser in total base commodity production costs
$CPCS^{TR}$	share of other tradable input in total base commodity production costs
$CPCS^{SD}$	share of seeds input in total base commodity production costs
GDPD	deflator for the gross domestic product
XP^{OIL}	world crude oil price
XP^{FT}	world fertiliser price
PP^{C}	producer price for crop product c

XR	nominal exchange rate with respect to the US Dollar
c	Crop product
r,t	region and time index, respectively
bas	base year (2000 or 2005 or 2008) value

The shares of the various cost categories are country specific. They were estimated based on historic cost structures in individual counties. Shares vary depending on the development stages of the countries and regions. Developed countries tend to have higher shares of energy, fertiliser and tradable inputs than developing nations.

The Fertilizer price is constructed by FAO fertiliser analysts as following:

XP^{FT} = 0.2*DAP + 0.16*MOP + 0.02*TSP + 0.62*Urea

With:

US Diammonium Phosphate (DAP)

Can Potassium Chloride (MOP)

Triple superphosphate (TSP)

Urea (Black Sea)

And is represented by an equation in the AGLINK-COSIMO model:

$$\log(XP_t^{FT}) = CON + elas_{FT}^{OIL} * \log(XP_t^{OIL}) + elas_{FT}^{crop} * \log(0.5 * XP_{t-1}^{CG} + 0.2 * XP_{t-1}^{WT} + 0.2 * XP_{t-1}^{OS} + 0.1 * XP_{t-1}^{RI})$$

With:

XP^{OIL}	world crude oil price
XP^{FT}	world fertiliser price
XP^{CG}	world coarse grain price
XP^{WT}	world wheat price
XP^{OS}	world oilseed price
XP^{RI}	world rice price

The methodology of stochastic simulations with AGLINK-COSIMO

The AGLINK-COSIMO model is a forward-looking medium term economic model which is used to perform simulations over a 10-year horizon. It is necessary to feed into the model a set of assumptions for exogenous variables. While a single set of assumptions is used for deterministic baseline, multiple sets of exogenous variables generated by random samplings, are fed into the model for stochastic experiments. The model is simulated for each set of assumptions and, thus, multiple sets of solutions are obtained. Implications of uncertainties for the baseline projections can be inferred from statistical information of the random outputs of the simulations.

Recently, analyses using the AGLINK-COSIMO stochastic model have been undertaken with an emphasis on risk and uncertainties in terms of price volatility (OECD, 2011, OECD, 2012). The methodology used in the present Outlook was developed and improved in the course of these works. The exogenous assumptions that are challenged in the stochastic framework relate to yields and macroeconomic variables. The methodology to obtain stochastic assumptions is detailed below.

Yields

Yields of coarse grains, rice and wheat for the modelled countries are made stochastic. Countries are sorted into eleven regional groups (Africa, East Asia, Europe, etc.) and the detrended yields in a region are assumed to follow multivariate normal distribution. Variance covariance matrices of the distributions, which reflect the magnitude of production shocks and the tendencies of crops in a region to be affected by a common risk factor, are calculated over historical data. Random samplings from the estimated distributions replicate the historically observed variation in yield projections.

Crude oil, fertiliser prices and macroeconomic variables

Crude oil prices are simulated using a truncated normal distribution that has been calibrated on past historical trends. The international fertiliser price is modeled as a function of the crude oil price calibrated on historical data. A simple macroeconomic model of GDP changes and consumer price index for leading economies (Brazil, China, European Union, India, Japan, the Russian Federation and the United States) was also developed and calibrated over historical data. The crude oil price being one of the variables of this simple model, random draws for macroeconomic variables are obtained by solving this macroeconomic model on random draws for the crude oil price.

Statistical Annex

Table A.1. Economic assumptions

Calendar year

		Avg 2009-11est.	2012	2013	2014	2015	2016	2017	2018	2019	2020	2021
REAL GDP[1]												
Australia	%	1.9	4.0	3.2	3.6	3.4	3.0	2.8	2.8	2.8	2.8	2.8
Canada	%	0.9	1.9	2.5	2.7	2.5	2.1	1.8	1.8	1.8	1.8	1.8
EU	%	-0.3	0.5	1.4	2.0	2.1	1.9	1.8	1.8	1.8	1.7	1.7
Japan	%	-0.8	2.0	1.6	2.5	2.7	1.9	1.4	1.5	1.5	1.4	1.4
Korea	%	3.4	3.8	4.3	3.4	3.3	3.2	3.1	2.9	2.8	2.5	2.3
Mexico	%	1.1	3.3	3.6	3.7	3.7	3.4	3.2	3.1	3.1	3.0	3.0
New Zealand	%	1.3	2.5	3.0	3.1	2.7	2.4	2.3	2.3	2.3	2.3	2.3
Norway	%	0.0	2.0	2.7	3.6	3.3	3.2	3.0	2.8	2.8	2.7	2.6
Switzerland	%	0.9	0.8	1.9	2.0	2.0	1.9	2.0	1.9	1.9	1.9	1.8
Turkey	%	3.6	2.2	3.4	3.8	4.1	4.3	4.0	4.0	4.0	4.0	4.0
United States	%	0.4	2.0	2.5	3.3	3.4	2.9	2.6	2.5	2.3	2.2	2.1
Argentina	%	6.3	4.6	6.1	2.7	2.7	2.5	2.5	2.5	2.5	2.5	2.5
Brazil	%	3.4	3.2	3.9	4.8	4.7	4.7	4.6	4.5	4.4	4.3	4.2
China	%	7.8	6.8	8.3	9.5	9.2	8.9	8.6	8.3	8.0	7.7	7.4
India	%	8.2	7.5	8.1	8.1	8.1	8.1	8.1	8.1	8.1	8.1	8.1
Russian Federation	%	0.1	4.1	4.1	5.1	5.0	5.0	4.9	4.9	4.9	4.9	4.8
South Africa	%	1.5	3.6	4.0	3.8	3.6	3.6	3.7	3.7	3.7	3.7	3.7
OECD[2,3]	%	0.2	1.6	2.1	2.7	2.8	2.5	2.2	2.2	2.1	2.0	2.0
PCE DEFLATOR[1]												
Australia	%	2.1	2.9	3.0	2.5	2.5	2.5	2.5	2.5	2.5	2.5	2.5
Canada	%	1.3	1.3	1.1	2.1	2.1	2.1	2.1	2.1	2.1	2.1	2.1
EU	%	1.8	1.9	1.7	1.9	2.0	1.9	1.9	1.9	1.9	1.9	1.9
Japan	%	-1.6	-0.9	-0.4	0.5	0.9	1.0	1.0	1.0	1.0	1.0	1.0
Korea	%	3.0	3.6	3.0	3.6	3.2	3.1	3.1	3.1	3.1	3.1	3.1
Mexico	%	4.7	3.6	3.4	3.0	3.0	3.0	3.0	3.0	3.0	3.0	3.0
New Zealand	%	2.2	2.2	2.3	2.1	2.0	2.0	2.0	2.0	2.0	2.0	2.0
Norway	%	1.9	2.0	2.0	2.3	2.5	2.5	2.5	2.5	2.5	2.5	2.5
Switzerland	%	0.3	0.2	0.3	1.5	1.9	2.0	2.0	2.0	2.0	2.0	2.0
Turkey	%	6.9	6.9	5.3	5.0	5.0	5.0	5.0	5.0	5.0	5.0	5.0
United States	%	1.5	1.9	1.4	1.8	2.0	2.0	2.0	2.0	2.0	2.0	2.0
Argentina	%	11.4	11.2	10.4	3.9	3.9	3.9	3.9	3.9	3.9	3.9	3.9
Brazil	%	6.0	6.6	5.3	4.3	4.3	4.3	4.3	4.3	4.3	4.3	4.3
China	%	3.6	5.8	4.1	2.7	2.7	2.7	2.7	2.7	2.8	2.8	2.8
India	%	11.1	8.6	7.1	5.4	5.0	4.1	4.8	4.8	4.8	4.8	4.8
Russian Federation	%	10.6	6.7	5.7	3.8	3.8	3.8	3.8	3.8	3.8	3.8	3.8
South Africa	%	5.8	5.0	5.0	4.9	4.8	4.7	4.8	4.8	4.8	4.8	4.8
OECD[2,3]	%	1.6	1.9	1.6	2.0	2.1	2.1	2.1	2.1	2.1	2.1	2.1

		2011est	2012	2013	2014	2015	2016	2017	2018	2019	2020	2021
POPULATION[1]												
Australia	%	1.5	1.4	1.3	1.2	1.2	1.2	1.2	1.2	1.2	1.1	1.1
Canada	%	1.0	0.9	0.9	0.9	0.9	0.9	0.9	0.9	0.8	0.8	0.8
EU	%	0.3	0.2	0.2	0.3	0.3	0.2	0.2	0.2	0.2	0.2	0.2
Japan	%	0.0	0.0	-0.1	-0.1	-0.1	-0.1	-0.2	-0.2	-0.2	-0.3	-0.3
Korea	%	0.4	0.4	0.4	0.4	0.3	0.3	0.3	0.3	0.3	0.2	0.2
Mexico	%	1.2	1.2	1.1	1.1	1.1	1.0	1.0	1.0	0.9	0.9	0.9
New Zealand	%	1.1	1.1	1.1	1.0	1.0	1.0	1.0	1.0	0.9	0.9	0.9
Norway	%	0.9	0.7	0.6	0.6	0.6	0.7	0.7	0.7	0.7	0.7	0.7
Switzerland	%	0.5	0.4	0.4	0.3	0.3	0.3	0.3	0.3	0.3	0.3	0.3
Turkey	%	1.2	1.2	1.1	1.1	1.1	1.0	1.0	1.0	0.9	0.9	0.9
United States	%	0.9	0.9	0.9	0.8	0.8	0.8	0.8	0.8	0.8	0.8	0.8
Argentina	%	0.9	0.9	0.9	0.9	0.8	0.8	0.8	0.8	0.8	0.7	0.7
Brazil	%	0.9	0.9	0.9	0.8	0.8	0.8	0.7	0.7	0.7	0.6	0.6
China	%	0.5	0.4	0.4	0.4	0.4	0.3	0.3	0.3	0.2	0.2	0.2
India	%	1.4	1.4	1.3	1.3	1.3	1.2	1.2	1.2	1.1	1.1	1.1
Russian Federation	%	-0.1	-0.1	-0.1	-0.1	-0.1	-0.1	-0.1	-0.2	-0.2	-0.2	-0.2
South Africa	%	0.7	0.6	0.5	0.4	0.4	0.4	0.4	0.4	0.4	0.4	0.5
OECD[3]	%	0.6	0.6	0.5	0.6	0.6	0.5	0.5	0.5	0.5	0.5	0.4
World	%	1.1	1.1	1.1	1.1	1.1	1.0	1.0	1.0	1.0	1.0	0.9

Table A.1. Economic assumptions *(cont.)*

Calendar year

		Avg 2009-11est.	2012	2013	2014	2015	2016	2017	2018	2019	2020	2021
GDP DEFLATOR[1]												
Australia	%	3.3	3.1	2.8	2.6	2.5	2.5	2.5	2.5	2.5	2.5	2.5
Canada	%	1.4	1.3	1.3	1.8	1.8	1.9	1.9	2.0	2.0	2.0	2.0
European Union	%	1.2	1.8	1.7	2.0	2.1	2.1	1.9	2.0	2.0	1.9	1.9
Japan	%	-1.5	-0.7	-0.3	0.5	0.9	1.0	1.0	1.0	1.0	1.0	1.0
Korea	%	3.0	2.5	2.1	3.2	2.9	2.9	2.9	2.9	2.9	2.9	2.9
Mexico	%	4.5	3.8	3.9	3.6	3.1	3.1	3.1	3.1	3.1	3.1	3.1
New Zealand	%	2.2	2.0	2.6	4.0	2.3	2.0	2.0	2.0	2.0	2.0	2.0
Norway	%	3.4	2.0	2.2	2.0	2.4	2.5	2.5	2.5	2.5	2.5	2.5
Switzerland	%	0.3	0.2	0.3	1.5	1.9	2.0	2.0	2.0	2.0	2.0	2.0
Turkey	%	6.1	7.1	5.1	4.9	4.9	4.9	4.9	4.9	4.9	4.9	4.9
United States	%	1.5	1.9	1.4	1.8	2.0	2.0	2.0	2.0	2.0	2.0	2.0
Argentina	%	14.2	11.2	10.4	3.9	3.9	3.9	3.9	3.9	3.9	3.9	3.9
Brazil	%	7.0	6.6	5.3	4.3	4.3	4.3	4.3	4.3	4.3	4.3	4.3
China	%	4.7	5.8	4.1	2.7	2.7	2.7	2.7	2.7	2.8	2.8	2.8
India	%	8.2	5.8	5.6	4.8	4.5	4.5	4.6	4.6	4.6	4.6	4.6
Russian Federation	%	9.0	6.7	5.7	3.8	3.8	3.8	3.8	3.8	3.8	3.8	3.8
South Africa	%	7.5	6.0	5.3	5.0	4.9	4.7	4.9	4.9	4.9	4.9	4.9
OECD[3]	%	1.4	1.8	1.6	2.0	2.1	2.1	2.1	2.1	2.1	2.1	2.1
WORLD OIL PRICE												
Brent crude oil price[4]	USD/barrel	84.1	110.0	112.1	115.5	119.0	122.6	126.3	130.1	134.1	138.2	142.4

		Avg 2009-11est.	2012	2013	2014	2015	2016	2017	2018	2019	2020	2021
EXCHANGE RATES												
Australia	AUD/USD	1.11	0.98	0.99	1.00	1.01	1.01	1.02	1.02	1.03	1.03	1.04
Canada	CAD/USD	1.05	0.98	0.98	0.98	0.98	0.98	0.98	0.98	0.98	0.98	0.98
European Union	EUR/USD	0.73	0.72	0.72	0.72	0.72	0.72	0.72	0.72	0.72	0.72	0.72
Japan	JPY/USD	86.99	77.60	76.31	75.35	74.58	73.84	73.12	72.40	71.69	70.98	70.28
Korea	000 KRW/USD	1.18	1.11	1.12	1.13	1.14	1.15	1.16	1.17	1.18	1.20	1.21
Mexico	MXN/USD	12.84	12.63	12.93	13.16	13.31	13.45	13.59	13.73	13.88	14.02	14.17
New Zealand	NZD/USD	1.42	1.26	1.28	1.31	1.31	1.31	1.31	1.31	1.31	1.31	1.31
Argentina	ARS/USD	3.88	4.38	4.77	4.87	4.96	5.05	5.15	5.24	5.34	5.44	5.54
Brazil	BRL/USD	1.81	1.74	1.81	1.85	1.90	1.94	1.98	2.03	2.07	2.12	2.17
China	CNY/USD	6.69	6.71	6.89	6.95	7.01	7.06	7.11	7.16	7.21	7.27	7.32
India	INR/USD	47.30	49.41	50.78	51.96	53.13	54.33	55.57	56.84	58.13	59.45	60.81
Russian Federation	RUB/USD	30.48	30.68	31.97	32.61	33.20	33.79	34.39	35.00	35.62	36.26	36.90
South Africa	ZAR/USD	7.59	7.32	7.64	7.94	8.23	8.52	8.84	9.17	9.51	9.86	10.23

Note: Calendar year: For OECD member countries (except Turkey, Chile and Israel), as well as Brazil, Argentina, China and Russia, historical data for real GDP, private consumption expenditure deflator and GDP deflator were obtained from the OECD Economic Outlook No. 90, December 2011. For other economies, historical macroeconomic data were obtained from the IMF, World Economic Outlook, September 2011. Assumptions for the projection period draw on the recent medium term macroeconomic projections of the OECD Economics Department, projections of the OECD Economic Outlook No. 89, projections of the IMF, and for population, projections from the United Nations World Population Prospects Database, 2010 Revision (medium variant). Data for the European Union are for the euro area aggregates.

1. Annual per cent change. The price index used is the private consumption expenditure deflator.
2. Annual weighted average real GDP and CPI growth rates in OECD countries are based on weights using purchasing power parities (PPPs).
3. Excludes Iceland but includes all 27 EU members.
4. Short term update for crude oil price from the OECD Economic Outlook No.90, December 2011.

Source: OECD and FAO Secretariats.

StatLink http://dx.doi.org/10.1787/888932642535

Table A.2. World prices

		Avg 09/10-11/12est	12/13	13/14	14/15	15/16	16/17	17/18	18/19	19/20	20/21	21/22
CEREALS												
Wheat[1]	USD/t	266.6	248.6	250.8	250.4	250.3	259.7	264.5	272.5	275.7	276.9	279.3
Coarse Grains[2]	USD/t	227.8	244.8	228.7	227.5	231.2	233.4	236.2	242.1	246.1	247.9	246.3
Rice[3]	USD/t	556.8	493.6	465.3	444.9	420.6	419.3	425.9	435.4	442.8	450.3	454.5
OILSEEDS												
Oilseeds[4]	USD/t	503.1	499.9	526.6	513.5	520.6	521.2	533.1	537.2	542.1	542.9	550.3
Protein meals[5]	USD/t	367.2	378.4	389.9	390.3	387.4	384.2	392.9	396.5	399.4	401.1	403.6
Vegetable oils[6]	USD/t	1 066.9	1 113.9	1 124.8	1 107.4	1 140.6	1 150.3	1 174.9	1 200.9	1 213.6	1 219.9	1 232.5
SWEETENERS												
Raw sugar[7]	USD/t rse	532.9	460.8	464.0	474.2	522.9	455.1	465.4	478.6	477.7	474.9	483.1
Refined sugar[8]	USD/t rse	632.5	537.5	545.6	557.2	607.8	537.3	546.2	561.6	560.3	559.0	565.7
HFCS[9]	USD/t	544.5	429.8	454.6	511.4	550.2	512.4	494.4	516.1	521.3	529.1	536.1
MEAT												
Beef and veal												
Price, EU[10]	USD/t dw	4 034.9	4 398.7	4 417.5	4 413.9	4 517.2	4 401.6	4 462.5	4 569.2	4 653.4	4 685.4	4 699.3
Price, United States[11]	USD/t dw	3 477.1	4 290.5	4 788.2	4 708.5	4 469.5	4 410.7	4 541.1	4 788.7	4 712.1	4 598.9	4 717.9
Price, Brazil[12]	USD/t pw	3 035.6	3 323.7	3 354.5	3 387.6	3 439.8	3 382.3	3 473.7	3 535.8	3 515.3	3 579.9	3 626.3
Pigmeat												
Price, EU[13]	USD/t dw	1 975.5	2 195.8	2 048.3	2 161.6	2 089.7	2 097.4	2 069.9	2 161.2	2 192.4	2 158.0	2 417.2
Price, United States[14]	USD/t dw	1 657.6	1 957.9	2 249.7	2 198.7	2 201.5	2 241.9	2 249.0	2 388.5	2 339.9	2 318.3	2 380.4
Price, Brazil[15]	USD/t dw	1 474.7	1 865.4	1 900.5	1 841.7	1 894.8	1 956.9	1 906.3	2 020.7	1 985.9	2 014.4	2 079.4
Poultry meat												
Price, EU[16]	USD/t rtc	2 569.4	2 685.9	2 575.8	2 539.5	2 667.5	2 805.1	2 887.3	2 946.8	2 992.3	3 079.9	3 014.2
Price, United States[17]	USD/t rtc	1 073.7	1 153.1	1 074.7	1 054.2	1 051.7	1 064.0	1 083.0	1 107.1	1 126.3	1 130.8	1 121.0
Price, Brazil[18]	USD/t rtc	1 175.4	1 354.1	1 308.5	1 280.2	1 292.7	1 310.1	1 340.3	1 368.1	1 390.6	1 400.2	1 419.1
Sheepmeat												
Price, New Zealand[19]	USD/t dw	3 789.5	4 257.3	4 121.8	4 027.0	4 168.5	4 277.7	4 381.5	4 462.2	4 784.8	4 742.3	4 812.1
FISH AND SEA FOOD												
Product traded[20]	USD/t	2 500.1	2 684.0	2 689.8	2 759.3	2 914.9	2 964.3	3 082.6	3 169.1	3 295.8	3 400.8	3 474.8
Aquaculture[21]	USD/t	1 904.7	1 985.8	2 001.5	2 048.8	2 175.4	2 233.5	2 361.1	2 471.5	2 578.2	2 763.3	2 818.0
Capture[22]	USD/t	1 288.7	1 362.3	1 391.2	1 437.7	1 510.2	1 550.5	1 611.2	1 665.0	1 730.9	1 792.6	1 843.1
Meal[23]	USD/t	1 484.8	1 565.1	1 702.4	1 868.1	1 995.5	2 016.7	2 047.8	2 146.6	2 246.5	2 347.2	2 360.8
DAIRY PRODUCTS												
Butter[24]	USD/t	3 626.0	3 859.9	3 821.4	3 764.5	3 774.4	3 794.5	3 911.9	4 031.6	4 086.5	4 136.2	4 214.0
Cheese[25]	USD/t	3 761.3	3 854.6	3 886.2	3 824.3	3 872.1	3 924.0	4 024.9	4 122.8	4 188.2	4 252.0	4 327.4
Skim milk powder[26]	USD/t	3 018.4	3 187.5	3 347.9	3 338.1	3 428.4	3 480.2	3 566.6	3 622.3	3 689.0	3 760.8	3 840.7
Whole milk powder[27]	USD/t	3 252.0	3 423.2	3 428.1	3 455.0	3 527.7	3 574.4	3 667.2	3 749.5	3 826.1	3 912.7	4 007.4
Whey powder wholesale price, United States[28]	USD/t	897.1	1 093.5	1 104.0	1 120.2	1 128.6	1 143.8	1 164.2	1 179.3	1 209.5	1 232.8	1 247.5
Casein[29]	USD/t	7 916.6	7 849.6	8 434.1	8 554.3	8 667.7	8 674.0	8 931.0	9 036.1	9 253.4	9 433.4	9 665.2
BIOFUEL												
Ethanol[30]	USD/hl	63.7	85.4	83.7	85.2	87.1	88.9	90.0	92.1	93.0	94.4	95.6
Biodiesel[31]	USD/hl	132.1	152.7	156.1	157.6	160.4	163.7	167.2	172.1	174.7	177.6	181.1

Note: This table is a compilation of price information presented in the detailed commodity tables further in this annex. Prices for crops are on marketing year basis and those for meat and dairy products on calendar year basis (e.g. 09/10 is calendar year 2009).

1. No.2 hard red winter wheat, ordinary protein, United States f.o.b. Gulf Ports (June/May), less EEP payments where applicable.
2. No.2 yellow corn, United States f.o.b. Gulf Ports (September/August).
3. Milled, 100%, grade b, Nominal Price Quote, NPQ, f.o.b. Bangkok (January/December).
4. Weighted average oilseed price, European port.
5. Weighted average meal price, European port.
6. Weighted average price of oilseed oils and palm oil, European port.
7. Raw sugar world price, ICE contract No11 nearby, October/September.
8. Refined sugar price,Euronext,Liffe,Contract No. 407 London,Europe,October/September.
9. United States wholesale list price HFCS-55 , October/September.
10. EU average beef producer price.
11. Choice steers, 1100-1300 lb lw, Nebraska - lw to dw conversion factor 0.63.
12. Brazil average beef producer price.
13. EU average pigmeat producer price.
14. Barrows and gilts, No. 1-3, 230-250 lb lw, Iowa/South Minnesota - lw to dw conversion factor 0.74.
15. Brazil average pigmeat producer price.
16. EU average producer price.
17. Wholesale weighted average broiler price 12 cities.
18. Brazil average chicken for slaughter producer price.
19. Lamb schedule price, all grade average.
20. World unit value of trade (sum of exports and imports).
21. World unit value of aquaculture fisheries production (live weight basis).
22. FAO estimated value of world ex vessel value of capture fisheries production excluding for reduction.
23. Fish meal, 64-65% protein, Hamburg, Germany.
24. F.o.b. export price, butter, 82% butterfat, Oceania.
25. F.o.b. export price, cheddar cheese, 39% moisture, Oceania.
26. F.o.b. export price, non-fat dry milk, 1.25% butterfat,Oceania.
27. F.o.b. export price, WMP 26% butterfat, Oceania.
28. Dry whey, West region, United States.
29. Export price, New Zealand.
30. Brazil, Sao Paulo (ex-distillery).
31. Producer price Germany net of biodiesel tariff.

Source: OECD and FAO Secretariats.

StatLink http://dx.doi.org/10.1787/888932642554

Table A.3.1. World trade projections, imports

		Avg 2009-11est.	2012	2013	2014	2015	2016	2017	2018	2019	2020	2021
Wheat												
World Trade	**kt**	**128 551**	**136 646**	**137 130**	**137 076**	**139 121**	**139 906**	**143 743**	**145 443**	**148 135**	**150 117**	**152 134**
OECD[1]	kt	29 336	28 820	29 266	29 384	29 349	28 940	28 851	28 695	28 564	28 510	28 364
Developing	kt	103 100	109 976	110 911	111 364	113 221	114 583	118 182	120 008	122 651	124 655	126 832
Least Developed Countries	kt	15 840	16 396	16 582	16 654	17 039	17 067	17 621	18 095	18 571	18 860	19 124
Coarse grains												
World Trade	**kt**	**119 562**	**123 666**	**126 612**	**127 337**	**128 793**	**131 739**	**135 170**	**138 130**	**140 770**	**143 185**	**146 058**
OECD[1]	kt	52 177	52 509	51 833	51 564	52 021	52 549	52 692	53 694	54 123	54 290	54 692
Developing	kt	87 037	90 820	94 876	96 451	97 592	100 319	103 917	106 270	108 708	111 151	113 964
Least Developed Countries	kt	2 690	3 300	3 631	3 701	3 605	3 601	3 615	3 693	3 698	3 828	4 030
Rice												
World Trade	**kt**	**32 871**	**35 755**	**36 687**	**37 413**	**38 764**	**39 542**	**40 159**	**40 748**	**41 202**	**41 806**	**42 775**
OECD[1]	kt	4 919	5 168	5 286	5 426	5 562	5 694	5 810	5 908	6 005	6 092	6 185
Developing	kt	27 959	30 715	31 547	32 179	33 424	34 105	34 631	35 144	35 522	36 086	36 967
Least Developed Countries	kt	6 552	6 692	7 624	8 105	8 732	9 192	9 489	9 790	9 817	9 897	10 181
Oilseeds												
World Trade	**kt**	**110 707**	**116 029**	**117 771**	**120 057**	**123 224**	**125 979**	**127 895**	**129 837**	**131 858**	**133 926**	**135 812**
OECD[1]	kt	35 830	36 896	36 837	37 375	37 819	38 377	38 708	38 900	39 114	39 282	39 547
Developing	kt	83 023	87 495	89 283	91 186	93 936	96 256	97 901	99 787	101 680	103 658	105 335
Least Developed Countries	kt	299	273	290	307	341	371	389	414	439	472	507
Protein Meals												
World Trade	**kt**	**71 895**	**75 592**	**77 379**	**78 948**	**80 289**	**81 413**	**82 545**	**83 957**	**85 211**	**86 817**	**87 935**
OECD[1]	kt	41 677	42 159	42 582	42 822	43 309	43 538	43 924	44 050	44 654	44 969	45 403
Developing	kt	33 054	36 283	37 843	39 345	40 322	41 227	42 010	43 325	44 042	45 444	46 236
Least Developed Countries	kt	484	534	566	605	649	691	742	785	826	878	912
Vegetable Oils												
World Trade	**kt**	**62 052**	**66 907**	**68 040**	**69 504**	**70 696**	**71 961**	**73 407**	**74 883**	**76 308**	**77 673**	**79 089**
OECD[1]	kt	17 791	19 976	19 941	20 221	20 578	20 710	21 048	21 332	21 739	22 150	22 475
Developing	kt	44 439	47 218	48 380	49 572	50 428	51 603	52 740	53 973	55 029	56 032	57 179
Least Developed Countries	kt	4 711	4 736	4 863	5 027	5 180	5 372	5 550	5 753	5 948	6 170	6 395
Sugar												
World Trade	**kt**	**49 129**	**47 052**	**47 697**	**48 235**	**51 543**	**50 912**	**52 362**	**54 960**	**55 640**	**56 617**	**57 978**
OECD[1]	kt	13 360	14 677	14 359	14 062	14 073	14 307	14 324	14 510	14 795	14 844	14 726
Developing	kt	33 252	30 635	31 696	32 574	36 179	35 165	36 798	39 256	39 642	40 599	42 149
Least Developed Countries	kt	5 229	4 730	4 794	4 835	4 835	5 306	5 254	5 328	5 537	5 595	5 629
Beef[2]												
World Trade	**kt**	**6 914**	**7 192**	**7 736**	**7 812**	**7 663**	**7 944**	**8 146**	**8 232**	**8 386**	**8 513**	**8 603**
OECD[1]	kt	3 260	3 335	3 757	3 804	3 584	3 779	3 887	3 854	3 930	3 882	3 861
Developing	kt	3 495	3 899	3 998	4 095	4 230	4 347	4 473	4 590	4 714	4 800	4 896
Least Developed Countries	kt	111	83	70	64	58	60	52	47	35	31	25
Pigmeat[2]												
World Trade	**kt**	**5 898**	**6 077**	**6 134**	**6 097**	**6 204**	**6 255**	**6 370**	**6 472**	**6 541**	**6 596**	**6 669**
OECD[1]	kt	2 964	3 222	3 214	3 293	3 353	3 405	3 481	3 513	3 553	3 523	3 554
Developing	kt	2 704	3 009	3 053	3 078	3 169	3 205	3 302	3 411	3 444	3 569	3 614
Least Developed Countries	kt	141	150	161	166	168	168	168	176	186	185	199
Poultry												
World Trade	**kt**	**11 128**	**11 608**	**11 881**	**12 249**	**12 552**	**12 693**	**12 838**	**12 964**	**13 160**	**13 231**	**13 461**
OECD[1]	kt	2 415	2 480	2 501	2 509	2 550	2 550	2 563	2 522	2 462	2 430	2 407
Developing	kt	7 672	8 325	8 627	8 951	9 219	9 383	9 543	9 728	9 927	10 227	10 523
Least Developed Countries	kt	775	864	895	903	922	922	916	947	982	1 006	1 041
Butter												
World Trade	**kt**	**855**	**858**	**875**	**897**	**919**	**935**	**950**	**966**	**978**	**991**	**1 001**
OECD[1]	kt	137	114	113	116	117	119	119	119	119	118	116
Developing	kt	569	603	620	643	664	680	694	713	725	738	751
Least Developed Countries	kt	1 218	1 016	892	836	745	693	630	706	567	333	177
Cheese												
World Trade	**kt**	**2 254**	**2 383**	**2 396**	**2 460**	**2 518**	**2 564**	**2 612**	**2 662**	**2 726**	**2 792**	**2 853**
OECD[1]	kt	738	776	796	807	822	838	858	877	895	905	920
Developing	kt	1 189	1 310	1 330	1 381	1 431	1 464	1 498	1 528	1 574	1 613	1 665
Least Developed Countries	kt	1 261	1 063	942	891	808	759	700	780	645	417	270
Whole Milk Powder												
World Trade	**kt**	**2 170**	**2 234**	**2 249**	**2 307**	**2 357**	**2 405**	**2 446**	**2 493**	**2 545**	**2 588**	**2 640**
OECD[1]	kt	74	73	75	77	78	80	80	81	81	81	81
Developing	kt	2 050	2 114	2 131	2 189	2 240	2 286	2 325	2 370	2 419	2 459	2 508
Least Developed Countries	kt	1 437	1 248	1 132	1 084	1 003	959	906	991	862	638	492
Skim Milk Powder												
World Trade	**kt**	**1 473**	**1 712**	**1 707**	**1 750**	**1 784**	**1 819**	**1 854**	**1 897**	**1 940**	**1 982**	**2 025**
OECD[1]	kt	207	211	210	216	219	222	223	226	229	232	235
Developing	kt	1 334	1 563	1 561	1 601	1 635	1 669	1 703	1 745	1 786	1 826	1 868
Least Developed Countries	kt	1 299	1 106	984	932	845	796	737	818	683	453	302

Note: The values do not add up to world trade due to double counting of certains countries and statistical differences (i.e. LDC are already included in the developping countries aggregate).

1. Excludes Iceland but includes all EU27 member countries.
2. Excludes trade of live animals.

Source: OECD and FAO Secretariats.

StatLink http://dx.doi.org/10.1787/888932642573

Table A.3.2. World trade projections, exports

		Avg 2009-11est.	2012	2013	2014	2015	2016	2017	2018	2019	2020	2021
Wheat												
OECD[1]	kt	87 566	82 111	79 695	79 480	80 447	80 012	80 650	81 203	81 940	81 606	82 546
Developing	kt	17 280	18 490	18 888	18 896	19 155	19 220	19 386	19 634	19 920	20 190	20 312
Least Developed Countries	kt	109	44	42	40	39	38	37	35	34	33	31
Coarse grains												
OECD[1]	kt	66 433	68 149	72 404	74 051	74 217	76 263	78 639	79 790	80 416	80 623	81 014
Developing	kt	38 079	35 611	35 556	35 222	36 520	37 224	38 448	39 561	40 785	42 001	43 542
Least Developed Countries	kt	3 763	3 195	2 779	2 636	2 311	2 435	2 447	2 034	1 542	1 541	2 299
Rice												
OECD[1]	kt	3 950	3 774	3 967	4 078	4 209	4 297	4 311	4 363	4 426	4 476	4 555
Developing	kt	28 652	31 649	32 422	33 029	34 257	34 946	35 552	36 087	36 488	37 086	37 985
Least Developed Countries	kt	1 940	2 592	3 192	3 725	4 252	4 818	5 359	5 858	6 370	6 890	7 381
Oilseeds												
OECD[1]	kt	52 120	50 192	50 697	50 349	52 070	53 517	54 932	55 945	57 223	58 745	60 253
Developing	kt	57 643	58 964	59 758	61 949	62 939	63 813	63 811	64 316	64 599	64 738	64 667
Least Developed Countries	kt	234	248	254	247	245	240	241	241	242	241	249
Protein Meals												
OECD[1]	kt	12 614	14 394	14 959	15 517	15 526	15 501	15 895	16 243	16 296	16 326	16 252
Developing	kt	57 345	60 060	61 266	62 275	63 586	64 772	65 514	66 540	67 725	69 258	70 397
Least Developed Countries	kt	219	212	215	207	188	176	162	153	150	134	134
Vegetable Oils												
OECD[1]	kt	5 133	5 305	5 457	5 598	5 715	5 820	5 930	6 001	6 030	6 041	6 008
Developing	kt	51 057	55 568	56 559	57 772	58 740	59 804	61 047	62 332	63 596	64 810	66 130
Least Developed Countries	kt	242	244	246	247	248	249	250	251	253	254	255
Sugar												
OECD[1]	kt	7 823	6 353	6 795	6 564	6 817	6 532	6 880	6 671	6 858	7 007	7 062
Developing	kt	46 850	46 535	46 784	47 575	50 722	50 529	51 574	54 450	54 927	55 813	57 054
Least Developed Countries	kt	938	888	872	891	946	1 028	1 060	1 225	1 191	1 162	1 254
Beef[2]												
OECD[1]	kt	3 509	3 743	4 137	4 024	3 782	3 994	4 082	4 036	4 072	4 027	4 008
Developing	kt	3 959	3 957	4 082	4 188	4 211	4 251	4 347	4 448	4 549	4 667	4 723
Least Developed Countries	kt	3	4	9	13	10	9	13	27	22	15	15
Pigmeat[2]												
OECD[1]	kt	5 300	5 676	5 736	5 726	5 821	5 853	5 956	6 017	6 063	6 120	6 170
Developing	kt	1 334	1 332	1 337	1 318	1 336	1 361	1 381	1 431	1 470	1 476	1 506
Least Developed Countries	kt	1	1	1	1	1	2	1	2	1	1	1
Poultry												
OECD[1]	kt	5 057	5 317	5 255	5 418	5 571	5 557	5 576	5 538	5 519	5 630	5 661
Developing	kt	6 387	6 633	6 947	7 140	7 261	7 399	7 506	7 667	7 872	7 828	8 023
Least Developed Countries	kt	2	1	1	1	1	1	11	42	96	126	154
Butter												
OECD[1]	kt	678	708	723	747	767	781	794	809	818	828	835
Developing	kt	86	84	85	85	86	88	89	90	93	95	97
Least Developed Countries	kt	5	3	3	3	3	3	3	3	3	3	3
Cheese												
OECD[1]	kt	1 421	1 551	1 524	1 580	1 623	1 655	1 680	1 707	1 747	1 793	1 830
Developing	kt	605	653	643	649	642	645	648	658	666	671	678
Least Developed Countries	kt	0	0	0	0	0	0	0	0	0	0	0
Whole Milk Powder												
OECD[1]	kt	1 495	1 721	1 729	1 779	1 817	1 855	1 881	1 916	1 955	1 989	2 030
Developing	kt	568	544	551	557	568	577	591	602	613	621	629
Least Developed Countries	kt	6	6	6	6	6	6	6	6	6	6	6
Skim Milk Powder												
OECD[1]	kt	1 294	1 527	1 511	1 553	1 582	1 612	1 640	1 676	1 711	1 747	1 780
Developing	kt	121	118	123	124	126	128	132	136	139	142	148
Least Developed Countries	kt	2	1	1	1	1	1	1	1	1	1	1
Biofuel[3]												
Ethanol World Trade	Mil l	3 983	3 726	3 123	2 963	5 478	7 692	8 891	10 187	9 955	10 158	11 863
Biodiesel World Trade	Mil l	2 147	2 830	2 973	3 059	3 175	3 304	3 403	3 502	3 585	3 669	3 728

1. Excludes Iceland but includes all EU27 member countries.
2. Excludes trade of live animals.
3. Sum of all positive net trade positions

Source: OECD and FAO Secretariats.

StatLink http://dx.doi.org/10.1787/888932642592

Table A.4.1. Biofuel projections: Ethanol

	PRODUCTION (MN L)		Growth (%)[1]	DOMESTIC USE (MN L)		Growth (%)[1]	FUEL USE (MN L)		Growth (%)[1]	SHARE IN GAZOLINE TYPE FUEL USE(%)				NET TRADE (MN L)[2]	
										Energy Shares		Volume Shares			
	Average 2009-11est.	2021	2012-21	Average 2009-11est.	2021	2012-21	Average 2009-11est.	2021	2012-21	Average 2009-11est.	2021	Average 2009-11est.	2021	Average 2009-11est.	2021
NORTH AMERICA															
Canada	1 565	1 992	1.25	1 759	2 356	-0.14	1 553	2 149	-0.16	2.6	3.4	3.8	5.0	-195	-364
United States	47 617	82 610	4.02	45 582	90 757	5.59	43 372	89 093	5.83	5.4	10.9	7.8	15.5	1 864	-8 269
of which second generation	27	16 353	..	..	..	..	..	..	..	..	..	..	..	..	..
WESTERN EUROPE															
EU(27)	6 424	15 747	6.97	7 877	19 388	7.14	5 477	16 938	8.66	2.7	8.3	4.0	12.0	-1 453	-3 641
of which second generation	29	440	..	..	..	..	..	..	..	..	..	..	..	..	..
OCEANIA DEVELOPED															
Australia	389	496	0.74	389	496	0.74	389	496	0.74	1.3	1.5	1.9	2.2	0	0
OTHER DEVELOPED															
Japan	102	104	0.28	832	1 016	0.20	225	416	0.50	0.0	0.0	0.0	0.0	-730	-912
of which second generation	98	99	..	..	..	..	..	..	..	..	..	..	..	..	..
South Africa	376	951	7.99	223	539	5.57	38	350	10.20	..	..	..	..	153	411
SUB-SAHARIAN AFRICA															
Mozambique	31	61	5.23	29	29	0.50	0	9	1.38	..	..	..	..	1	32
Tanzania	32	67	6.77	42	52	5.78	0	19	..	..	..	..	..	-10	15
LATIN AMERICA AND CARRIBBEAN															
Argentina	330	815	3.95	359	449	0.87	190	276	1.46	2.7	3.4	4.0	5.0	-29	367
Brazil	25 331	51 305	7.58	23 347	39 808	5.14	21 715	37 424	5.25	47.1	64.3	57.0	72.9	1 984	11 496
Columbia	316	769	8.32	369	707	5.74	314	650	6.41	..	..	..	..	-52	62
Mexico	199	243	0.98	336	413	0.98	0	0	..	0.0	0.0	0.0	0.0	-137	-170
Peru	126	246	4.31	92	248	3.81	66	225	4.27	..	..	..	..	35	-2
ASIA AND PACIFIC															
China	8 094	10 058	0.80	7 994	10 242	1.15	2 211	2 748	2.04	1.8	1.3	2.7	2.0	100	-184
India	1 976	4 194	5.32	2 254	4 384	5.23	264	2 300	13.15	..	..	..	..	-278	-190
Indonesia	199	295	0.60	153	239	0.09	28	100	0.00	..	..	..	..	46	56
Malaysia	59	67	0.23	75	85	0.11	0	0	5.89	..	..	..	..	-16	-18
Philippines	120	345	8.96	361	518	1.84	184	350	2.76	..	..	..	..	-241	-173
Thailand	777	2 102	8.51	723	1 714	7.86	473	1 488	9.67	..	..	..	..	54	388
Turkey	74	118	1.71	123	139	1.69	50	69	3.59	..	..	..	..	-48	-21
Viet Nam	209	493	4.15	166	419	9.89	23	255	23.81	..	..	..	..	43	74
TOTAL	**98 219**	**180 402**	**5.02**	**97 220**	**179 919**	**5.03**	**77 178**	**155 954**	**5.82**	**5.9**	**10.8**	**8.6**	**15.4**	**3 983**	**11 863**

Note: .. : Not available.

1. Least-squares growth rate (see glossary).
2. For total net trade exports are shown.

Source: OECD and FAO Secretariats.

StatLink http://dx.doi.org/10.1787/888932642611

Table A.4.2. Biofuel projections: Biodiesel

	PRODUCTION (MN L)		Growth (%)[1]	DOMESTIC USE (MN L)		Growth (%)[1]	SHARE IN DIESEL TYPE FUEL USE(%)				NET TRADE (MN L)[2]	
							Energy Shares		Volume Shares			
	Average 2009-11est.	2021	2012-21	Average 2009-11est.	2021	2012-21	Average 2009-11est.	2021	Average 2009-11est.	2021	Average 2009-11est.	2021
NORTH AMERICA												
Canada	147	522	8.79	193	576	0.43	0.7	1.6	0.8	2.0	-46	-54
United States	2 834	5 083	0.43	2 546	4 979	0.55	0.9	1.5	1.1	1.9	288	104
WESTERN EUROPE												
European Union	10 436	19 864	5.04	12 467	22 046	4.99	5.1	8.5	6.3	10.4	-2 030	-2 182
of which second generation	47	552	..	..	..	..	..	..	..	..	..	..
OCEANIA DEVELOPED												
Australia	641	727	1.11	641	727	1.11	3.1	3.1	3.9	3.8	0	0
OTHER DEVELOPED												
South Africa	62	108	4.14	62	94	1.16	..	..	..	..	0	14
Sub-Saharian Africa												
Mozambique	54	72	0.40	9	32	1.37	..	..	..	..	45	41
Tanzania	51	61	-0.19	0	58	110.33	..	..	..	..	51	3
LATIN AMERICA AND CARIBBEAN												
Argentina	2 231	4 204	3.18	372	581	2.06	3.2	4.0	4.0	5.0	1 859	3 623
Brazil	2 015	3 205	2.66	2 100	3 205	2.66	4.0	4.6	5.0	5.7	-85	0
Columbia	431	917	5.92	358	800	6.27	..	..	..	..	73	117
Peru	138	137	3.93	179	309	3.16	..	..	..	..	-41	-172
ASIA AND PACIFIC												
India	330	1 297	10.98	391	2 300	14.84	..	..	..	..	-60	-1 003
Indonesia	397	1 384	10.44	294	1 250	12.22	..	..	..	..	104	134
Malaysia	563	956	11.78	147	650	19.24	..	..	..	..	416	306
Philippines	209	406	10.72	203	307	6.57	..	..	..	..	6	99
Thailand	664	1 339	6.42	626	1 139	3.81	..	..	..	..	38	200
Turkey	63	58	5.95	99	179	1.05	..	..	..	..	-35	-121
Viet Nam	11	109	14.77	6	100	14.35	..	..	..	..	5	9
TOTAL	**21 322**	**41 595**	**4.64**	**20 749**	**40 461**	**4.81**	**2.5**	**3.8**	**3.1**	**4.8**	**2 147**	**3 728**

Note: .. : Not available.

1. Least-squares growth rate (see glossary).
2. For total net trade exports are shown.

Source: OECD and FAO Secretariats.

StatLink http://dx.doi.org/10.1787/888932642630

Table A.5. Main policy assumptions for biofuel markets

		11/12est.	12/13	13/14	14/15	15/16	16/17	17/18	18/19	19/20	20/21	21/22
BRAZIL												
Ethanol												
Import tariffs	%	0.0	0.0	0.0	0.0	0.0	0.0	0.0	0.0	0.0	0.0	0.0
Incorporation mandate[1]	%	18.3	18.3	18.3	18.3	18.3	18.3	18.3	18.3	18.3	18.3	18.3
Biodiesel												
Tax concessions[2]	BRL/hl	6.7	6.7	6.7	6.7	6.7	6.7	6.7	6.7	6.7	6.7	6.7
Import tariffs	%	4.6	4.6	4.6	4.6	4.6	4.6	4.6	4.6	4.6	4.6	4.6
CANADA												
Ethanol												
Tax concessions[2]	CAD/hl	6.4	6.4	6.4	6.4	6.4	6.4	6.4	6.4	6.4	6.4	6.4
Import tariffs	CAD/hl	5.0	5.0	5.0	5.0	5.0	5.0	5.0	5.0	5.0	5.0	5.0
Incorporation mandate[1]	%	3.4	3.4	3.4	3.4	3.4	3.4	3.4	3.4	3.4	3.4	3.4
Direct support												
Federal	CAD/hl	8.0	7.0	6.0	5.0	4.0	3.0	0.0	0.0	0.0	0.0	0.0
Provincial	CAD/hl	4.3	4.3	4.3	4.3	4.3	4.3	0.0	0.0	0.0	0.0	0.0
Biodiesel												
Tax concessions[2]	CAD/hl	7.7	7.7	7.7	7.7	7.7	7.7	7.7	7.7	7.7	7.7	7.7
Incorporation mandate[1]	%	0.9	1.6	1.6	1.6	1.6	1.6	1.6	1.6	1.6	1.6	1.6
Direct support												
Federal	CAD/hl	16.0	14.0	12.0	10.0	8.0	6.0	0.0	0.0	0.0	0.0	0.0
Provincial	CAD/hl	0.0	0.0	0.0	0.0	0.0	0.0	0.0	0.0	0.0	0.0	0.0
COLOMBIA												
Ethanol												
Import tariffs	%	9.6	9.6	9.6	9.6	9.6	9.6	9.6	9.6	9.6	9.6	9.6
Blending target[3,4]	%	9.0	10.0	10.0	10.0	10.0	10.0	10.0	10.0	10.0	10.0	10.0
Biodiesel												
Blending target[4]	%	5.0	5.0	5.0	5.0	5.0	5.0	5.0	5.0	5.0	5.0	5.0
EUROPEAN UNION												
Ethanol												
Tax concessions[2]	EUR/hl	20.0	23.0	23.0	23.0	23.0	23.0	23.0	23.0	23.0	23.0	25.0
Import tariffs	EUR/hl	19.2	19.2	19.2	19.2	19.2	19.2	19.2	19.2	19.2	19.2	19.2
Biodiesel												
Tax concessions[2]	EUR/hl	23.0	23.0	23.0	23.0	23.0	23.0	23.0	23.0	23.0	23.0	23.0
Import tariffs	%	6.5	6.5	6.5	6.5	6.5	6.5	6.5	6.5	6.5	6.5	6.5
INDONESIA												
Ethanol												
Import tariffs	%	26.3	26.3	26.3	26.3	26.3	26.3	26.3	26.3	26.3	26.3	26.3
Blending target[4]	%	0.0	0.0	0.0	0.0	0.0	0.0	0.0	0.0	0.0	0.0	0.0
Biodiesel												
Blending target[4]	%	3.0	3.0	3.0	3.0	7.0	7.0	7.0	7.0	7.0	7.0	7.0
MALAYSIA												
Ethanol												
Import tariffs	%	0.0	0.0	0.0	0.0	0.0	0.0	0.0	0.0	0.0	0.0	0.0
Blending target[4]	%	0.0	0.0	0.0	0.0	0.0	0.0	0.0	0.0	0.0	0.0	0.0
Biodiesel												
Blending target[4]	%	5.0	5.0	5.0	5.0	5.0	5.0	5.0	5.0	5.0	5.0	5.0
PERU												
Ethanol												
Import tariffs	%	0.0	0.0	0.0	0.0	0.0	0.0	0.0	0.0	0.0	0.0	0.0
Blending target[4]	%	7.8	7.8	7.8	7.8	7.8	7.8	7.8	7.8	7.8	7.8	7.8
Biodiesel												
Import tariffs	%	0.0	0.0	0.0	0.0	0.0	0.0	0.0	0.0	0.0	0.0	0.0
Blending target[4]	%	5.0	5.0	5.0	5.0	5.0	5.0	5.0	5.0	5.0	5.0	5.0
THAILAND												
Ethanol												
Import tariffs	%	0.0	0.0	0.0	0.0	0.0	0.0	0.0	0.0	0.0	0.0	0.0
Blending target[4]	%	5.2	5.2	5.2	5.2	5.2	5.2	5.2	5.2	5.2	5.2	5.2
Biodiesel												
Blending target[4]	%	5.0	5.0	5.0	5.0	5.0	5.0	5.0	5.0	5.0	5.0	5.0

Table A.5. Main policy assumptions for biofuel markets *(cont.)*

		11/12est.	12/13	13/14	14/15	15/16	16/17	17/18	18/19	19/20	20/21	21/22
UNITED STATES												
Renewable Fuel Standard[5]												
Total	MN l	57 538	61 760	65 673	70 913	74 019	77 201	80 828	84 639	87 179	90 932	96 615
advanced mandate	MN l	7 571	9 522	11 163	14 132	17 238	20 420	24 047	27 858	30 398	34 151	39 834
cellulosic ethanol	MN l	33	232	560	1 224	2 106	3 208	4 698	6 450	8 889	12 613	16 353
biodiesel	MN l	3 785	4 845	4 845	4 845	4 845	4 845	4 845	4 845	4 845	4 845	4 845
Ethanol												
Import surcharge	USD/hl	14.27	0.00	0.00	0.00	0.00	0.00	0.00	0.00	0.00	0.00	0.00
Import tariffs (undenatured)	%	2.40	2.40	2.40	2.40	2.40	2.40	2.40	2.40	2.40	2.40	2.40
Import tariffs (denatured)	%	1.90	1.90	1.90	1.90	1.90	1.90	1.90	1.90	1.90	1.90	1.90
Blenders tax credit	USD/hl	11.89	0.00	0.00	0.00	0.00	0.00	0.00	0.00	0.00	0.00	0.00
Biodiesel												
Import tariffs	%	4.60	4.60	4.60	4.60	4.60	4.60	4.60	4.60	4.60	4.60	4.60
Blenders tax credit	USD/hl	26.42	0.00	0.00	0.00	0.00	0.00	0.00	0.00	0.00	0.00	0.00

Note that for many countries, shares for ethanol and biodiesel are not specified individually in the legislation.

Figures are based on a combination of the EU mandate in the context of the Renewable Energy Directive and the National Renewable Energy Action Plan (NREAP) in the EU member states.

1. Share in respective fuel type, energy equivalent.
2. Difference between tax rates applying to fossil and biogen fuels.
3. Applies to cities with more than 500 000 inhabitants.
4. Expressed in volume share.
5. The total, advanced and cellulosic mandates are not at the levels defined in EISA. As those mandates are subject to uncertainties regarding EPA implementation, the following assumptions were taken to construct the baseline: The total and advanced mandates have been reduced by about 90% of the shortfall in cellulosic production. More details are provided in the annex of the biofuel chapter.

Source: OECD and FAO Secretariats.

StatLink http://dx.doi.org/10.1787/888932642649

Table A.6. World cereal projections

Crop year

		Avg 09/10-11/12est	12/13	13/14	14/15	15/16	16/17	17/18	18/19	19/20	20/21	21/22
WHEAT												
OECD[1]												
Production	mt	273.6	275.8	273.7	272.6	276.2	277.4	281.6	284.0	287.2	289.7	291.0
Consumption	mt	216.0	220.6	222.3	224.4	226.3	227.2	229.7	231.4	233.2	234.9	236.1
Closing stocks	mt	57.2	56.7	57.7	55.8	54.7	53.8	53.9	54.0	54.6	56.3	57.0
Non-OECD												
Production	mt	403.2	426.8	432.4	436.9	440.8	441.9	448.6	452.8	460.3	466.5	469.9
Consumption	mt	452.2	473.4	477.9	484.7	490.7	494.5	500.1	505.3	511.2	517.2	522.5
Closing stocks	mt	142.4	151.8	156.4	158.3	159.1	157.2	157.1	156.8	158.9	161.0	162.2
World												
Production	mt	676.8	702.6	706.1	709.4	716.9	719.2	730.2	736.8	747.5	756.3	760.9
Area	mha	220.8	223.5	223.3	222.9	223.7	223.1	224.2	224.6	225.8	227.1	226.9
Yield	t/ha	3.1	3.1	3.2	3.2	3.2	3.2	3.3	3.3	3.3	3.3	3.4
Consumption	mt	668.2	694.0	700.2	709.1	716.9	721.7	729.8	736.7	744.4	752.2	758.6
Feed use	mt	128.3	137.2	138.8	141.5	144.3	144.5	146.7	148.0	150.5	152.8	154.3
Food use	mt	462.5	475.5	477.6	481.9	485.1	487.9	491.7	495.2	498.8	502.5	505.9
Biofuel use	mt	6.3	8.2	9.5	10.6	11.1	12.0	13.1	14.2	14.8	15.3	15.7
Other use	mt	71.4	73.0	74.3	75.1	76.4	77.2	78.3	79.2	80.4	81.6	82.7
Exports	mt	130.0	137.0	137.5	137.4	139.5	140.3	144.1	145.8	148.5	150.5	152.5
Closing stocks	mt	199.6	208.6	214.2	214.1	213.8	211.0	211.0	210.9	213.5	217.3	219.2
Price[2]	USD/t	266.6	248.6	250.8	250.4	250.3	259.7	264.5	272.5	275.7	276.9	279.3
COARSE GRAINS												
OECD[1]												
Production	mt	561.5	603.4	595.7	598.5	607.9	619.0	628.1	633.8	643.2	651.6	662.0
Consumption	mt	560.9	568.0	572.1	577.9	588.1	595.1	600.8	607.9	615.6	622.9	631.1
Closing stocks	mt	94.0	94.4	97.4	95.4	93.0	93.2	94.5	94.3	95.7	98.1	102.6
Non-OECD												
Production	mt	572.7	602.0	614.2	620.3	630.4	640.8	651.5	662.1	675.5	689.6	700.6
Consumption	mt	582.4	617.0	630.2	641.7	653.1	664.3	676.7	688.4	701.3	714.3	724.4
Closing stocks	mt	111.0	115.3	119.9	121.1	120.6	121.0	121.8	121.8	122.4	124.2	126.8
World												
Production	mt	1 134.1	1 205.4	1 209.9	1 218.7	1 238.3	1 259.8	1 279.6	1 295.9	1 318.7	1 341.2	1 362.6
Area	mha	321.8	328.3	329.8	330.0	331.7	333.5	335.2	336.4	339.3	342.4	344.4
Yield	t/ha	3.5	3.7	3.7	3.7	3.7	3.8	3.8	3.9	3.9	3.9	4.0
Consumption	mt	1 143.4	1 185.0	1 202.4	1 219.7	1 241.3	1 259.4	1 277.5	1 296.3	1 316.9	1 337.1	1 355.6
Feed use	mt	601.3	624.1	636.7	644.9	654.2	666.0	676.9	688.1	699.9	711.4	721.6
Food use	mt	205.9	212.7	216.8	219.9	223.5	226.2	229.4	232.2	236.5	240.5	243.0
Biofuel use	mt	137.9	161.5	165.9	172.5	182.4	183.2	183.9	184.6	184.0	184.6	185.3
Other use	mt	160.1	148.0	143.6	142.5	141.2	143.6	146.4	150.1	154.6	158.4	162.9
Exports	mt	121.8	123.5	126.5	127.2	128.7	131.6	135.0	138.0	140.6	143.1	145.9
Closing stocks	mt	205.0	209.7	217.3	216.5	213.6	214.2	216.4	216.1	218.1	222.3	229.4
Price[3]	USD/t	227.8	244.8	228.7	227.5	231.2	233.4	236.2	242.1	246.1	247.9	246.3
RICE												
OECD[1]												
Production	mt	22.0	21.7	21.7	21.6	21.7	22.0	22.2	22.2	22.2	22.2	22.3
Consumption	mt	22.8	23.0	23.1	23.2	23.4	23.4	23.6	23.7	23.8	23.8	23.9
Closing stocks	mt	6.3	6.1	6.0	5.7	5.4	5.4	5.5	5.7	5.7	5.7	5.8
Non-OECD												
Production	mt	445.6	466.4	471.3	477.5	482.4	486.8	492.8	499.6	506.7	513.0	519.8
Consumption	mt	437.3	461.0	469.5	476.0	482.5	488.2	494.3	500.4	506.6	512.8	518.5
Closing stocks	mt	137.2	150.2	150.7	150.9	149.6	146.9	143.9	141.7	140.3	139.0	138.8
World												
Production	mt	467.6	488.0	493.0	499.1	504.1	508.8	515.0	521.9	528.9	535.2	542.1
Area	mha	161.3	166.0	165.8	165.8	165.5	165.0	164.9	165.0	165.3	165.5	165.6
Yield	t/ha	2.9	2.9	3.0	3.0	3.0	3.1	3.1	3.2	3.2	3.2	3.3
Consumption	mt	460.2	483.9	492.7	499.3	505.8	511.7	517.9	524.1	530.4	536.6	542.4
Feed use	mt	15.7	17.0	17.6	18.1	18.4	18.7	18.9	19.3	19.5	20.0	20.3
Food use	mt	391.3	409.0	416.1	422.1	428.6	434.8	440.5	446.1	452.0	457.4	463.0
Exports	mt	32.8	35.7	36.6	37.3	38.7	39.4	40.1	40.6	41.1	41.7	42.7
Closing stocks	mt	143.5	156.3	156.7	156.6	155.0	152.2	149.5	147.4	146.0	144.7	144.5
Price[4]	USD/t	556.8	493.6	465.3	444.9	420.6	419.3	425.9	435.4	442.8	450.3	454.5

Note: Crop year: Beginning crop marketing year - see Glossary of Terms for definitions.

1. Excludes Iceland but includes all EU27 member countries.
2. No.2 hard red winter wheat, ordinary protein, United States f.o.b. Gulf Ports (June/May), less EEP payments where applicable.
3. No.2 yellow corn, United States f.o.b. Gulf Ports (September/August).
4. Milled, 100%, grade b, Nominal Price Quote, NPQ, f.o.b. Bangkok (January/December).

Source: OECD and FAO Secretariats.

StatLink http://dx.doi.org/10.1787/888932642668

Table A.7. Wheat projections

Crop year

	PRODUCTION (Kt)		Growth (%)[4]		IMPORTS (Kt)		Growth (%)[4]		EXPORTS (Kt)		Growth (%)[4]	
	Avg 2009-11est	2021	2002-2011	2012-2021	Avg 2009-11est	2021	2002-2011	2012-2021	Avg 2009-11est	2021	2002-2011	2012-2021
WORLD	**676 759**	**760 926**	**2.10**	**0.94**	**128 551**	**152 134**	**2.92**	**1.30**	**129 996**	**152 492**	**3.15**	**1.30**
DEVELOPED	**356 770**	**406 456**	**1.95**	**1.06**	**25 451**	**25 302**	**0.26**	**-0.47**	**112 716**	**132 179**	**4.08**	**1.34**
NORTH AMERICA	**83 370**	**84 241**	**1.78**	**0.39**	**3 123**	**3 599**	**5.77**	**1.97**	**45 890**	**44 072**	**2.11**	**0.39**
Canada	25 092	28 613	2.82	1.15	75	44	3.80	2.09	17 358	19 695	4.46	1.78
United States	58 279	55 628	1.37	0.02	3 048	3 555	6.00	1.97	28 531	24 377	0.82	-0.60
EUROPE	**217 620**	**254 708**	**1.93**	**1.27**	**7 921**	**6 993**	**-5.90**	**-2.17**	**41 106**	**58 531**	**7.42**	**3.05**
EU(27)	136 571	149 306	0.90	0.88	5 919	5 677	-3.99	-1.41	19 618	17 391	5.09	0.33
Russian Federation	54 535	69 566	3.31	1.49	36	57	-66.34	0.41	13 954	26 114	6.88	3.43
Ukraine	19 629	27 174	8.04	2.74	21	3	-53.63	-6.18	6 952	13 815	21.84	6.18
OCEANIA DEVELOPED	**25 518**	**25 181**	**5.13**	**0.57**	**320**	**348**	**-0.75**	**-0.09**	**17 809**	**16 271**	**4.44**	**-0.74**
Australia	25 142	24 753	5.19	0.57	0	0	..	..	17 804	16 268	4.43	-0.74
New Zealand	376	428	3.00	0.32	320	348	-0.75	-0.09	5	4	43.94	0.00
OTHER DEVELOPED[1]	**30 262**	**42 326**	**1.96**	**1.49**	**14 087**	**14 362**	**4.28**	**-0.12**	**7 911**	**13 305**	**6.71**	**0.53**
Japan	662	985	-3.00	2.38	5 735	5 513	1.19	-0.50	0	1	..	..
South Africa	1 753	2 053	-1.15	1.24	1 462	1 754	4.75	2.15	213	171	4.30	-2.15
DEVELOPING	**319 989**	**354 469**	**2.25**	**0.81**	**103 100**	**126 832**	**3.68**	**1.69**	**17 280**	**20 312**	**-1.74**	**1.01**
AFRICA	**22 730**	**30 447**	**2.89**	**2.63**	**37 024**	**43 205**	**4.12**	**1.43**	**756**	**301**	**0.73**	**-2.59**
NORTH AFRICA	**18 249**	**22 957**	**2.24**	**2.11**	**22 363**	**24 083**	**4.42**	**0.76**	**317**	**53**	**0.20**	**-0.79**
Algeria	2 934	4 486	2.36	4.82	5 800	5 445	2.54	-1.99	0	0	0.00	0.14
Egypt	8 021	9 968	1.89	1.62	10 230	11 987	5.86	2.19	0	0	-37.73	-0.16
SUB-SAHARAN AFRICA	**4 481**	**7 490**	**5.78**	**4.40**	**14 660**	**19 122**	**3.66**	**2.32**	**439**	**248**	**1.46**	**-2.93**
LATIN AMERICA and CARIBBEAN	**26 888**	**32 565**	**1.28**	**1.50**	**19 348**	**21 816**	**0.70**	**1.04**	**9 752**	**13 507**	**-0.95**	**1.98**
Argentina	12 264	16 302	-2.21	1.55	0	0	..	..	6 721	9 334	-4.95	1.57
Brazil	5 441	5 600	3.06	1.10	5 906	6 910	-0.02	1.74	500	729	54.47	2.51
Chile	1 565	1 675	-2.38	1.71	700	943	8.86	0.46	0	0	0.00	-0.03
Mexico	4 134	4 615	5.58	1.26	3 275	3 153	-0.41	-0.50	694	998	5.80	4.10
Uruguay	1 595	2 408	23.02	2.40	8	0	-99.65	-2.95	1 059	1 758	97.07	2.95
ASIA and PACIFIC	**270 371**	**291 457**	**2.31**	**0.57**	**46 728**	**61 811**	**4.76**	**2.12**	**6 772**	**6 504**	**-2.94**	**-0.61**
Bangladesh	973	981	-5.38	-0.40	3 351	3 549	7.50	2.59	0	0	-67.08	-0.18
China[2]	115 695	119 468	3.57	0.11	1 387	5 322	-5.05	-1.04	67	0	..	..
India	82 470	88 739	2.42	0.44	180	4 200	17.60	26.46	667	1 279	-83.20	-3.78
Indonesia	0	0	0.00	1.13	5 481	6 720	3.12	2.04	30	29	3.93	-0.29
Iran, Islamic Republic of	13 333	15 330	-0.55	1.27	2 033	2 418	14.26	-0.49	567	133	107.93	0.41
Korea	24	18	13.10	-0.04	4 733	4 011	3.23	0.44	50	55	-6.55	0.00
Malaysia	0	0	0.00	1.71	1 194	1 411	-2.04	1.67	57	78	-13.37	-0.24
Pakistan	23 875	26 717	3.02	1.03	200	1 342	-2.49	16.47	700	193	7.64	-9.81
Saudi Arabia	1 100	1 004	-11.11	0.09	1 900	2 622	44.89	3.19	0	0	-118.10	-0.23
Turkey	20 687	23 797	0.26	1.27	3 017	2 519	19.21	-1.82	3 500	3 752	16.29	1.82
LEAST DEVELOPED COUNTRIES (LDC)	**11 369**	**16 283**	**3.43**	**3.72**	**15 840**	**19 124**	**4.61**	**1.84**	**109**	**31**	**-7.05**	**-3.60**
OECD[3]	**273 588**	**291 029**	**1.36**	**0.75**	**29 336**	**28 364**	**1.66**	**-0.33**	**87 566**	**82 546**	**3.40**	**0.24**
NON-OECD	**403 171**	**469 897**	**2.64**	**1.06**	**99 215**	**123 770**	**3.34**	**1.71**	**42 430**	**69 946**	**2.58**	**2.68**

Table A.7. Wheat projections *(cont.)*

Crop year

	CONSUMPTION (Kt)		Growth (%)[4]		FOOD USE (Kt)		Growth (%)[4]		PER CAPITA (Kg)		Growth (%)[4]	
	Avg 2009-11est	2021	2002-2011	2012-2021	Avg 2009-11est	2021	2002-2011	2012-2021	Avg 2009-11est	2021	2002-2011	2012-2021
WORLD	**668 216**	**758 648**	**1.53**	**0.99**	**462 529**	**505 925**	**0.93**	**0.70**	**67.1**	**65.4**	**-0.24**	**-0.32**
DEVELOPED	**267 788**	**298 389**	**0.78**	**0.85**	**129 385**	**137 652**	**-0.12**	**0.45**	**94.3**	**96.4**	**-0.58**	**0.09**
NORTH AMERICA	**38 992**	**43 423**	**0.03**	**0.44**	**28 023**	**30 698**	**0.13**	**0.79**	**81.4**	**81.4**	**-0.79**	**-0.03**
Canada	7 757	8 554	0.17	-0.28	2 775	2 900	-0.96	0.49	81.6	77.4	-2.01	-0.37
United States	31 235	34 869	0.00	0.62	25 247	27 798	0.25	0.82	81.3	81.8	-0.66	0.01
EUROPE	**187 777**	**202 754**	**0.91**	**0.69**	**78 567**	**79 288**	**-0.57**	**-0.03**	**106.2**	**106.0**	**-0.76**	**-0.14**
EU(27)	127 230	137 291	0.58	0.82	55 671	57 041	0.35	0.21	111.0	110.8	-0.05	-0.03
Russian Federation	39 711	43 503	0.94	0.39	11 967	11 740	-4.92	-0.82	83.7	83.4	-4.73	-0.66
Ukraine	12 417	13 276	0.83	0.61	5 547	5 107	-0.40	-0.94	122.0	119.3	0.27	-0.40
OCEANIA DEVELOPED	**6 320**	**9 257**	**0.08**	**3.29**	**1 798**	**2 602**	**-1.38**	**1.05**	**67.5**	**85.6**	**-2.94**	**-0.10**
Australia	5 629	8 486	0.06	3.62	1 508	2 264	-1.24	1.25	67.7	88.7	-2.88	0.05
New Zealand	691	771	0.21	0.13	290	338	-2.17	-0.16	66.4	69.5	-3.35	-1.13
OTHER DEVELOPED[1]	**34 699**	**42 955**	**1.63**	**1.58**	**20 997**	**25 064**	**1.44**	**1.59**	**80.4**	**92.2**	**0.90**	**1.24**
Japan	6 516	6 448	0.60	-0.06	5 490	5 478	0.63	-0.05	43.4	44.0	0.59	0.12
South Africa	3 033	3 622	0.91	1.72	2 891	3 475	1.26	1.78	57.7	65.8	0.23	1.34
DEVELOPING	**400 428**	**460 260**	**1.89**	**1.09**	**333 144**	**368 273**	**1.35**	**0.79**	**60.3**	**58.4**	**0.00**	**-0.38**
AFRICA	**57 713**	**72 863**	**3.64**	**1.97**	**49 736**	**63 207**	**2.87**	**2.08**	**51.2**	**50.5**	**0.50**	**-0.21**
NORTH AFRICA	**38 846**	**46 670**	**3.18**	**1.44**	**32 438**	**38 964**	**2.18**	**1.53**	**196.1**	**204.2**	**0.64**	**0.26**
Algeria	8 468	9 860	2.59	0.92	7 278	8 472	1.76	1.02	205.1	208.8	0.26	-0.16
Egypt	17 451	21 797	3.35	1.88	15 450	19 258	2.40	1.97	190.4	200.4	0.59	0.47
SUB-SAHARAN AFRICA	**18 867**	**26 193**	**4.65**	**2.98**	**17 298**	**24 243**	**4.28**	**3.03**	**21.4**	**22.8**	**1.72**	**0.54**
LATIN AMERICA and CARIBBEAN	**35 041**	**40 763**	**1.51**	**0.97**	**30 624**	**36 042**	**1.61**	**1.06**	**51.8**	**54.8**	**0.42**	**0.10**
Argentina	4 504	6 933	0.66	0.64	3 611	5 958	1.00	0.76	89.2	134.9	0.12	-0.04
Brazil	10 557	11 765	1.33	1.40	10 118	11 273	1.52	1.42	51.9	53.2	0.50	0.70
Chile	2 250	2 613	0.19	1.21	2 063	2 352	1.08	1.10	120.5	126.1	0.08	0.34
Mexico	6 715	6 769	2.12	0.07	5 073	5 230	2.32	0.34	44.7	41.2	1.07	-0.65
Uruguay	630	643	5.34	0.37	388	405	1.50	0.47	115.0	115.6	1.31	0.09
ASIA and PACIFIC	**307 673**	**346 634**	**1.64**	**0.93**	**252 784**	**269 023**	**1.04**	**0.48**	**63.8**	**61.3**	**-0.10**	**-0.43**
Bangladesh	3 524	4 460	1.33	1.83	3 325	4 217	1.94	1.95	22.3	25.0	0.70	0.80
China[2]	116 027	125 311	1.54	0.43	86 653	75 722	-0.52	-1.31	64.6	54.5	-1.05	-1.61
India	82 249	91 602	1.76	0.76	75 750	84 439	1.84	0.77	61.9	60.2	0.38	-0.43
Indonesia	5 284	6 593	3.08	1.96	4 498	5 538	2.65	1.85	18.8	20.9	1.53	0.98
Iran, Islamic Republic of	15 433	17 568	0.46	1.13	12 350	13 833	1.17	1.00	167.0	169.6	-0.03	0.16
Korea	4 474	3 977	2.62	-0.27	2 367	2 416	0.73	0.30	49.1	48.4	0.26	0.00
Malaysia	1 044	1 322	-1.90	1.90	841	1 074	-1.73	2.06	29.6	32.1	-3.53	0.60
Pakistan	23 309	27 814	1.82	1.66	21 001	25 661	1.75	1.77	121.0	123.1	-0.05	0.14
Saudi Arabia	2 767	3 567	2.54	2.22	2 540	3 327	5.36	2.46	92.6	97.6	2.41	0.54
Turkey	19 970	22 515	0.10	0.86	14 902	16 643	1.24	1.03	204.9	204.4	-0.08	0.04
LEAST DEVELOPED COUNTRIES (LDC)	**26 356**	**35 113**	**4.27**	**2.62**	**23 518**	**32 163**	**4.01**	**2.71**	**28.2**	**30.4**	**1.76**	**0.55**
OECD[3]	**216 016**	**236 100**	**0.56**	**0.77**	**117 416**	**124 689**	**0.51**	**0.50**	**92.4**	**92.7**	**-0.17**	**-0.01**
NON-OECD	**452 199**	**522 549**	**2.02**	**1.10**	**345 112**	**381 236**	**1.07**	**0.77**	**61.3**	**59.7**	**-0.21**	**-0.37**

Note: Crop year: Beginning crop marketing year - see Glossary of Terms for definitions.

1. Includes Israel and also transition economies: Kazakhstan, Kyrgyzstan, Tajikistan, Turkmenistan, Uzbekistan, Armenia, Azerbaijan and Georgia.
2. Refers to mainland only. The economies of Chinese Taipei, Hong Kong (China) and Macau (China) are included in the Other Asia Pacific aggregate.
3. Excludes Iceland but includes all EU27 member countries.
4. Least-squares growth rate (see glossary).

Source: OECD and FAO Secretariats.

StatLink http://dx.doi.org/10.1787/888932642687

Table A.8. Coarse grain projections

Crop year

	PRODUCTION (Kt)		Growth (%)[4]		IMPORTS (Kt)		Growth (%)[4]		EXPORTS (Kt)		Growth (%)[4]	
	Avg 2009-11est	2021	2002-2011	2012-2021	Avg 2009-11est	2021	2002-2011	2012-2021	Avg 2009-11est	2021	2002-2011	2012-2021
WORLD	**1 134 118**	**1 362 560**	**2.74**	**1.44**	**119 562**	**146 058**	**1.83**	**1.88**	**121 800**	**145 925**	**1.54**	**1.88**
DEVELOPED	**606 555**	**717 602**	**1.67**	**1.15**	**32 524**	**32 094**	**-1.11**	**0.06**	**83 721**	**102 384**	**1.10**	**1.66**
NORTH AMERICA	**356 246**	**438 926**	**2.63**	**1.27**	**3 838**	**3 568**	**-3.19**	**-1.41**	**54 571**	**68 884**	**-0.10**	**1.97**
Canada	22 186	28 160	-0.25	1.38	1 814	1 159	-6.83	-4.03	4 900	6 336	4.87	1.45
United States	334 060	410 766	2.85	1.26	2 024	2 409	0.93	0.25	49 671	62 548	-0.51	2.04
EUROPE	**219 182**	**244 341**	**0.25**	**1.04**	**6 441**	**7 994**	**-0.87**	**3.01**	**22 441**	**25 387**	**4.09**	**1.46**
EU(27)	147 869	153 776	-0.27	0.75	5 297	6 912	1.06	3.30	7 056	5 458	-4.59	0.98
Russian Federation	26 643	37 221	-1.92	2.26	212	288	-22.69	8.82	1 679	434	-4.30	1.21
Ukraine	25 368	32 018	4.82	1.25	46	35	-12.31	-1.89	11 847	17 073	14.48	1.89
OCEANIA DEVELOPED	**12 902**	**15 339**	**2.27**	**0.88**	**27**	**1**	**-5.70**	**-8.57**	**4 118**	**6 212**	**0.66**	**1.18**
Australia	12 291	14 746	2.38	0.92	0	0	..	..	4 113	6 206	0.64	1.18
New Zealand	611	593	1.58	-0.10	27	1	-5.01	-8.57	5	6	..	0.00
OTHER DEVELOPED[1]	**18 226**	**18 996**	**2.12**	**0.35**	**22 218**	**20 530**	**-0.81**	**-0.65**	**2 592**	**1 901**	**8.05**	**-3.18**
Japan	191	204	-2.56	0.69	20 045	17 661	-0.68	-1.10	0	0	..	..
South Africa	12 792	13 164	2.56	0.36	133	216	-22.65	1.41	2 190	1 758	9.84	-1.41
DEVELOPING	**527 563**	**644 958**	**4.06**	**1.76**	**87 037**	**113 964**	**3.11**	**2.45**	**38 079**	**43 542**	**2.81**	**2.41**
AFRICA	**104 323**	**125 602**	**4.30**	**1.99**	**16 273**	**22 065**	**2.13**	**2.50**	**4 106**	**3 603**	**11.78**	**-1.87**
NORTH AFRICA	**13 936**	**15 193**	**2.05**	**1.33**	**12 369**	**15 850**	**3.42**	**2.46**	**50**	**45**	**-4.86**	**-0.19**
Algeria	1 810	2 143	7.23	3.78	2 498	2 327	3.16	-0.18	0	0	0.00	0.01
Egypt	8 241	8 028	1.20	-0.16	5 745	9 180	2.60	4.17	0	0	0.00	-0.30
SUB-SAHARAN AFRICA	**90 388**	**110 409**	**4.66**	**2.08**	**3 904**	**6 216**	**-1.38**	**2.61**	**4 056**	**3 558**	**12.19**	**-1.90**
LATIN AMERICA and CARIBBEAN	**132 344**	**162 936**	**3.60**	**1.98**	**26 343**	**33 083**	**4.09**	**2.68**	**29 398**	**38 299**	**7.82**	**3.59**
Argentina	29 524	37 073	4.94	2.68	0	0	..	..	18 310	24 845	4.72	4.01
Brazil	56 668	69 548	4.56	1.73	1 026	1 057	4.67	2.00	9 135	11 341	15.83	3.48
Chile	1 803	2 047	1.04	0.65	2 112	2 468	8.24	2.17	57	56	-5.84	-1.40
Mexico	28 513	35 085	0.95	2.26	10 214	13 308	1.32	3.05	310	214	37.87	-4.23
Uruguay	852	1 538	5.55	3.21	117	22	29.94	-8.06	100	541	5.25	8.06
ASIA and PACIFIC	**290 895**	**356 420**	**4.20**	**1.59**	**44 421**	**58 815**	**2.94**	**2.30**	**4 575**	**1 640**	**-13.48**	**-8.11**
Bangladesh	1 069	1 552	26.44	2.36	50	1	-15.40	-53.99	0	0	0.00	3.85
China[2]	184 374	227 804	4.59	1.59	4 604	11 036	9.89	6.53	166	40	-59.67	-8.12
India	39 110	44 184	3.40	0.78	38	780	-23.78	26.45	2 625	175	33.22	-26.06
Indonesia	17 729	20 893	7.30	1.69	1 691	4 816	4.37	5.69	88	91	18.50	-0.63
Iran, Islamic Republic of	4 405	4 948	-1.21	1.73	4 183	4 505	9.94	-0.19	0	0	0.00	0.01
Korea	335	345	-1.70	0.35	7 929	7 578	-2.00	-0.33	0	0	..	..
Malaysia	96	116	4.27	1.89	2 863	3 399	1.93	1.74	8	7	-0.38	-1.73
Pakistan	3 941	4 653	6.04	1.24	26	127	2.94	15.62	0	0	-60.36	-2.88
Saudi Arabia	461	420	1.64	-0.55	8 909	11 101	2.37	1.46	0	0	0.00	-0.10
Turkey	12 314	14 912	0.25	1.65	500	948	-4.58	2.11	321	190	-11.09	-1.08
LEAST DEVELOPED COUNTRIES (LDC)	**67 610**	**85 308**	**6.01**	**2.41**	**2 690**	**4 030**	**0.05**	**1.36**	**3 763**	**2 299**	**13.79**	**-6.15**
OECD[3]	**561 462**	**661 988**	**1.61**	**1.19**	**52 177**	**54 692**	**-0.31**	**0.62**	**66 433**	**81 014**	**-0.69**	**1.80**
NON-OECD	**572 656**	**700 572**	**3.92**	**1.68**	**67 385**	**91 366**	**3.75**	**2.70**	**55 367**	**64 911**	**4.83**	**1.99**

Table A.8. Coarse grain projections *(cont.)*

Crop year

	CONSUMPTION (Kt)		Growth (%)[4]		FEED USE (Kt)		Growth (%)[4]		PER CAPITA (Kg)		Growth (%)[4]	
	Avg 2009-11est	2021	2002-2011	2012-2021	Avg 2009-11est	2021	2002-2011	2012-2021	Avg 2009-11est	2021	2002-2011	2012-2021
WORLD	**1 143 371**	**1 355 589**	**2.71**	**1.50**	**601 332**	**721 613**	**0.64**	**0.02**	**29.9**	**31.4**	**1.24**	**0.44**
DEVELOPED	**570 150**	**642 660**	**1.88**	**1.14**	**306 323**	**336 137**	**-1.37**	**0.01**	**24.5**	**25.2**	**1.28**	**-0.16**
NORTH AMERICA	**314 397**	**370 785**	**3.34**	**1.43**	**118 666**	**145 290**	**-3.46**	**0.01**	**32.4**	**33.5**	**3.18**	**-0.01**
Canada	20 238	22 491	-1.24	1.38	14 745	16 097	-3.35	0.02	58.5	55.1	-1.78	-0.66
United States	294 160	348 294	3.72	1.43	103 921	129 192	-3.47	0.01	29.6	31.1	4.51	0.12
EUROPE	**209 810**	**225 135**	**0.17**	**0.90**	**156 294**	**160 101**	**0.02**	**0.01**	**22.5**	**23.1**	**0.76**	**-0.32**
EU(27)	151 090	153 607	0.50	0.63	115 171	107 363	0.17	0.00	19.0	18.6	0.34	-0.23
Russian Federation	26 502	36 992	-1.71	2.30	17 487	27 747	-1.55	0.03	26.9	33.0	0.91	-0.36
Ukraine	13 708	14 952	-0.26	0.56	8 702	9 116	-0.53	0.00	46.5	43.6	3.59	-0.27
OCEANIA DEVELOPED	**8 500**	**9 128**	**2.81**	**0.57**	**5 854**	**6 163**	**0.07**	**0.01**	**14.7**	**14.9**	**-0.89**	**-2.52**
Australia	7 867	8 540	2.93	0.62	5 305	5 665	-0.15	0.01	13.8	14.2	0.57	-2.81
New Zealand	633	588	1.71	-0.13	549	498	2.80	0.00	19.2	18.5	-4.55	-1.29
OTHER DEVELOPED[1]	**37 443**	**37 611**	**0.34**	**-0.02**	**25 510**	**24 584**	**0.77**	**0.00**	**20.7**	**20.4**	**-0.54**	**-0.08**
Japan	20 153	17 972	-0.54	-1.20	15 445	13 175	-0.28	-0.02	2.3	2.3	-2.52	0.00
South Africa	10 626	11 567	1.64	0.78	4 802	5 187	3.08	0.01	94.5	90.8	-0.67	-0.30
DEVELOPING	**573 221**	**712 929**	**3.59**	**1.84**	**295 009**	**385 477**	**3.06**	**0.02**	**31.2**	**32.9**	**1.19**	**0.52**
AFRICA	**115 688**	**143 411**	**3.74**	**2.16**	**24 108**	**30 074**	**4.37**	**0.02**	**78.2**	**77.2**	**0.91**	**0.05**
NORTH AFRICA	**25 835**	**30 877**	**2.71**	**1.96**	**18 124**	**21 771**	**3.23**	**0.02**	**39.5**	**41.2**	**0.10**	**0.49**
Algeria	4 224	4 470	5.55	1.53	3 378	3 463	6.99	0.02	19.9	21.0	-0.05	0.56
Egypt	13 951	17 191	1.40	1.93	9 472	11 628	1.35	0.02	48.3	51.6	-0.13	0.80
SUB-SAHARAN AFRICA	**89 854**	**112 534**	**4.05**	**2.22**	**5 985**	**8 303**	**8.53**	**0.03**	**86.2**	**83.7**	**0.89**	**-0.09**
LATIN AMERICA and CARIBBEAN	**129 562**	**157 325**	**2.81**	**1.79**	**77 809**	**92 066**	**3.26**	**0.01**	**62.8**	**73.0**	**0.50**	**1.52**
Argentina	10 551	12 550	4.88	0.79	5 874	8 404	6.48	0.02	78.3	56.7	2.46	-3.72
Brazil	48 992	58 840	2.91	1.38	38 060	44 622	3.78	0.01	26.7	36.5	-4.35	1.26
Chile	3 876	4 449	5.05	1.56	3 234	3 712	5.72	0.02	19.0	20.9	1.19	1.06
Mexico	38 622	48 114	1.03	2.54	15 570	16 527	-1.11	0.01	172.6	212.4	1.71	2.65
Uruguay	878	1 002	6.18	1.47	290	315	4.30	0.03	21.2	21.6	-1.55	0.67
ASIA and PACIFIC	**327 970**	**412 193**	**3.85**	**1.75**	**193 091**	**263 336**	**2.83**	**0.03**	**14.9**	**14.2**	**1.04**	**-0.67**
Bangladesh	1 173	1 528	18.17	2.06	690	846	23.94	0.01	2.6	3.2	9.13	2.29
China[2]	186 056	238 195	4.51	1.78	112 850	159 747	2.77	0.03	11.0	7.7	1.43	-3.61
India	36 370	44 822	2.18	1.28	6 829	13 194	-0.52	0.05	20.2	18.9	0.01	-1.09
Indonesia	19 033	25 463	6.98	2.42	7 256	11 436	10.39	0.03	33.5	36.7	3.24	1.00
Iran, Islamic Republic of	8 722	9 422	3.44	0.84	8 286	8 914	3.75	0.01	1.4	1.4	-0.79	0.56
Korea	8 188	7 905	-1.78	-0.16	6 094	6 043	-1.59	0.00	4.4	4.2	0.14	-0.28
Malaysia	2 988	3 501	2.26	1.76	2 801	3 297	2.44	0.02	1.6	1.7	-0.73	0.55
Pakistan	4 034	4 777	6.67	1.44	1 625	1 969	11.13	0.02	9.2	9.4	0.71	0.39
Saudi Arabia	9 637	11 469	3.49	1.49	9 393	11 203	3.82	0.02	3.5	2.9	-2.61	-1.45
Turkey	12 610	15 576	0.12	1.78	10 148	12 878	0.31	0.02	17.6	16.8	-0.05	-0.26
LEAST DEVELOPED COUNTRIES (LDC)	**65 741**	**86 307**	**5.29**	**2.59**	**6 987**	**9 593**	**12.03**	**0.03**	**58.0**	**61.6**	**1.98**	**0.72**
OECD[3]	**560 944**	**631 140**	**2.02**	**1.19**	**293 317**	**314 382**	**-1.48**	**0.01**	**33.8**	**38.7**	**1.84**	**1.48**
NON-OECD	**582 427**	**724 450**	**3.41**	**1.78**	**308 015**	**407 232**	**2.98**	**0.02**	**29.0**	**29.9**	**1.10**	**0.20**

Note: Crop year: Beginning crop marketing year - see Glossary of Terms for definitions.

1. Includes Israel and also transition economies: Kazakhstan, Kyrgyzstan, Tajikistan, Turkmenistan, Uzbekistan, Armenia, Azerbaijan and Georgia.
2. Refers to mainland only. The economies of Chinese Taipei, Hong Kong (China) and Macau (China) are included in the Other Asia Pacific aggregate.
3. Excludes Iceland but includes all EU27 member countries.
4. Least-squares growth rate (see glossary).

Source: OECD and FAO Secretariats.

StatLink http://dx.doi.org/10.1787/888932642706

Table A.9. Rice projections

Crop year

	PRODUCTION (Kt)		Growth (%)[5]		IMPORTS (Kt)		Growth (%)[5]		EXPORTS (Kt)		Growth (%)[5]	
	Avg 2009-11est	2021	2002-2011	2012-2021	Avg 2009-11est	2021	2002-2011	2012-2021	Avg 2009-11est	2021	2002-2011	2012-2021
WORLD	**467 554**	**542 072**	**2.49**	**1.17**	**32 871**	**42 775**	**1.75**	**1.92**	**32 802**	**42 679**	**2.10**	**1.93**
DEVELOPED	**18 101**	**18 941**	**0.54**	**0.61**	**4 912**	**5 808**	**1.63**	**1.57**	**4 151**	**4 694**	**-0.62**	**1.61**
NORTH AMERICA	**6 819**	**7 285**	**-0.21**	**1.29**	**968**	**1 291**	**3.83**	**2.39**	**3 261**	**3 736**	**-1.70**	**2.29**
Canada	0	0	..	..	375	421	4.27	1.01	0	0	..	..
United States	6 819	7 285	-0.21	1.29	592	870	3.61	3.13	3 261	3 736	-1.70	2.29
EUROPE	**2 632**	**2 918**	**4.00**	**1.04**	**1 772**	**2 066**	**0.55**	**1.21**	**424**	**396**	**7.61**	**-1.47**
EU(27)	1 810	1 742	1.74	-0.03	1 222	1 444	2.28	1.50	234	220	0.99	-0.14
Russian Federation	705	1 050	11.06	3.13	244	249	-5.62	-1.28	180	169	27.60	-2.92
Ukraine	101	111	8.58	0.74	100	115	-0.68	1.90	7	6	55.83	-1.86
OCEANIA DEVELOPED	**438**	**744**	**-2.95**	**1.03**	**195**	**156**	**5.28**	**0.94**	**236**	**375**	**-0.54**	**0.37**
Australia	438	744	-2.95	1.03	151	112	5.98	1.33	236	375	-0.54	0.37
New Zealand	0	0	..	..	44	44	2.65	0.00	0	0	..	..
OTHER DEVELOPED[1]	**8 212**	**7 994**	**0.09**	**-0.17**	**1 978**	**2 295**	**1.32**	**1.50**	**229**	**187**	**-0.94**	**-0.82**
Japan	7 680	7 391	-0.04	-0.29	764	753	-0.61	0.00	193	181	-1.18	0.00
South Africa	2	18	0.00	19.12	867	1 030	2.17	1.77	0	0	0.00	-0.25
DEVELOPING	**449 453**	**523 131**	**2.57**	**1.19**	**27 959**	**36 967**	**1.76**	**1.98**	**28 652**	**37 985**	**2.54**	**1.97**
AFRICA	**16 441**	**28 516**	**4.31**	**4.81**	**9 261**	**10 499**	**2.07**	**0.84**	**345**	**188**	**-13.98**	**-6.51**
NORTH AFRICA	**3 821**	**4 529**	**-1.00**	**0.60**	**468**	**772**	**8.97**	**6.97**	**183**	**40**	**-26.21**	**-16.48**
Algeria	1	5	0.00	15.33	78	107	2.27	2.84	0	0	0.00	-0.41
Egypt	3 791	4 497	-1.05	0.60	153	278	51.48	18.27	183	40	-26.21	-16.48
SUB-SAHARAN AFRICA	**12 620**	**23 987**	**6.47**	**5.81**	**8 792**	**9 726**	**1.78**	**0.49**	**162**	**149**	**13.53**	**-0.64**
LATIN AMERICA and CARIBBEAN	**18 321**	**21 242**	**1.64**	**1.28**	**3 593**	**5 154**	**0.37**	**2.87**	**2 807**	**2 537**	**10.74**	**-1.03**
Argentina	970	1 179	6.37	1.51	0	0	..	..	600	648	11.09	0.28
Brazil	7 978	7 850	0.45	-0.39	675	1 322	-3.99	3.51	695	597	41.85	-3.72
Chile	79	88	-2.10	-0.09	115	216	2.78	4.88	1	1	38.84	-1.16
Mexico	156	177	-1.57	1.71	775	1 223	0.96	3.57	6	28	31.00	-0.90
Uruguay	952	1 069	4.21	-0.09	0	0	-48.79	0.12	872	983	4.13	-0.12
ASIA and PACIFIC	**414 692**	**473 372**	**2.55**	**1.00**	**15 104**	**21 314**	**1.98**	**2.38**	**25 500**	**35 259**	**2.26**	**2.28**
Bangladesh	33 459	37 671	3.87	1.04	910	3 782	-6.16	9.35	6	13	12.87	-0.68
China[2]	135 092	128 860	1.99	-0.51	486	857	-0.23	5.96	624	216	-10.48	-1.88
India	95 807	118 507	2.65	1.39	83	29	32.86	13.27	3 581	5 155	-3.52	-2.66
Indonesia	41 213	52 720	3.09	2.39	1 217	1 931	-9.75	-3.34	2	2	42.52	0.24
Iran, Islamic Republic of	1 458	1 606	-2.82	0.33	1 200	2 434	2.57	6.43	0	0	0.00	-0.46
Korea	4 476	4 285	-1.06	0.01	341	433	9.37	0.79	3	3	-66.83	0.00
Pakistan	6 296	9 179	3.93	0.88	30	11	32.10	0.01	3 208	4 399	4.26	-0.01
Philippines	10 747	15 368	2.65	2.60	1 576	1 252	2.77	-1.70	0	0	0.00	0.12
Thailand	21 620	23 145	2.17	1.04	367	367	61.56	-1.63	9 177	8 717	1.39	1.63
Turkey	502	583	10.21	1.23	303	425	1.33	3.68	15	11	35.84	-2.18
Viet Nam	26 961	32 732	2.11	1.46	567	447	48.44	-2.67	6 993	9 404	6.98	2.67
LDC Asia[3]	29 828	40 499	4.27	2.57	1 027	313	4.30	2.00	1 774	7 221	21.51	11.71
LEAST DEVELOPED COUNTRIES (LDC)	**72 619**	**95 504**	**4.44**	**2.40**	**6 552**	**10 181**	**0.76**	**4.20**	**1 940**	**7 381**	**20.46**	**11.25**
OECD[4]	**21 960**	**22 321**	**0.03**	**0.39**	**4 919**	**6 185**	**2.22**	**2.01**	**3 950**	**4 555**	**-2.02**	**1.85**
NON-OECD	**445 594**	**519 752**	**2.62**	**1.20**	**27 952**	**36 590**	**1.66**	**1.91**	**28 852**	**38 125**	**2.77**	**1.94**

Note: Crop year: Beginning crop marketing year - see Glossary of Terms for definitions.

1. Includes Israel and also transition economies: Kazakhstan, Kyrgyzstan, Tajikistan, Turkmenistan, Uzbekistan, Armenia, Azerbaijan and Georgia.
2. Refers to mainland only. The economies of Chinese Taipei, Hong Kong (China) and Macau (China) are included in the Other Asia Pacific aggregate.
3. LDC Asia includes Afghanistan, Bhutan, Myanmar, Cambodia, Lao Peoples' Democratic Republic, Nepal, Yemen, Timor Meste, Maldives.
4. Excludes Iceland but includes all EU27 member countries.
5. Least-squares growth rate (see glossary).

Source: OECD and FAO Secretariats.

Table A.9. Rice projections *(cont.)*

Crop year

	CONSUMPTION (Kt)		Growth (%)[3]		PER CAPITA (Kg)		Growth (%)[3]	
	Avg 2009-11est	2021	2002-2011	2012-2021	Avg 2009-11est	2021	2002-2011	2012-2021
WORLD	**460 150**	**542 359**	**1.69**	**1.24**	**56.7**	**59.9**	**0.15**	**0.34**
DEVELOPED	**18 822**	**19 983**	**0.69**	**0.53**	**13.0**	**13.3**	**0.09**	**0.19**
NORTH AMERICA	**4 440**	**4 907**	**1.61**	**1.01**	**12.9**	**13.0**	**0.68**	**0.19**
Canada	375	421	4.27	1.01	11.0	11.2	3.22	0.15
United States	4 065	4 485	1.39	1.01	13.1	13.2	0.48	0.19
EUROPE	**3 917**	**4 583**	**1.34**	**1.38**	**5.3**	**6.1**	**1.15**	**1.28**
EU(27)	2 740	2 965	1.35	0.69	5.5	5.8	0.95	0.44
Russian Federation	766	1 128	0.82	3.26	5.4	8.0	1.02	3.42
Ukraine	194	220	2.68	1.45	4.2	5.0	3.33	1.95
OCEANIA DEVELOPED	**369**	**518**	**-1.77**	**2.12**	**13.8**	**17.0**	**-3.33**	**0.96**
Australia	324	474	-2.28	2.34	14.6	18.6	-3.91	1.14
New Zealand	44	44	2.65	0.00	10.1	9.1	1.46	-0.97
OTHER DEVELOPED[1]	**10 096**	**9 975**	**0.15**	**-0.13**	**35.0**	**32.9**	**-0.68**	**-0.53**
Japan	8 394	7 837	-0.31	-0.68	59.4	55.4	-0.78	-0.62
South Africa	862	1 046	2.27	1.91	16.4	19.0	1.62	1.52
DEVELOPING	**441 328**	**522 377**	**1.73**	**1.27**	**67.6**	**70.5**	**0.01**	**0.22**
AFRICA	**25 512**	**38 662**	**3.87**	**3.72**	**22.8**	**27.6**	**1.45**	**1.70**
NORTH AFRICA	**4 210**	**5 189**	**1.90**	**1.82**	**22.1**	**23.6**	**1.52**	**0.66**
Algeria	79	111	2.20	3.20	2.2	2.7	0.70	2.01
Egypt	3 861	4 664	1.68	1.64	40.7	41.3	1.10	0.22
SUB-SAHARAN AFRICA	**21 303**	**33 472**	**4.30**	**4.05**	**23.0**	**28.4**	**1.43**	**1.84**
LATIN AMERICA and CARIBBEAN	**19 041**	**23 621**	**0.58**	**1.87**	**30.4**	**33.8**	**-0.59**	**0.87**
Argentina	370	530	1.10	3.25	9.2	12.0	0.22	2.46
Brazil	7 795	8 387	-1.10	0.32	40.0	39.6	-2.12	-0.40
Chile	201	303	0.93	3.25	11.5	15.9	0.04	2.49
Mexico	931	1 365	-0.29	3.62	8.2	10.7	-1.54	2.63
Uruguay	92	85	10.84	-2.24	8.0	7.9	-1.09	-0.78
ASIA and PACIFIC	**396 775**	**460 094**	**1.66**	**1.05**	**84.1**	**88.2**	**0.12**	**0.26**
Bangladesh	33 863	41 249	3.57	1.53	171.7	179.3	1.15	0.20
China[2]	128 754	131 420	0.38	-0.20	77.8	75.5	-0.25	-0.34
India	92 243	113 176	1.64	1.38	71.4	77.1	-0.28	0.28
Indonesia	41 595	54 455	2.24	2.26	156.5	184.7	0.81	1.41
Iran, Islamic Republic of	2 634	4 019	-0.36	3.68	31.3	44.7	-1.18	3.16
Korea	4 702	4 722	-0.41	0.07	69.5	62.6	-1.21	-1.24
Pakistan	3 215	4 751	3.96	2.38	14.7	18.2	2.47	1.27
Philippines	12 382	16 520	2.86	2.52	119.9	131.9	1.65	0.93
Thailand	12 210	14 583	2.80	1.53	128.8	139.8	1.07	0.70
Turkey	799	996	5.67	2.32	10.3	11.5	4.32	1.31
Viet Nam	20 851	23 703	1.36	1.06	189.2	186.7	0.31	-0.12
LDC Asia[4]	29 489	33 411	4.13	1.24	128.6	131.8	0.45	0.25
LEAST DEVELOPED COUNTRIES (LDC)	**77 171**	**97 862**	**4.03**	**2.09**	**68.8**	**72.0**	**0.51**	**0.32**
OECD[5]	**22 809**	**23 882**	**0.41**	**0.43**	**16.2**	**15.8**	**-0.43**	**-0.21**
NON-OECD	**437 341**	**518 478**	**1.76**	**1.28**	**65.9**	**69.2**	**0.10**	**0.29**

Note: Crop year: Beginning crop marketing year - see Glossary of Terms for definitions.

1. Includes Israel and also transition economies: Kazakhstan, Kyrgyzstan, Tajikistan, Turkmenistan, Uzbekistan, Armenia, Azerbaijan and Georgia.
2. Refers to mainland only. The economies of Chinese Taipei, Hong Kong (China) and Macau (China) are included in the Other Asia Pacific aggregate.
3. Least-squares growth rate (see glossary).
4. LDC Asia includes Afghanistan, Bhutan, Myanmar, Cambodia, Lao Peoples' Democratic Republic, Nepal, Yemen, Timor Meste, Maldives.
5. Excludes Iceland but includes all EU27 member countries.

Source: OECD and FAO Secretariats.

StatLink http://dx.doi.org/10.1787/888932642744

Table A.10. Main policy assumptions for cereal markets

Crop year

		Avg 2009-11est.	2012	2013	2014	2015	2016	2017	2018	2019	2020	2021
Argentina												
Crops export tax	%	21.0	21.0	21.0	21.0	21.0	21.0	21.0	21.0	21.0	21.0	21.0
Rice export tax	%	5.5	5.5	5.5	5.5	5.5	5.5	5.5	5.5	5.5	5.5	5.5
Canada												
Tariff-quotas[1]												
Wheat	kt	350.0	350.0	350.0	350.0	350.0	350.0	350.0	350.0	350.0	350.0	350.0
In-quota tariff	%	1.1	1.1	1.1	1.1	1.1	1.1	1.1	1.1	1.1	1.1	1.1
Out-of-quota tariff	%	61.7	61.7	61.7	61.7	61.7	61.7	61.7	61.7	61.7	61.7	61.7
Barley	kt	399.0	399.0	399.0	399.0	399.0	399.0	399.0	399.0	399.0	399.0	399.0
In-quota tariff	%	0.6	0.6	0.6	0.6	0.6	0.6	0.6	0.6	0.6	0.6	0.6
Out-of-quota tariff	%	58.0	58.0	58.0	58.0	58.0	58.0	58.0	58.0	58.0	58.0	58.0
European Union[2]												
Cereal support price[3]	EUR/t	101.3	101.3	101.3	101.3	101.3	101.3	101.3	101.3	101.3	101.3	102.3
Single farm payment[4]	EUR/ha	187.9	187.9	187.9	187.9	187.9	187.9	187.9	187.9	187.9	187.9	188.9
Rice support price[5]	EUR/t	150.0	150.0	150.0	150.0	150.0	150.0	150.0	150.0	150.0	150.0	150.0
Compulsory set-aside rate	%	0.0	0.0	0.0	0.0	0.0	0.0	0.0	0.0	0.0	0.0	0.0
Direct payment for rice[6]	EUR/ha	261.1	11.9	11.9	11.9	11.9	11.9	11.9	11.9	11.9	11.9	11.9
Wheat tariff-quota[1]	kt	3 779.9	3 779.9	3 779.9	3 779.9	3 779.9	3 779.9	3 779.9	3 779.9	3 779.9	3 779.9	3 780.9
Coarse grain tariff-quota[1]	kt	3 469.4	3 469.4	3 469.4	3 469.4	3 469.4	3 469.4	3 469.4	3 469.4	3 469.4	3 469.4	3 470.4
Subsidised export limits[1]												
Wheat	mt	15.4	15.4	15.4	15.4	15.4	15.4	15.4	15.4	15.4	15.4	15.4
Coarse grains[7]	mt	10.8	10.8	10.8	10.8	10.8	10.8	10.8	10.8	10.8	10.8	10.8
Japan												
Wheat tariff-quota	kt	5 740.0	5 740.0	5 740.0	5 740.0	5 740.0	5 740.0	5 740.0	5 740.0	5 740.0	5 740.0	5 740.0
In-quota tariff	'000 JPY/t	0.0	0.0	0.0	0.0	0.0	0.0	0.0	0.0	0.0	0.0	0.0
Out-of-quota tariff	'000 JPY/t	55.0	55.0	55.0	55.0	55.0	55.0	55.0	55.0	55.0	55.0	55.0
Barley tariff-quota	kt	1 369.0	1 369.0	1 369.0	1 369.0	1 369.0	1 369.0	1 369.0	1 369.0	1 369.0	1 369.0	1 369.0
In-quota tariff	'000 JPY/t	0.0	0.0	0.0	0.0	0.0	0.0	0.0	0.0	0.0	0.0	0.0
Out-of-quota tariff	'000 JPY/t	39.0	39.0	39.0	39.0	39.0	39.0	39.0	39.0	39.0	39.0	39.0
Rice tariff-quota	kt	682.2	682.2	682.2	682.2	682.2	682.2	682.2	682.2	682.2	682.2	682.2
In-quota tariff	'000 JPY/t	0.0	0.0	0.0	0.0	0.0	0.0	0.0	0.0	0.0	0.0	0.0
Out-of-quota tariff	'000 JPY/t	341.0	341.0	341.0	341.0	341.0	341.0	341.0	341.0	341.0	341.0	341.0
Korea												
Wheat tariff	%	5.4	5.4	5.4	5.4	5.4	5.4	5.4	5.4	5.4	5.4	5.4
Maize tariff-quota	kt	6 102.0	6 102.0	6 102.0	6 102.0	6 102.0	6 102.0	6 102.0	6 102.0	6 102.0	6 102.0	6 102.0
In-quota tariff	%	1.6	1.6	1.6	1.6	1.6	1.6	1.6	1.6	1.6	1.6	1.6
Out-of-quota tariff	%	403.5	403.5	403.5	403.5	403.5	403.5	403.5	403.5	403.5	403.5	403.5
Barley tariff-quota	kt	0.4	0.4	0.4	0.4	0.4	0.4	0.4	0.4	0.4	0.4	0.4
In-quota tariff	%	40.0	40.0	40.0	40.0	40.0	40.0	40.0	40.0	40.0	40.0	40.0
Out-of-quota tariff	%	89.0	89.0	89.0	89.0	89.0	89.0	89.0	89.0	89.0	89.0	89.0
Rice quota[8]	kt	205.2	205.2	205.2	205.2	205.2	205.2	205.2	205.2	205.2	205.2	205.2
In-quota tariff	%	5.0	5.0	5.0	5.0	5.0	5.0	5.0	5.0	5.0	5.0	5.0
MERCOSUR												
Wheat tariff	%	10.0	10.0	10.0	10.0	10.0	10.0	10.0	10.0	10.0	10.0	10.0
Coarse grain tariff	%	8.0	8.0	8.0	8.0	8.0	8.0	8.0	8.0	8.0	8.0	8.0
Rice tariff	%	10.0	10.0	10.0	10.0	10.0	10.0	10.0	10.0	10.0	10.0	10.0
Mexico												
Wheat NAFTA tariff	%	0.0	0.0	0.0	0.0	0.0	0.0	0.0	0.0	0.0	0.0	0.0
Maize tariff-quota	kt	2 501.0	2 501.0	2 501.0	2 501.0	2 501.0	2 501.0	2 501.0	2 501.0	2 501.0	2 501.0	2 501.0
In-tariff-quota	%	50.0	50.0	50.0	50.0	50.0	50.0	50.0	50.0	50.0	50.0	50.0
Out-of-tariff-quota	%	194.0	194.0	194.0	194.0	194.0	194.0	194.0	194.0	194.0	194.0	194.0
Barley tariff-quota	kt	4.7	4.7	4.7	4.7	4.7	4.7	4.7	4.7	4.7	4.7	4.7
In-tariff-quota	%	50.0	50.0	50.0	50.0	50.0	50.0	50.0	50.0	50.0	50.0	50.0
Out-of-tariff-quota	%	115.2	115.2	115.2	115.2	115.2	115.2	115.2	115.2	115.2	115.2	115.2

Table A.10. Main policy assumptions for cereal markets *(cont.)*

Crop year

		Avg 2009-11est.	2012	2013	2014	2015	2016	2017	2018	2019	2020	2021
United States												
ACRE participation rate												
Wheat	%	10.0	10.0	10.0	10.0	10.0	10.0	10.0	10.0	10.0	10.0	10.0
Coarse grains	%	10.0	10.0	10.0	10.0	10.0	10.0	10.0	10.0	10.0	10.0	10.0
Rice	%	0.1	0.1	0.1	0.1	0.1	0.1	0.1	0.1	0.1	0.1	0.1
Wheat loan rate	USD/t	105.7	108.0	108.0	108.0	108.0	108.0	108.0	108.0	108.0	108.0	108.0
Maize loan rate	USD/t	76.8	76.8	76.8	76.8	76.8	76.8	76.8	76.8	76.8	76.8	76.8
Prod. flex. contract payment												
Wheat	USD/t	16.9	16.9	16.9	16.9	16.9	16.9	16.9	16.9	16.9	16.9	16.9
Maize	USD/t	10.3	10.3	10.3	10.3	10.3	10.3	10.3	10.3	10.3	10.3	10.3
CRP areas[9]												
Wheat	mha	3.3	3.1	3.1	3.2	3.3	3.3	3.3	3.3	3.3	3.3	3.3
Coarse grains	mha	3.0	2.8	2.8	2.9	3.0	3.0	3.0	3.0	3.0	3.0	2.6
Subsidised export limits[1]												
Wheat	mt	14.5	14.5	14.5	14.5	14.5	14.5	14.5	14.5	14.5	14.5	14.5
Coarse grains	mt	1.6	1.6	1.6	1.6	1.6	1.6	1.6	1.6	1.6	1.6	1.6
Wheat EEP payment[10]	USD/t	0.0	0.0	0.0	0.0	0.0	0.0	0.0	0.0	0.0	0.0	0.0
China												
Wheat tariff-quota	kt	9 636	9 636	9 636	9 636	9 636	9 636	9 636	9 636	9 636	9 636	9 636
In quota tariff	%	2.2	2.2	2.2	2.2	2.2	2.2	2.2	2.2	2.2	2.2	2.2
Out of quota tariff	%	65.0	65.0	65.0	65.0	65.0	65.0	65.0	65.0	65.0	65.0	65.0
Coarse grains tariff	%	2.0	2.0	2.0	2.0	2.0	2.0	2.0	2.0	2.0	2.0	2.0
Maize tariff-quota	kt	7 200	7 200	7 200	7 200	7 200	7 200	7 200	7 200	7 200	7 200	7 200
In quota tariff	%	3.7	3.7	3.7	3.7	3.7	3.7	3.7	3.7	3.7	3.7	3.7
Out of quota tariff	%	41.7	41.7	41.7	41.7	41.7	41.7	41.7	41.7	41.7	41.7	41.7
Rice tariff-quota	kt	5 320	5 320	5 320	5 320	5 320	5 320	5 320	5 320	5 320	5 320	5 320
In quota tariff	%	2.3	2.3	2.3	2.3	2.3	2.3	2.3	2.3	2.3	2.3	2.3
Out of quota tariff	%	51.7	51.7	51.7	51.7	51.7	51.7	51.7	51.7	51.7	51.7	51.7
India												
Input subsidy coarse grains[11]	INR/ha	6 877	7 055	7 022	7 063	7 090	7 101	7 062	7 071	6 990	6 945	6 964
Input subsidy rice[11]	INR/ha	6 877	7 055	7 022	7 063	7 090	7 101	7 062	7 071	6 990	6 945	6 964
Input subsidy wheat[11]	INR/ha	6 877	7 055	7 022	7 063	7 090	7 101	7 062	7 071	6 990	6 945	6 964
Minimum support price												
Maize	INR/t	8 784	9 518	9 899	10 295	10 707	11 135	11 580	12 043	12 525	13 026	13 514
Rice	INR/t	10 067	10 816	11 249	11 699	12 167	12 653	13 159	13 686	14 233	14 802	15 357
Wheat	INR/t	11 216	12 114	12 598	13 102	13 627	14 172	14 738	15 328	15 941	16 579	17 200
Wheat Export subsidy	INR/t	1 941	1 941	1 941	1 941	1 941	1 941	1 941	1 941	1 941	1 941	1 941
Wheat tariff	%	87.5	87.5	87.5	87.5	87.5	87.5	87.5	87.5	87.5	87.5	87.5
Maize tariff	%	50.0	50.0	50.0	50.0	50.0	50.0	50.0	50.0	50.0	50.0	50.0
Rice tariff	%	30.0	30.0	30.0	30.0	30.0	30.0	30.0	30.0	30.0	30.0	30.0
Barley tariff	%	100.0	100.0	100.0	100.0	100.0	100.0	100.0	100.0	100.0	100.0	100.0
Russian Federation												
Wheat ad valorem import tax	%	3.9	5.0	5.0	5.0	5.0	5.0	5.0	5.0	5.0	5.0	5.0
Rice tariff equivalent of import barriers	%	0.0	15.0	13.3	11.7	10.0	10.0	10.0	10.0	10.0	10.0	10.0
Coarse grains tariff equivalent of import barriers	%	2.7	2.7	2.7	2.7	2.7	2.7	2.7	2.7	2.7	2.7	2.7
Coarse grain specific tariff	RUB/t	0.0	5.0	4.5	4.5	4.5	4.5	4.5	4.5	4.5	4.5	4.5
Coarse grain ad valorem import tax	%	2.7	2.7	2.7	2.7	2.7	2.7	2.7	2.7	2.7	2.7	2.7

Note: Crop year: Beginning crop marketing year - see Glossary of Terms for definitions.

The source for tariffs and Tariff Rate Quotas is AMAD (Agricultural market access database). The tariff and TRQ data are based on Most Favoured Nation rates scheduled with the WTO and exclude those under preferential or regional agreements, which may be substantially different. Tariffs are simple averages of several product lines. Specific rates are converted to ad valorem rates using world prices in the Outlook. Import quotas are based on global commitments scheduled in the WTO rather than those allocated to preferential partners under regional or other agreements. For Mexico, the NAFTA in-quota tariff on maize and barley is zero, while the tariff-rate quota becomes unlimited in 2003 for barley and 2008 for maize.

1. Year beginning 1 July.
2. EU farmers also benefit from the Single Farm Payment (SFP) Scheme, which provides flat-rate payments independent from current production decisions and market developments. For the new member states, payments are phased in with the assumption of maximum top-ups from national budgets up to 2013 through the Single Area Payment (SAP), and through the (SFP) from 2014. Due to modulation, an increasing share of the total SFP will go to rural development spending rather than directly to farmers.
3. Common intervention price for soft wheat, barley, maize and sorghum.
4. Actual payments are made per eligible hectare (all agricultural land) based on historical reference or regional average.
5. Subject to a purchase limit of 75 000 tonnes per year.
6. Actual payments made per hectare based on program yields; will be integrated in the single farm payment as of 2012.
7. The export volume excludes 0.4mt of exported potato starch. The original limit on subsidised exports is 10.8 mt.
8. Husked rice basis.
9. Includes wheat, barley, maize, oats and sorghum.
10. Average per tonne of total exports.
11. Indian input subsidies consist of those for electricty, fertiliser and irrigation.

Source: OECD and FAO Secretariats.

StatLink http://dx.doi.org/10.1787/888932642763

Table A.11. World oilseed projections

		Avg 09/10-11/12est	12/13	13/14	14/15	15/16	16/17	17/18	18/19	19/20	20/21	21/22
OILSEED (crop year)[1]												
OECD[2]												
Production	mt	150.1	151.9	154.9	158.2	160.8	163.9	166.4	169.5	172.0	175.0	177.6
Consumption	mt	132.9	140.2	141.6	144.4	146.3	148.6	150.5	152.3	153.8	155.4	156.7
Crush	mt	117.8	124.9	126.1	128.7	130.6	132.7	134.6	136.3	137.7	139.2	140.5
Closing stocks	mt	16.3	15.9	15.4	16.2	16.5	16.6	16.2	16.3	16.4	16.6	16.7
Non-OECD												
Production	mt	286.3	299.4	305.2	312.6	318.7	324.9	329.5	336.0	342.3	348.3	352.2
Consumption	mt	300.7	315.1	321.6	327.8	335.3	342.2	348.4	355.2	362.4	369.7	375.7
Crush	mt	245.8	258.6	264.1	269.3	275.6	281.4	286.5	292.2	298.2	304.2	309.2
Closing stocks	mt	22.2	18.6	18.7	19.1	19.4	19.8	19.8	20.2	20.8	21.5	21.3
WORLD												
Production	mt	436.5	451.3	460.2	470.8	479.5	488.8	495.9	505.4	514.4	523.3	529.7
Area	mha	205.3	221.4	223.6	226.0	228.2	230.4	231.4	233.6	236.0	238.3	238.9
Yield	t/ha	1.9	2.0	2.1	2.1	2.1	2.1	2.1	2.2	2.2	2.2	2.2
Consumption	mt	433.6	455.3	463.2	472.2	481.6	490.9	498.9	507.6	516.3	525.1	532.5
Crush	mt	363.7	383.5	390.2	398.0	406.2	414.2	421.1	428.5	435.9	443.4	449.7
Exports	mt	96.5	113.4	115.2	117.4	120.6	123.4	125.3	127.2	129.2	131.3	133.2
Closing stocks	mt	38.5	34.5	34.1	35.3	35.9	36.4	36.0	36.5	37.2	38.1	38.0
Price[3]	USD/t	503.1	499.9	526.6	513.5	520.6	521.2	533.1	537.2	542.1	542.9	550.3
PROTEIN MEALS (marketing year)												
OECD[2]												
Production	mt	80.9	85.4	86.1	87.9	89.0	90.5	91.7	92.8	93.8	94.7	95.6
Consumption	mt	109.8	113.2	113.8	115.2	116.8	118.5	119.7	120.6	122.1	123.4	124.7
Closing stocks	mt	1.5	1.5	1.5	1.5	1.5	1.6	1.5	1.5	1.5	1.5	1.5
Non-OECD												
Production	mt	174.9	183.3	187.2	190.9	195.4	199.6	203.3	207.3	211.5	215.7	219.3
Consumption	mt	142.6	152.7	156.3	160.3	164.2	168.2	172.1	176.2	179.8	183.7	186.8
Closing stocks	mt	13.5	14.1	14.4	14.8	15.2	15.5	15.7	16.0	16.3	16.6	16.9
WORLD												
Production	mt	255.8	268.7	273.3	278.8	284.4	290.0	294.9	300.1	305.2	310.5	314.9
Consumption	mt	252.4	265.8	270.1	275.5	281.0	286.7	291.8	296.8	301.9	307.1	311.6
Closing stocks	mt	15.0	15.7	15.9	16.3	16.7	17.1	17.2	17.5	17.8	18.1	18.4
Price[4]	USD/t	367.2	378.4	389.9	390.3	387.4	384.2	392.9	396.5	399.4	401.1	403.6
VEGETABLE OILS (marketing year)												
OECD[2]												
Production	mt	32.1	33.6	34.0	34.9	35.5	36.2	36.8	37.3	37.7	38.1	38.5
Consumption	mt	44.9	48.2	48.5	49.5	50.4	51.1	51.9	52.6	53.4	54.3	55.0
Closing stocks	mt	3.1	2.9	3.0	2.9	2.9	2.9	2.9	2.9	2.9	2.8	2.8
Non-OECD												
Production	mt	112.2	119.8	122.8	125.7	128.7	131.6	134.3	137.2	140.2	143.2	146.0
Consumption	mt	99.5	107.7	109.7	112.6	115.5	118.3	121.0	123.7	126.3	128.8	131.4
Closing stocks	mt	17.5	18.0	18.6	19.0	19.3	19.6	19.7	19.8	20.0	20.3	20.4
WORLD												
Production	mt	144.3	153.4	156.8	160.5	164.2	167.7	171.1	174.5	177.9	181.3	184.5
Of which palm oil	mt	48.1	51.9	53.4	54.9	56.3	57.7	59.1	60.5	61.9	63.3	64.6
Consumption	mt	144.4	155.9	158.2	162.1	165.8	169.4	172.9	176.3	179.7	183.1	186.4
Food	mt	113.2	126.9	128.4	130.9	133.1	135.7	138.1	140.5	142.8	145.1	147.4
Biofuel	mt	19.9	20.7	21.3	22.6	24.0	25.0	25.9	26.8	27.7	28.8	29.6
Exports	mt	55.2	65.0	66.1	67.6	68.8	70.0	71.5	72.9	74.4	75.7	77.1
Closing stocks	mt	20.7	20.9	21.5	21.9	22.2	22.5	22.6	22.7	22.9	23.1	23.2
Price[5]	USD/t	1 066.9	1 113.9	1 124.8	1 107.4	1 140.6	1 150.3	1 174.9	1 200.9	1 213.6	1 219.9	1 232.5

1. Beginning crop marketing year - see Glossary of Terms for definitions.
2. Excludes Iceland but includes all EU27 member countries.
3. Weighted average oilseed price, European port.
4. Weighted average protein meal, European port.
5. Weighted average price of oilseed oils and palm oil, European port.

Source: OECD and FAO Secretariats.

StatLink http://dx.doi.org/10.1787/888932642782

Table A.12. Oilseed projections

Crop year

	PRODUCTION (Kt)		Growth (%)[4]		IMPORTS (Kt)		Growth (%)[4]		EXPORTS (Kt)		Growth (%)[4]	
	Avg 2009-11est	2021	2002-2011	2012-2021	Avg 2009-11est	2021	2002-2011	2012-2021	Avg 2009-11est	2021	2002-2011	2012-2021
WORLD	**436 466**	**529 707**	**3.56**	**1.79**	**110 707**	**135 812**	**6.61**	**1.79**	**113 148**	**133 199**	**6.38**	**1.83**
DEVELOPED	**174 412**	**213 898**	**3.78**	**2.02**	**27 684**	**30 477**	**0.37**	**0.80**	**55 505**	**68 532**	**6.36**	**2.67**
NORTH AMERICA	**114 212**	**135 960**	**2.46**	**1.82**	**1 723**	**1 961**	**2.38**	**-0.05**	**49 640**	**56 997**	**5.91**	**2.19**
Canada	17 372	22 168	9.64	1.26	608	730	-5.99	2.28	9 969	12 542	12.41	1.15
United States	96 841	113 792	1.47	1.94	1 115	1 231	9.95	-1.17	39 671	44 456	4.66	2.50
EUROPE	**51 294**	**66 015**	**6.94**	**2.41**	**18 900**	**21 726**	**1.22**	**1.30**	**3 983**	**8 375**	**13.09**	**6.37**
EU(27)	29 223	33 508	4.83	1.58	17 247	19 742	0.50	1.21	810	631	1.21	-1.23
Russian Federation	8 587	12 052	7.47	2.99	1 019	1 323	49.69	3.00	241	974	-2.03	8.03
Ukraine	11 323	17 520	14.38	3.69	21	10	1.64	-7.30	2 690	6 226	26.37	7.30
OCEANIA DEVELOPED	**3 378**	**4 699**	**7.53**	**1.98**	**86**	**69**	**1.12**	**0.00**	**1 607**	**2 564**	**6.37**	**2.47**
Australia	3 368	4 689	7.51	1.98	72	62	1.72	0.00	1 606	2 563	6.37	2.47
New Zealand	10	10	13.88	0.00	14	7	1.01	-0.05	0	0	-0.17	-1.44
OTHER DEVELOPED[1]	**5 527**	**7 224**	**3.76**	**2.28**	**6 975**	**6 720**	**-2.11**	**-0.43**	**275**	**596**	**11.44**	**7.06**
Japan	253	251	0.26	-0.01	5 963	5 767	-2.78	-0.38	0	0	-4.07	0.00
South Africa	1 583	2 145	6.70	2.54	149	148	3.81	-0.51	177	390	24.74	9.62
DEVELOPING	**262 053**	**315 809**	**3.41**	**1.64**	**83 023**	**105 335**	**9.58**	**2.10**	**57 643**	**64 667**	**6.30**	**1.01**
AFRICA	**12 726**	**16 129**	**1.01**	**2.00**	**2 913**	**3 731**	**12.55**	**2.58**	**524**	**408**	**3.01**	**-2.55**
NORTH AFRICA	**762**	**922**	**-2.23**	**0.40**	**2 774**	**3 590**	**13.65**	**2.60**	**37**	**50**	**-3.14**	**-2.96**
Algeria	105	121	0.75	-0.70	206	232	13.55	0.70	0	0	0.00	-0.05
Egypt	488	586	-4.01	0.18	1 729	2 407	20.73	3.20	31	47	9.38	-3.09
SUB-SAHARAN AFRICA	**11 963**	**15 207**	**1.24**	**2.11**	**139**	**141**	**-0.18**	**2.13**	**487**	**358**	**3.45**	**-2.49**
LATIN AMERICA and CARIBBEAN	**145 912**	**178 357**	**4.86**	**1.81**	**8 474**	**9 254**	**1.85**	**0.58**	**55 668**	**63 224**	**6.67**	**1.14**
Argentina	56 971	68 288	4.43	1.20	1 294	1 686	15.13	1.16	14 454	14 566	4.70	-0.65
Brazil	76 253	91 380	4.86	1.89	79	58	-28.50	1.50	34 298	37 225	6.48	0.90
Chile	83	89	1.11	0.33	295	350	4.06	2.98	7	5	-2.79	-2.96
Mexico	300	328	0.08	2.11	5 297	5 739	1.10	0.51	7	5	-5.32	0.33
Uruguay	2 028	4 300	18.87	8.06	10	5	-2.99	-8.28	1 767	3 752	18.83	8.28
ASIA and PACIFIC	**103 415**	**121 324**	**1.90**	**1.35**	**71 636**	**92 351**	**10.77**	**2.24**	**1 451**	**1 034**	**-2.36**	**-4.04**
Bangladesh	320	438	0.89	2.48	164	347	3.07	9.16	0	0	0.00	-0.65
China[2]	56 375	64 373	0.74	1.21	55 078	72 754	13.35	2.41	866	559	-5.14	-5.51
India	33 032	40 679	4.77	1.68	227	265	50.58	-3.91	316	243	7.46	-2.80
Indonesia	2 070	2 554	1.57	2.13	2 076	2 585	5.78	1.76	6	5	-1.79	-0.17
Iran, Islamic Republic of	744	870	4.07	0.67	828	1 334	3.75	4.71	8	12	3.20	-0.84
Korea	135	135	0.63	0.00	1 587	1 955	-0.39	1.45	0	0	-7.78	0.00
Malaysia	6	7	3.26	1.18	662	615	0.24	0.12	21	27	-7.90	-0.02
Pakistan	4 915	5 817	1.67	1.45	1 372	1 422	6.25	0.34	46	50	77.90	-0.05
Saudi Arabia	4	4	0.00	0.09	52	78	5.04	2.21	0	0	0.00	-0.32
Turkey	2 365	2 393	0.69	-1.22	2 520	2 785	9.31	1.96	28	30	19.19	-0.39
LEAST DEVELOPED COUNTRIES (LDC)	**8 595**	**10 729**	**1.14**	**1.92**	**299**	**507**	**1.13**	**6.93**	**234**	**249**	**-0.04**	**-0.29**
OECD[3]	**150 141**	**177 553**	**2.94**	**1.73**	**35 830**	**39 547**	**0.48**	**0.85**	**52 120**	**60 253**	**5.82**	**2.16**
NON-OECD	**286 325**	**352 155**	**3.90**	**1.82**	**74 878**	**96 265**	**10.95**	**2.21**	**61 028**	**72 946**	**6.82**	**1.57**

Note: Crop year: Beginning crop marketing year - see Glossary of Terms for definitions.

1. Includes Israel and also transition economies: Kazakhstan, Kyrgyzstan, Tajikistan, Turkmenistan, Uzbekistan, Armenia, Azerbaijan and Georgia.
2. Refers to mainland only. The economies of Chinese Taipei, Hong Kong (China) and Macau (China) are included in the Other Asia Pacific aggregate.
3. Excludes Iceland but includes all EU27 member countries.
4. Least-squares growth rate (see glossary).

Source: OECD and FAO Secretariats.

Table A.12. Oilseed projections *(cont.)*

Crop year

	CONSUMPTION (Kt)		Growth (%)[1]		DOMESTIC CRUSH (kt)		Growth (%)[1]	
	Avg 2009-11est	2021	2002-2011	2012-2021	Avg 2009-11est	2021	2002-2011	2012-2021
WORLD	**433 620**	**532 451**	**3.79**	**1.76**	**363 651**	**449 699**	**4.10**	**1.79**
DEVELOPED	**145 954**	**175 739**	**2.31**	**1.50**	**129 865**	**158 231**	**2.70**	**1.59**
NORTH AMERICA	**65 425**	**80 855**	**0.44**	**1.39**	**58 071**	**73 135**	**0.89**	**1.49**
Canada	8 110	10 338	5.43	1.46	7 224	9 336	6.54	1.44
United States	57 315	70 517	-0.15	1.38	50 847	63 799	0.23	1.50
EUROPE	**66 632**	**79 340**	**4.86**	**1.76**	**60 470**	**72 463**	**5.25**	**1.84**
EU(27)	45 900	52 587	3.13	1.48	41 957	48 301	3.37	1.55
Russian Federation	9 543	12 415	9.55	2.67	8 917	11 643	9.71	2.72
Ukraine	8 648	11 298	12.15	2.14	7 697	10 177	14.27	2.21
OCEANIA DEVELOPED	**1 651**	**2 204**	**6.80**	**1.37**	**1 447**	**1 948**	**7.33**	**1.57**
Australia	1 628	2 188	6.82	1.38	1 434	1 934	7.33	1.58
New Zealand	23	17	5.33	0.02	13	14	7.74	0.01
OTHER DEVELOPED[2]	**12 247**	**13 340**	**-0.06**	**0.71**	**9 876**	**10 685**	**-0.39**	**0.64**
Japan	6 225	6 019	-2.67	-0.36	5 273	5 123	-3.03	-0.29
South Africa	1 591	1 902	4.30	1.22	1 400	1 663	4.46	1.15
DEVELOPING	**287 666**	**356 712**	**4.61**	**1.89**	**233 786**	**291 468**	**4.94**	**1.91**
AFRICA	**15 111**	**19 446**	**2.54**	**2.23**	**9 330**	**11 624**	**3.30**	**1.87**
NORTH AFRICA	**3 500**	**4 457**	**8.65**	**2.19**	**2 869**	**3 711**	**9.27**	**2.29**
Algeria	306	353	7.16	0.25	185	228	5.51	0.71
Egypt	2 186	2 943	10.65	2.65	1 846	2 534	12.05	2.76
SUB-SAHARAN AFRICA	**11 611**	**14 989**	**1.14**	**2.25**	**6 461**	**7 913**	**1.31**	**1.68**
LATIN AMERICA and CARIBBEAN	**97 764**	**124 230**	**3.97**	**2.05**	**90 435**	**116 161**	**3.83**	**2.14**
Argentina	43 348	55 281	4.39	1.71	42 355	54 235	4.47	1.73
Brazil	41 597	54 214	3.96	2.61	36 907	49 053	3.35	2.84
Chile	372	434	3.55	2.47	330	390	3.58	2.50
Mexico	5 589	6 062	1.01	0.59	5 098	5 571	1.63	0.64
Uruguay	275	543	19.18	6.62	236	499	20.34	7.22
ASIA and PACIFIC	**174 791**	**213 036**	**5.18**	**1.77**	**134 021**	**163 683**	**5.89**	**1.74**
Bangladesh	486	785	1.71	4.95	415	689	1.81	5.24
China[3]	112 263	136 520	6.04	1.87	89 876	110 906	7.07	1.97
India	32 666	41 150	4.86	1.71	25 962	32 138	4.58	1.48
Indonesia	4 139	5 132	3.53	1.96	119	139	2.57	1.36
Iran, Islamic Republic of	1 555	2 191	3.90	2.97	1 357	1 933	4.03	3.09
Korea	1 721	2 090	0.03	1.35	1 247	1 415	-0.25	0.82
Malaysia	637	595	0.56	0.14	546	483	0.67	-0.34
Pakistan	6 229	7 187	2.52	1.25	4 728	5 068	2.97	0.59
Saudi Arabia	55	82	4.43	2.13	38	56	4.98	1.64
Turkey	4 754	5 148	4.18	0.44	3 676	3 796	3.64	-0.20
LEAST DEVELOPED COUNTRIES (LDC)	**8 655**	**10 986**	**1.17**	**2.16**	**5 873**	**7 433**	**1.33**	**2.09**
OECD[4]	**132 906**	**156 746**	**1.36**	**1.28**	**117 831**	**140 468**	**1.69**	**1.35**
NON-OECD	**300 714**	**375 704**	**5.01**	**1.97**	**245 820**	**309 231**	**5.43**	**2.00**

Note: Crop year: Beginning crop marketing year - see Glossary of Terms for definitions.

1. Least-squares growth rate (see glossary).
2. Includes Israel and also transition economies: Kazakhstan, Kyrgyzstan, Tajikistan, Turkmenistan, Uzbekistan, Armenia, Azerbaijan and Georgia.
3. Refers to mainland only. The economies of Chinese Taipei, Hong Kong (China) and Macau (China) are included in the Other Asia Pacific aggregate.
4. Excludes Iceland but includes all EU27 member countries.

Source: OECD and FAO Secretariats.

StatLink http://dx.doi.org/10.1787/888932642820

Table A.13. Protein meal projections

Marketing year

	PRODUCTION (Kt)		Growth (%)[4]		IMPORTS (Kt)		Growth (%)[4]		EXPORTS (Kt)		Growth (%)[4]	
	Avg 2009-11est	2021	2002-2011	2012-2021	Avg 2009-11est	2021	2002-2011	2012-2021	Avg 2009-11est	2021	2002-2011	2012-2021
WORLD	**255 808**	**314 887**	**4.16**	**1.79**	**71 895**	**87 935**	**3.98**	**1.63**	**73 930**	**90 956**	**3.45**	**1.59**
DEVELOPED	**84 980**	**102 343**	**2.12**	**1.49**	**38 840**	**41 698**	**2.03**	**0.67**	**16 585**	**20 559**	**7.94**	**1.12**
NORTH AMERICA	**43 079**	**53 499**	**0.72**	**1.50**	**3 115**	**3 678**	**3.16**	**1.84**	**11 710**	**15 418**	**6.81**	**1.30**
Canada	4 369	5 702	4.98	1.49	1 186	1 397	1.07	0.57	2 794	4 078	10.42	1.57
United States	38 710	47 797	0.30	1.50	1 929	2 282	4.64	2.70	8 916	11 339	5.81	1.20
EUROPE	**35 202**	**41 455**	**4.61**	**1.61**	**30 029**	**31 562**	**0.72**	**0.52**	**4 604**	**4 836**	**11.11**	**0.68**
EU(27)	26 269	29 807	2.61	1.33	27 887	28 836	0.31	0.37	753	681	2.82	1.38
Russian Federation	4 464	5 816	12.17	2.60	461	727	1.90	5.26	865	490	12.32	-5.41
Ukraine	3 636	4 814	16.48	2.21	68	51	-3.58	-1.90	2 724	3 417	14.90	1.90
OCEANIA DEVELOPED	**822**	**1 094**	**7.30**	**1.57**	**1 896**	**2 606**	**17.55**	**1.63**	**22**	**22**	**5.93**	**0.00**
Australia	815	1 086	7.29	1.58	627	851	5.39	3.51	22	22	5.94	0.00
New Zealand	8	8	8.29	0.00	1 269	1 756	33.60	0.81	0	0	..	..
OTHER DEVELOPED[1]	**5 877**	**6 295**	**-1.04**	**0.54**	**3 800**	**3 852**	**8.61**	**0.20**	**248**	**283**	**11.60**	**-0.85**
Japan	3 627	3 524	-3.38	-0.29	2 196	2 058	9.21	-0.80	0	0	..	..
South Africa	757	920	5.23	1.14	1 144	1 397	7.52	2.40	84	76	38.33	-2.40
DEVELOPING	**170 828**	**212 544**	**5.29**	**1.94**	**33 054**	**46 236**	**6.70**	**2.58**	**57 345**	**70 397**	**2.40**	**1.73**
AFRICA	**5 589**	**6 997**	**4.18**	**1.96**	**2 727**	**4 367**	**3.94**	**4.08**	**418**	**254**	**-0.15**	**-6.76**
NORTH AFRICA	**2 107**	**2 710**	**10.86**	**2.31**	**2 410**	**3 794**	**3.67**	**4.01**	**5**	**6**	**4.17**	**-0.49**
Algeria	119	148	6.87	0.71	836	1 084	8.47	2.26	0	0	0.00	2.06
Egypt	1 370	1 867	14.28	2.76	566	1 012	-6.83	5.13	2	2	0.00	-0.42
SUB-SAHARAN AFRICA	**3 482**	**4 287**	**1.32**	**1.75**	**316**	**572**	**6.08**	**4.61**	**412**	**248**	**-0.20**	**-6.87**
LATIN AMERICA and CARIBBEAN	**68 196**	**86 828**	**3.90**	**2.14**	**6 504**	**8 279**	**4.36**	**2.48**	**43 665**	**55 274**	**1.52**	**1.95**
Argentina	32 063	40 496	4.64	1.69	0	0	0.00	0.00	28 679	35 978	3.64	1.53
Brazil	27 694	36 584	3.34	2.89	30	7	-29.77	-7.72	12 755	16 623	-2.30	3.15
Chile	239	277	3.86	2.50	914	1 032	5.23	0.82	10	9	92.67	-0.67
Mexico	3 669	3 987	1.74	0.66	637	1 326	16.97	6.87	15	15	48.47	0.07
Uruguay	128	269	19.94	7.22	256	280	16.50	1.14	6	4	3.11	-0.94
ASIA and PACIFIC	**97 043**	**118 719**	**6.44**	**1.79**	**23 824**	**33 590**	**7.79**	**2.43**	**13 263**	**14 869**	**5.92**	**1.15**
Bangladesh	257	422	2.20	5.13	317	478	11.34	3.54	0	0	0.00	-0.25
China[2]	61 931	75 995	7.87	1.96	2 205	6 339	31.61	3.27	883	327	-2.60	2.48
India	16 419	20 343	5.03	1.48	81	84	3.94	-0.30	4 943	6 253	8.01	1.85
Indonesia	3 633	4 758	8.46	2.13	3 086	4 047	7.33	2.16	3 142	3 991	8.17	1.52
Iran, Islamic Republic of	969	1 371	4.23	3.09	2 086	2 983	20.11	3.01	290	229	27.90	-3.01
Korea	942	1 070	-0.27	0.87	3 209	3 432	3.25	0.49	0	0	..	..
Malaysia	3 040	3 765	2.63	1.90	1 111	1 082	7.46	-0.22	2 359	2 520	3.34	0.22
Pakistan	2 771	2 975	2.93	0.59	531	695	19.56	3.07	164	119	21.24	-2.96
Saudi Arabia	20	29	5.21	1.57	478	571	-3.18	1.87	4	0	17.68	0.00
Turkey	2 009	2 056	4.07	-0.20	989	1 616	4.37	5.39	43	45	-2.44	-2.08
LEAST DEVELOPED COUNTRIES (LDC)	**2 853**	**3 626**	**1.12**	**2.11**	**484**	**912**	**7.74**	**6.11**	**219**	**134**	**-0.11**	**-5.89**
OECD[3]	**80 905**	**95 597**	**1.26**	**1.30**	**41 677**	**45 403**	**2.05**	**0.80**	**12 614**	**16 252**	**6.50**	**1.28**
NON-OECD	**174 903**	**219 290**	**5.73**	**2.01**	**30 218**	**42 532**	**7.19**	**2.60**	**61 316**	**74 704**	**2.91**	**1.66**

1. Includes Israel and also transition economies: Kazakhstan, Kyrgyzstan, Tajikistan, Turkmenistan, Uzbekistan, Armenia, Azerbaijan and Georgia.
2. Refers to mainland only. The economies of Chinese Taipei, Hong Kong (China) and Macau (China) are included in the Other Asia Pacific aggregate.
3. Excludes Iceland but includes all EU27 member countries.
4. Least-squares growth rate (see glossary).

Source: OECD and FAO Secretariats.

Table A.13. Protein meal projections *(cont.)*

Marketing year

	CONSUMPTION (Kt)		Growth (%)[1]	
	Avg 2009-11est	2021	2002-2011	2012-2021
WORLD	**252 395**	**311 571**	**4.30**	**1.80**
DEVELOPED	**107 139**	**123 474**	**1.38**	**1.27**
NORTH AMERICA	**34 409**	**41 765**	**-0.64**	**1.61**
Canada	2 716	3 025	-0.41	0.98
United States	31 693	38 740	-0.66	1.66
EUROPE	**60 632**	**68 170**	**2.22**	**1.16**
EU(27)	53 402	57 958	1.36	0.84
Russian Federation	4 047	6 052	10.77	3.84
Ukraine	999	1 447	18.62	2.82
OCEANIA DEVELOPED	**2 700**	**3 678**	**13.73**	**1.62**
Australia	1 420	1 915	6.40	2.42
New Zealand	1 280	1 763	33.20	0.81
OTHER DEVELOPED[2]	**9 398**	**9 861**	**1.62**	**0.43**
Japan	5 805	5 582	-0.10	-0.48
South Africa	1 803	2 238	5.79	2.00
DEVELOPING	**145 256**	**188 096**	**6.95**	**2.16**
AFRICA	**7 870**	**11 103**	**4.33**	**3.09**
NORTH AFRICA	**4 478**	**6 494**	**6.47**	**3.27**
Algeria	939	1 232	8.10	2.09
Egypt	1 920	2 872	4.48	3.51
SUB-SAHARAN AFRICA	**3 392**	**4 609**	**1.96**	**2.85**
LATIN AMERICA and CARIBBEAN	**30 512**	**39 768**	**8.28**	**2.47**
Argentina	3 035	4 500	19.07	3.21
Brazil	14 757	19 927	10.84	2.63
Chile	1 143	1 299	4.90	1.12
Mexico	4 291	5 297	3.19	1.92
Uruguay	380	545	17.15	3.74
ASIA and PACIFIC	**106 873**	**137 225**	**6.80**	**2.00**
Bangladesh	573	900	6.46	4.25
China[3]	62 657	81 872	8.44	2.05
India	11 670	14 149	4.23	1.32
Indonesia	3 471	4 796	7.54	2.65
Iran, Islamic Republic of	2 748	4 117	12.01	3.49
Korea	4 152	4 502	2.22	0.58
Malaysia	1 816	2 324	4.53	2.78
Pakistan	3 131	3 550	4.19	1.18
Saudi Arabia	494	600	-3.10	1.85
Turkey	2 918	3 622	4.25	2.00
LEAST DEVELOPED COUNTRIES (LDC)	**3 118**	**4 403**	**2.06**	**3.21**
OECD[4]	**109 843**	**124 744**	**1.05**	**1.12**
NON-OECD	**142 552**	**186 826**	**7.48**	**2.28**

1. Least-squares growth rate (see glossary).
2. Includes Israel and also transition economies: Kazakhstan, Kyrgyzstan, Tajikistan, Turkmenistan, Uzbekistan, Armenia, Azerbaijan and Georgia.
3. Refers to mainland only. The economies of Chinese Taipei, Hong Kong (China) and Macau (China) are included in the Other Asia Pacific aggregate.
4. Excludes Iceland but includes all EU27 member countries.

Source: OECD and FAO Secretariats.

StatLink http://dx.doi.org/10.1787/888932642858

Table A.14. Vegetable oil projections

Marketing year

	PRODUCTION (Kt)		Growth (%)[4]		IMPORTS (Kt)		Growth (%)[4]		EXPORTS (Kt)		Growth (%)[4]	
	Avg 2009-11est	2021	2002-2011	2012-2021	Avg 2009-11est	2021	2002-2011	2012-2021	Avg 2009-11est	2021	2002-2011	2012-2021
WOLRD	**144 313**	**184 530**	**5.07**	**2.06**	**62 052**	**79 089**	**5.99**	**1.87**	**59 918**	**77 145**	**5.30**	**1.93**
DEVELOPED	**36 962**	**45 481**	**4.41**	**1.82**	**17 612**	**21 910**	**6.42**	**1.25**	**8 861**	**11 015**	**10.52**	**1.85**
NORTH AMERICA	**12 831**	**16 093**	**2.02**	**1.52**	**3 849**	**4 169**	**11.21**	**0.50**	**3 902**	**4 908**	**10.60**	**1.69**
Canada	2 796	3 747	9.01	1.60	444	632	8.12	0.13	2 315	3 131	14.80	1.30
United States	10 035	12 346	0.56	1.50	3 405	3 538	11.51	0.56	1 587	1 777	5.91	2.42
EUROPE	**21 201**	**26 085**	**6.63**	**2.14**	**11 526**	**15 324**	**5.70**	**1.50**	**4 739**	**5 925**	**10.40**	**2.01**
EU(27)	14 374	17 113	5.09	1.96	9 448	12 940	5.62	1.65	893	745	-2.11	0.48
Russian Federation	3 118	4 114	8.27	2.86	993	1 221	2.80	1.02	886	1 821	23.81	5.38
Ukraine	3 278	4 325	13.79	2.21	426	406	13.74	-0.92	2 779	3 193	15.49	0.92
OCEANIA DEVELOPED	**451**	**633**	**8.21**	**2.29**	**392**	**420**	**3.70**	**1.02**	**80**	**39**	**3.75**	**4.27**
Australia	447	629	8.21	2.31	288	312	5.52	1.36	80	38	3.76	4.32
New Zealand	5	5	8.99	0.00	104	108	0.01	0.09	0	0	-0.27	-0.23
OTHER DEVELOPED[1]	**2 479**	**2 669**	**0.36**	**0.57**	**1 845**	**1 996**	**3.50**	**1.06**	**140**	**143**	**11.65**	**0.52**
Japan	1 455	1 422	-1.58	-0.29	757	746	3.08	-0.28	2	36	30.19	11.13
South Africa	421	489	3.74	1.13	755	807	4.08	1.55	113	91	17.05	-1.55
DEVELOPING	**107 351**	**139 049**	**5.31**	**2.14**	**44 439**	**57 179**	**5.82**	**2.12**	**51 057**	**66 130**	**4.58**	**1.94**
AFRICA	**5 055**	**6 355**	**2.59**	**2.06**	**6 859**	**9 740**	**6.25**	**3.37**	**719**	**551**	**6.37**	**-1.80**
NORTH AFRICA	**625**	**800**	**7.93**	**2.16**	**2 792**	**3 819**	**3.25**	**2.49**	**149**	**134**	**21.12**	**-0.59**
Algeria	55	67	3.33	0.71	563	779	1.78	2.70	20	12	9.75	-2.70
Egypt	346	474	11.41	2.76	1 542	2 081	4.59	2.16	57	42	25.54	-2.16
SUB-SAHARAN AFRICA	**4 430**	**5 555**	**1.97**	**2.05**	**4 067**	**5 921**	**8.76**	**3.98**	**571**	**418**	**4.84**	**-2.15**
LATIN AMERICA and CARIBBEAN	**20 242**	**27 080**	**3.88**	**2.19**	**4 005**	**5 012**	**4.53**	**1.94**	**7 853**	**11 597**	**-2.19**	**2.19**
Argentina	8 488	11 523	3.95	1.81	13	13	4.28	-0.10	4 728	7 086	-2.17	1.80
Brazil	6 617	9 028	3.35	2.80	399	515	13.96	2.31	1 664	3 008	-6.53	4.55
Chile	76	94	2.91	2.50	271	485	2.15	4.70	2	1	-7.24	-2.44
Mexico	1 489	1 675	2.25	0.84	1 029	1 376	5.32	2.39	4	0	..	..
Uruguay	67	141	20.87	7.21	82	61	11.93	-2.54	2	2	-3.88	1.32
ASIA and PACIFIC	**82 054**	**105 614**	**5.87**	**2.13**	**33 575**	**42 428**	**5.90**	**1.87**	**42 485**	**53 982**	**6.26**	**1.93**
Bangladesh	175	282	2.89	4.92	1 393	1 590	4.75	1.26	0	0	0.00	-0.09
China[2]	20 001	24 704	6.05	2.11	9 229	10 447	5.54	0.25	200	55	4.72	-3.48
India	6 436	7 856	3.33	1.48	8 873	11 791	7.92	2.87	63	48	-4.95	-0.49
Indonesia	27 206	36 888	9.16	2.42	73	36	1.88	-2.38	19 935	27 225	9.57	2.38
Iran, Islamic Republic of	288	407	3.81	3.09	1 333	1 688	0.62	1.67	204	236	3.25	-1.67
Korea	240	270	-0.11	0.82	744	782	5.43	0.57	8	10	4.14	0.00
Malaysia	20 045	26 014	3.54	2.17	2 001	2 496	9.04	1.07	18 645	23 240	4.19	1.90
Pakistan	932	991	4.44	0.59	2 198	3 374	3.85	3.81	80	48	46.84	-3.80
Saudi Arabia	7	10	4.29	1.64	306	388	-1.99	2.26	11	1	-9.18	-0.32
Turkey	1 092	1 129	4.87	-0.20	942	1 166	3.91	1.45	184	214	20.64	-1.45
LEAST DEVELOPED COUNTRIES (LDC)	**2 619**	**3 205**	**1.69**	**1.75**	**4 711**	**6 395**	**5.68**	**3.37**	**242**	**255**	**4.88**	**0.48**
OECD[3]	**32 088**	**38 520**	**3.27**	**1.57**	**17 791**	**22 475**	**6.37**	**1.38**	**5 133**	**6 008**	**7.39**	**1.43**
NON-OECD	**112 225**	**146 010**	**5.63**	**2.19**	**44 261**	**56 614**	**5.85**	**2.07**	**54 785**	**71 136**	**5.13**	**1.97**

1. Includes Israel and also transition economies: Kazakhstan, Kyrgyzstan, Tajikistan, Turkmenistan, Uzbekistan, Armenia, Azerbaijan and Georgia.
2. Refers to mainland only. The economies of Chinese Taipei, Hong Kong (China) and Macau (China) are included in the Other Asia Pacific aggregate.
3. Excludes Iceland but includes all EU27 member countries.
4. Least-squares growth rate (see glossary).

Source: OECD and FAO Secretariats.

Table A.14. Vegetable oil projections *(cont.)*

Marketing year

	CONSUMPTION (Kt)		Growth (%)[1]		FOOD VEGETABLE USE PER CAPITA (Kt)		Growth (%)[1]	
	Avg 2009-11est	2021	2002-2011	2012-2021	Avg 2009-11est	2021	2002-2011	2012-2021
WORLD	**144 396**	**186 384**	**5.23**	**2.03**	**17.4**	**19.1**	**2.58**	**0.69**
DEVELOPED	**45 750**	**56 399**	**4.21**	**1.60**	**24.2**	**25.6**	**0.90**	**0.50**
NORTH AMERICA	**12 907**	**15 432**	**2.24**	**1.19**	**33.7**	**35.1**	**-0.02**	**0.25**
Canada	945	1 249	1.34	1.59	27.0	28.4	-0.06	-0.50
United States	11 961	14 183	2.30	1.15	34.4	35.8	-0.03	0.32
EUROPE	**27 904**	**35 437**	**5.68**	**1.90**	**22.7**	**24.1**	**1.44**	**0.60**
EU(27)	22 984	29 272	5.73	1.89	23.8	24.2	-0.10	0.34
Russian Federation	3 080	3 511	3.64	1.02	21.5	25.0	3.83	1.18
Ukraine	943	1 532	11.45	4.37	20.6	21.9	12.13	0.26
OCEANIA DEVELOPED	**767**	**1 011**	**6.04**	**1.69**	**25.3**	**29.5**	**2.73**	**0.52**
Australia	659	899	7.31	1.90	25.4	30.8	3.63	0.73
New Zealand	108	112	0.28	0.09	24.8	23.0	-0.90	-0.88
OTHER DEVELOPED[2]	**4 172**	**4 519**	**1.43**	**0.78**	**15.7**	**16.1**	**0.67**	**0.36**
Japan	2 209	2 132	-0.05	-0.41	17.5	17.1	-0.09	-0.24
South Africa	1 052	1 202	3.04	1.60	19.6	20.6	1.10	0.94
DEVELOPING	**98 647**	**129 986**	**5.73**	**2.22**	**15.8**	**17.6**	**3.43**	**0.83**
AFRICA	**11 196**	**15 533**	**4.57**	**3.06**	**11.3**	**12.2**	**2.11**	**0.77**
NORTH AFRICA	**3 277**	**4 481**	**3.99**	**2.56**	**19.6**	**23.3**	**2.45**	**1.30**
Algeria	606	833	2.90	2.65	16.9	20.4	1.40	1.49
Egypt	1 834	2 512	5.64	2.38	22.4	25.9	3.84	0.90
SUB-SAHARAN AFRICA	**7 919**	**11 052**	**4.82**	**3.27**	**9.6**	**10.2**	**2.15**	**0.79**
LATIN AMERICA and CARIBBEAN	**15 320**	**20 621**	**7.67**	**2.31**	**18.9**	**20.3**	**2.23**	**0.94**
Argentina	2 813	4 577	18.13	2.64	23.3	23.8	1.31	0.20
Brazil	5 192	6 526	9.19	2.01	18.7	19.7	3.24	1.40
Chile	345	578	2.22	4.33	19.8	25.5	1.08	2.01
Mexico	2 536	3 051	3.52	1.51	22.4	24.0	2.27	0.52
Uruguay	146	200	15.83	3.35	15.6	17.7	1.89	1.02
ASIA and PACIFIC	**72 130**	**93 832**	**5.53**	**2.07**	**16.4**	**18.7**	**3.94**	**0.93**
Bangladesh	1 545	1 869	4.60	1.74	10.3	11.0	3.37	0.61
China[3]	28 304	35 042	5.58	1.37	20.9	25.0	5.06	1.06
India	15 295	19 562	5.74	2.34	12.2	13.6	4.07	1.16
Indonesia	6 976	9 704	7.67	2.52	18.6	20.0	2.71	0.50
Iran, Islamic Republic of	1 399	1 855	0.84	2.53	18.7	22.6	-0.35	1.70
Korea	974	1 042	3.72	0.64	20.2	20.9	3.25	0.34
Malaysia	3 542	5 166	3.35	2.84	23.0	27.0	3.21	1.39
Pakistan	3 036	4 318	4.27	3.10	17.3	20.5	2.48	1.48
Saudi Arabia	305	397	-1.22	2.27	11.0	11.5	-4.17	0.36
Turkey	1 800	2 076	3.11	0.92	23.6	24.5	1.88	-0.20
LEAST DEVELOPED COUNTRIES (LDC)	**7 052**	**9 339**	**4.08**	**2.87**	**8.3**	**8.7**	**1.71**	**0.76**
OECD[4]	**44 904**	**55 018**	**4.04**	**1.52**	**25.5**	**26.6**	**0.45**	**0.33**
NON-OECD	**99 493**	**131 366**	**5.81**	**2.25**	**15.6**	**17.5**	**3.57**	**0.87**

1. Least-squares growth rate (see glossary).
2. Includes Israel and also transition economies: Kazakhstan, Kyrgyzstan, Tajikistan, Turkmenistan, Uzbekistan, Armenia, Azerbaijan and Georgia.
3. Refers to mainland only. The economies of Chinese Taipei, Hong Kong (China) and Macau (China) are included in the Other Asia Pacific aggregate.
4. Excludes Iceland but includes all EU27 member countries.

Source: OECD and FAO Secretariats.

StatLink http://dx.doi.org/10.1787/888932642896

Table A.15. Main policy assumptions for oilseed markets

Crop year

		Avg 09/10-11/12est.	12/13	13/14	14/15	15/16	16/17	17/18	18/19	19/20	20/21	21/22
ARGENTINA												
Oilseed export tax	%	33.5	33.5	33.5	33.5	33.5	33.5	33.5	33.5	33.5	33.5	33.5
Protein meal export tax	%	31.0	31.0	31.0	31.0	31.0	31.0	31.0	31.0	31.0	31.0	31.0
Oilseed oil export tax	%	31.0	31.0	31.0	31.0	31.0	31.0	31.0	31.0	31.0	31.0	31.0
AUSTRALIA												
Tariffs												
Soybean oil	%	8.0	8.0	8.0	8.0	8.0	8.0	8.0	8.0	8.0	8.0	8.0
Rapeseed oil	%	8.0	8.0	8.0	8.0	8.0	8.0	8.0	8.0	8.0	8.0	8.0
CANADA												
Tariffs												
Rapeseed oil	%	6.4	6.4	6.4	6.4	6.4	6.4	6.4	6.4	6.4	6.4	6.4
EUROPEAN UNION												
Single farm payment[1]	EUR/ha	187.9	187.9	187.9	187.9	187.9	187.9	187.9	187.9	187.9	187.9	188.9
Compulsory set-aside rate	%	0.0	0.0	0.0	0.0	0.0	0.0	0.0	0.0	0.0	0.0	0.0
Tariffs												
Soybean oil	%	6.00	6.00	6.00	6.00	6.00	6.00	6.00	6.00	6.00	6.00	6.00
Rapeseed oil	%	6.00	6.00	6.00	6.00	6.00	6.00	6.00	6.00	6.00	6.00	6.00
JAPAN												
New output payments												
Soybeans	JPY/kg	..	188.5	188.5	188.5	188.5	188.5	188.5	188.5	188.5	188.5	188.5
Tariffs												
Soybean oil	JPY/kg	10.9	10.9	10.9	10.9	10.9	10.9	10.9	10.9	10.9	10.9	10.9
Rapeseed oil	JPY/kg	10.9	10.9	10.9	10.9	10.9	10.9	10.9	10.9	10.9	10.9	10.9
KOREA												
Soybean tariff-quota	kt	1 032	1 032	1 032	1 032	1 032	1 032	1 032	1 032	1 032	1 032	1 032
In-quota tariff	%	5	5	5	5	5	5	5	5	5	5	5
Out-of-quota tariff	%	487	487	487	487	487	487	487	487	487	487	487
Soybean (for food) mark up	'000 KRW/t	161	150	146	141	138	134	130	127	123	119	115
MEXICO												
Tariffs												
Soybeans	%	33	33	33	33	33	33	33	33	33	33	33
Soybeans meal	%	23.8	23.8	23.8	23.8	23.8	23.8	23.8	23.8	23.8	23.8	23.8
Soybeans oil	%	45	45	45	45	45	45	45	45	45	45	45
UNITED STATES												
ACRE participation rate												
Soybeans	%	10.0	10.0	10.0	10.0	10.0	10.0	10.0	10.0	10.0	10.0	10.0
Soybeans loan rate	USD/t	183.7	183.7	183.7	183.7	183.7	183.7	183.7	183.7	183.7	183.7	183.7
CRP area												
Soybeans	mha	1.9	1.8	1.7	1.8	1.9	1.9	1.9	1.9	1.9	1.8	1.8
Tariffs												
Rapeseed	%	3	3	3	3	3	3	3	3	3	3	3
Soybean meal	%	2.2	2.2	2.2	2.2	2.2	2.2	2.2	2.2	2.2	2.2	2.2
Rapeseed meal	%	1.2	1.2	1.2	1.2	1.2	1.2	1.2	1.2	1.2	1.2	1.2
Soybean oil	%	12.7	12.7	12.7	12.7	12.7	12.7	12.7	12.7	12.7	12.7	12.7
Rapeseed oil	%	3.2	3.2	3.2	3.2	3.2	3.2	3.2	3.2	3.2	3.2	3.2
Subsidised export limits												
Oilseed oils	kt	141	141	141	141	141	141	141	141	141	141	142
CHINA												
Soybeans support price	CNY/t	1 643.7	1 815.3	1 885.7	1 951.7	2 014.0	2 079.0	2 149.3	2 219.8	2 291.2	2 364.8	2 440.9
Tariffs												
Soybeans	%	2.4	2.4	2.4	2.4	2.4	2.4	2.4	2.4	2.4	2.4	2.4
Soybean meal	%	6.3	6.3	6.3	6.3	6.3	6.3	6.3	6.3	6.3	6.3	6.3
Soybean oil in-quota tariff	%	9.0	9.0	9.0	9.0	9.0	9.0	9.0	9.0	9.0	9.0	9.0
Vegetable oil tariff-quota	kt	7 998.1	7 998.1	7 998.1	7 998.1	7 998.1	7 998.1	7 998.1	7 998.1	7 999.1	7 999.1	8 000.1
INDIA												
Input subsidy rate, oilseeds[2]	INR/t	4 888.3	4 888.3	4 888.3	4 888.3	4 888.3	4 888.3	4 888.3	4 888.3	4 888.3	4 888.3	4 888.3
Soybean tariff	%	30.0	30.0	30.0	30.0	30.0	30.0	30.0	30.0	30.0	30.0	30.0
Rapeseed tariff	%	30.0	30.0	30.0	30.0	30.0	30.0	30.0	30.0	30.0	30.0	30.0
Sunflower tariff	%	30.0	30.0	30.0	30.0	30.0	30.0	30.0	30.0	30.0	30.0	30.0
Oilseed tariff	%	30.0	30.0	30.0	30.0	30.0	30.0	30.0	30.0	30.0	30.0	30.0
Soybean meal tariff	%	100.0	100.0	100.0	100.0	100.0	100.0	100.0	100.0	100.0	100.0	100.0
Rapeseed meal tariff	%	100.0	100.0	100.0	100.0	100.0	100.0	100.0	100.0	100.0	100.0	100.0
Sunflower meal tariff	%	100.0	100.0	100.0	100.0	100.0	100.0	100.0	100.0	100.0	100.0	100.0
Soybean oil tariff	%	0.0	0.0	0.0	0.0	0.0	0.0	0.0	0.0	0.0	0.0	0.0
Rapeseed oil tariff	%	7.5	7.5	7.5	7.5	7.5	7.5	7.5	7.5	7.5	7.5	7.5
Sunflower oil tariff	%	0.0	0.0	0.0	0.0	0.0	0.0	0.0	0.0	0.0	0.0	0.0
Palm oil tariff	%	0.0	0.0	0.0	0.0	0.0	0.0	0.0	0.0	0.0	0.0	0.0

Note: Beginning crop marketing year - see Glossary of Terms for definitions.

The source for tariffs and Tariff Rate Quotas is AMAD (Agricultural market access database). The tariff and TRQ data are based on Most Favoured Nation rates scheduled with the WTO and exclude those under preferential or regional agreements, which may be substantially different. Tariffs are simple averages of several product lines. Specific rates are converted to ad valorem rates using world prices in the Outlook. Import quotas are based on global commitments scheduled in the WTO rather than those allocated to preferential partners under regional or other agreements. For Mexico, the NAFTA tariffs on soybeans, oil meals and soybean oil are zero after 2003.

1. EU farmers benefit from the Single Farm Payment (SFP) Scheme, which provides flat-rate payments independent from current production decisions and market developments. For the new member states, payments are phased in with the assumption of maximum top-ups from national budgets up to 2013 through the Single Area Payment (SAP), and through the (SFP) from 2014. Due to modulation, an increasing share of the total SFP will go to rural development spending rather than directly to farmers.
2. Indian input subsidies consist of those for electricty, fertiliser and irrigation.

Source: OECD and FAO Secretariats.

StatLink http://dx.doi.org/10.1787/888932642915

Table A.16. World sugar projections

Crop year

		Avg 09/10-11/12est	12/13	13/14	14/15	15/16	16/17	17/18	18/19	19/20	20/21	21/22
OECD[1]												
SUGAR BEET												
Production	mt	163.5	160.0	163.7	164.6	166.4	168.7	168.6	167.7	169.0	168.6	169.4
Biofuel use	mt	18.6	23.1	27.0	28.1	28.0	27.6	28.1	28.2	27.5	26.4	26.0
SUGAR CANE												
Production	mt	104.0	107.3	108.8	109.7	110.1	108.7	109.9	109.9	110.9	110.8	111.1
SUGAR												
Production	mt rse	37.4	36.6	37.0	36.9	37.5	37.8	38.0	37.7	38.4	38.5	39.0
Consumption	mt rse	43.2	43.6	44.0	44.4	44.7	45.1	45.5	45.7	46.1	46.4	46.7
Closing stocks	mt rse	9.6	11.7	12.2	12.2	12.2	12.6	12.6	12.4	12.6	12.6	12.6
HFCS												
Production	mt	12.8	13.0	13.1	13.1	13.2	13.3	13.4	13.5	13.6	13.8	13.9
Consumption	mt	12.4	12.5	12.6	12.7	12.7	12.8	12.9	13.0	13.2	13.3	13.5
NON-OECD												
SUGAR BEET												
Production	mt	76.8	89.3	90.7	91.8	93.7	95.9	96.4	97.0	98.1	98.5	99.5
SUGAR CANE												
Production	mt	1 517.6	1 561.2	1 610.3	1 627.0	1 648.8	1 738.8	1 854.4	1 858.0	1 941.8	1 981.5	2 052.4
Biofuel use	mt	331.8	344.6	383.7	427.5	474.0	521.5	581.5	608.3	642.4	683.5	733.3
SUGAR												
Production	mt rse	127.7	139.8	142.3	143.8	145.1	152.5	155.0	156.7	162.3	165.4	168.9
Consumption	mt rse	117.4	123.6	127.1	130.3	132.4	136.3	139.8	143.0	147.4	151.3	155.5
HFCS												
Production	mt	1.7	1.9	2.0	2.0	2.0	2.1	2.1	2.2	2.2	2.3	2.3
Consumption	mt	1.6	1.9	2.0	2.0	2.1	2.1	2.2	2.2	2.3	2.3	2.4
WORLD												
SUGAR BEET												
Production	mt	240.3	249.3	254.4	256.4	260.0	264.6	264.9	264.7	267.1	267.1	268.9
Area	mha	4.6	4.8	4.9	4.9	5.0	5.0	5.0	5.0	5.0	5.0	5.0
Yield	t/ha	52.0	51.7	52.2	52.5	52.1	52.6	52.7	52.9	53.2	53.3	53.6
Biofuel use	mt	18.6	23.1	27.0	28.1	28.0	27.6	28.1	28.2	27.5	26.4	26.0
SUGAR CANE												
Production	mt	1 621.7	1 668.5	1 719.1	1 736.7	1 759.0	1 847.5	1 964.3	1 968.0	2 052.7	2 092.2	2 163.4
Area	mha	23.5	24.4	24.4	24.4	24.4	25.6	27.4	27.1	27.9	28.5	29.2
Yield	t/ha	68.9	68.4	70.3	71.2	72.0	72.3	71.8	72.6	73.6	73.5	74.1
Biofuel use	mt	331.8	344.6	383.7	427.5	474.0	521.5	581.5	608.3	642.4	683.5	733.3
SUGAR												
Production	mt rse	165.1	176.4	179.3	180.7	182.5	190.3	193.0	194.3	200.7	204.0	207.9
Consumption	mt rse	160.6	167.3	171.1	174.7	177.1	181.4	185.3	188.6	193.5	197.7	202.2
Price, raw sugar[2]	USD/t	532.9	460.8	464.0	474.2	522.9	455.1	465.4	478.6	477.7	474.9	483.1
Price, white sugar[3]	USD/t	632.5	537.5	545.6	557.2	607.8	537.3	546.2	561.6	560.3	559.0	565.7
Price, HFCS[4]	USD/t	544.5	429.8	454.6	511.4	550.2	512.4	494.4	516.1	521.3	529.1	536.1

Note: Crop year: Beginning crop marketing year (Oct/Sept)- see the Glossary of Terms for definitions.
rse : raw sugar equivalent.
HFCS: High fructose corn syrup

1. Excludes Iceland but includes all EU27 member countries.
2. Raw sugar world price, ICE contract No11 nearby, October/September.
3. Refined sugar price, White Sugar Futures Contract No. 407, Euronext market, Liffe, London, Europe, October/September.
4. United States wholesale list price HFCS-55 , October/September.

Source: OECD and FAO Secretariats.

StatLink http://dx.doi.org/10.1787/888932642934

Table A.17. Sugar projections (in raw sugar equivalent)

Crop year

	PRODUCTION (Kt)		Growth (%)[4]		IMPORTS (Kt)		Growth (%)[4]		EXPORTS (Kt)		Growth (%)[4]	
	Avg 2009-11est	2021	2002-2011	2012-2021	Avg 2009-11est	2021	2002-2011	2012-2021	Avg 2009-11est	2021	2002-2011	2012-2021
WORLD	**165 086**	**207 915**	**1.71**	**1.88**	**49 129**	**57 978**	**2.09**	**2.43**	**54 571**	**63 653**	**2.01**	**2.19**
DEVELOPED	**38 154**	**41 891**	**-1.83**	**0.74**	**15 877**	**15 828**	**-0.68**	**-0.10**	**7 721**	**7 286**	**-8.47**	**1.45**
NORTH AMERICA	**7 256**	**7 949**	**-0.79**	**0.66**	**4 511**	**5 287**	**6.02**	**1.21**	**293**	**270**	**9.16**	**-0.17**
Canada	95	91	2.08	0.33	1 256	1 579	-0.34	0.76	72	90	17.25	-0.50
United States	7 161	7 858	-0.82	0.67	3 254	3 708	9.73	1.41	221	180	7.25	0.00
EUROPE	**24 088**	**25 373**	**-1.69**	**0.59**	**7 074**	**5 773**	**-3.93**	**-1.95**	**3 566**	**2 703**	**-11.38**	**2.21**
EU(27)	17 073	16 128	-4.01	0.43	3 894	3 944	5.69	-1.60	2 776	1 633	-12.95	2.28
Russian Federation	3 949	5 402	8.45	0.96	1 921	709	-13.55	-6.00	101	220	-7.25	0.00
Ukraine	1 867	2 211	1.15	-1.15	300	260	-9.51	3.04	102	289	-35.31	-1.84
OCEANIA DEVELOPED	**3 768**	**5 141**	**-4.20**	**1.57**	**270**	**276**	**0.32**	**0.81**	**3 201**	**4 060**	**-4.22**	**1.58**
Australia	3 768	5 141	-4.20	1.57	34	10	10.09	0.00	3 184	4 030	-4.23	1.59
New Zealand	0	0	..	..	236	266	-0.85	0.84	17	30	-4.47	0.00
OTHER DEVELOPED[1]	**3 042**	**3 428**	**-2.01**	**0.80**	**4 022**	**4 492**	**-0.60**	**1.07**	**662**	**253**	**-14.03**	**-4.35**
Japan	798	945	-2.32	0.60	1 363	1 243	-1.10	-1.95	2	5	-2.70	0.00
South Africa	2 159	2 277	-1.79	0.40	331	401	3.45	5.51	494	211	-12.91	-4.73
DEVELOPING	**126 932**	**166 753**	**2.97**	**2.24**	**33 252**	**42 149**	**3.66**	**3.56**	**46 850**	**57 054**	**4.70**	**2.44**
AFRICA	**8 326**	**13 430**	**2.45**	**3.69**	**8 753**	**9 222**	**1.12**	**1.58**	**2 196**	**2 688**	**-2.26**	**1.93**
NORTH AFRICA	**2 531**	**3 332**	**2.86**	**2.91**	**3 745**	**4 012**	**-0.12**	**1.24**	**98**	**7**	**-12.67**	**-10.10**
Algeria	0	0	..	..	1 400	1 771	0.34	2.80	0	0	..	..
Egypt	2 058	2 636	4.09	2.85	1 014	795	-1.65	-1.37	98	7	-7.34	-10.13
SUB-SAHARAN AFRICA	**5 795**	**10 098**	**2.26**	**3.96**	**5 008**	**5 211**	**2.14**	**1.86**	**2 098**	**2 681**	**-2.10**	**2.02**
LATIN AMERICA and CARIBBEAN	**58 208**	**71 920**	**3.01**	**2.37**	**2 632**	**2 899**	**5.94**	**1.59**	**32 453**	**39 266**	**5.22**	**3.04**
Argentina	2 070	2 642	1.19	1.43	13	5	14.26	0.00	326	465	-0.49	3.63
Brazil	38 469	46 838	4.95	2.77	0	0	..	..	25 809	30 623	6.76	3.39
Chile	337	405	-3.12	2.28	473	582	13.62	2.01	0	0	0.00	-11.29
Mexico	5 325	5 882	-0.47	1.05	448	571	28.98	3.17	1 197	1 428	26.17	0.99
Uruguay	13	21	7.27	1.76	114	147	-1.22	1.97	3	8	-44.60	-1.97
ASIA and PACIFIC	**60 398**	**81 404**	**3.01**	**1.92**	**21 867**	**30 028**	**4.61**	**4.48**	**12 200**	**15 100**	**4.92**	**1.18**
Bangladesh	130	184	-2.12	-0.01	1 299	1 839	11.08	3.73	1	0	5.26	0.36
China[2]	11 533	15 182	1.57	1.86	2 323	4 799	10.29	7.83	73	32	-10.71	-3.28
India	24 869	29 348	4.43	0.90	1 626	2 475	-8.68	86.44	2 161	15	17.05	-61.56
Indonesia	2 617	4 444	3.81	3.61	2 558	4 190	4.39	4.11	0	0	-9.53	0.51
Iran, Islamic Republic of	925	1 105	-5.07	3.35	1 626	2 067	9.11	2.78	1	0	-59.36	1.12
Korea	0	0	..	..	1 605	1 973	0.33	3.39	294	350	-1.76	17.02
Malaysia	30	51	-15.95	1.84	1 576	1 783	1.14	1.81	152	96	-8.46	3.29
Pakistan	4 233	7 437	1.83	3.67	554	123	30.17	-4.31	103	129	5.53	1.18
Saudi Arabia	0	0	..	..	1 214	1 857	5.70	3.97	249	330	19.83	4.42
Thailand	9 064	14 442	4.64	2.82	13	10	52.81	-0.06	6 474	10 968	6.20	2.93
Turkey	2 558	2 869	2.35	2.05	8	11	22.76	0.71	49	0	-19.96	-134.54
LEAST DEVELOPED COUNTRIES (LDC)	**3 950**	**7 959**	**4.07**	**5.04**	**5 229**	**5 629**	**5.19**	**2.19**	**938**	**1 254**	**0.06**	**4.47**
OECD[3]	**37 386**	**38 994**	**-2.51**	**0.65**	**13 360**	**14 726**	**3.78**	**0.37**	**7 823**	**7 062**	**-5.96**	**0.83**
NON-OECD	**127 699**	**168 922**	**3.22**	**2.19**	**35 769**	**43 252**	**1.47**	**3.24**	**46 748**	**56 591**	**3.89**	**2.37**

Note: Crop year: Beginning crop marketing year (Oct/Sept) - see Glossary of Terms for definitions.

1. Includes Israel and also transition economies: Kazakhstan, Kyrgyzstan, Tajikistan, Turkmenistan, Uzbekistan, Armenia, Azerbaijan and Georgia.
2. Refers to mainland only. The economies of Chinese Taipei, Hong Kong (China) and Macau (China) are included in the Other Asia Pacific aggregate.
3. Excludes Iceland but includes all EU27 member countries.
4. Least-squares growth rate (see glossary).

Source: OECD and FAO Secretariats.

Table A.17. Sugar projections (in raw sugar equivalent) *(cont.)*

Crop year

	CONSUMPTION (Kt)		Growth (%)[1]		PER CAPITA (Kg)		Growth (%)[1]	
	Avg 2009-11est	2021	2002-2011	2012-2021	Avg 2009-11est	2021	2002-2011	2012-2021
WORLD	**160 565**	**202 205**	**1.90**	**2.09**	**23.3**	**26.2**	**0.72**	**1.06**
DEVELOPED	**46 937**	**50 416**	**0.10**	**0.61**	**34.2**	**35.3**	**-0.36**	**0.25**
NORTH AMERICA	**11 634**	**12 897**	**1.60**	**0.97**	**33.8**	**34.2**	**0.67**	**0.15**
Canada	1 385	1 536	0.91	0.75	40.7	41.0	-0.14	-0.11
United States	10 249	11 361	1.69	0.99	33.0	33.4	0.78	0.18
EUROPE	**27 552**	**28 417**	**-0.72**	**0.17**	**37.2**	**38.0**	**-0.91**	**0.07**
EU(27)	17 966	18 483	-0.36	0.20	35.8	35.9	-0.76	-0.05
Russian Federation	5 882	5 908	-1.75	-0.13	41.1	42.0	-1.56	0.02
Ukraine	2 084	2 089	-1.33	-0.22	45.9	48.8	-0.65	0.32
OCEANIA DEVELOPED	**1 245**	**1 342**	**-0.38**	**0.71**	**46.8**	**44.2**	**-1.94**	**-0.45**
Australia	1 025	1 106	-0.35	0.70	46.1	43.3	-1.99	-0.49
New Zealand	220	236	-0.50	0.73	50.4	48.5	-1.69	-0.23
OTHER DEVELOPED[2]	**6 506**	**7 760**	**1.28**	**1.68**	**24.9**	**28.5**	**0.75**	**1.33**
Japan	2 257	2 186	-0.97	-0.21	17.8	17.6	-1.00	-0.04
South Africa	1 954	2 559	3.83	2.38	39.0	48.5	2.80	1.93
DEVELOPING	**113 627**	**151 789**	**2.71**	**2.62**	**20.6**	**24.1**	**1.36**	**1.45**
AFRICA	**14 598**	**19 978**	**3.05**	**2.94**	**15.0**	**15.9**	**0.67**	**0.65**
NORTH AFRICA	**6 029**	**7 344**	**1.53**	**2.01**	**36.5**	**38.5**	**-0.01**	**0.74**
Algeria	1 325	1 765	2.35	2.82	37.4	43.5	0.85	1.63
Egypt	2 895	3 426	1.68	1.77	35.7	35.7	-0.12	0.28
SUB-SAHARAN AFRICA	**8 569**	**12 634**	**4.22**	**3.53**	**10.6**	**11.9**	**1.66**	**1.04**
LATIN AMERICA and CARIBBEAN	**28 738**	**35 682**	**1.50**	**1.78**	**48.7**	**54.2**	**0.32**	**0.81**
Argentina	1 861	2 206	2.18	1.53	46.0	49.9	1.30	0.73
Brazil	12 990	16 154	2.63	1.90	66.6	76.3	1.61	1.17
Chile	803	1 043	2.32	2.33	46.9	55.9	1.32	1.57
Mexico	4 407	5 007	-2.23	1.33	38.9	39.4	-3.48	0.33
Uruguay	128	159	1.68	2.17	37.9	45.4	1.48	1.79
ASIA and PACIFIC	**70 291**	**96 129**	**3.16**	**2.89**	**17.7**	**21.9**	**2.03**	**1.98**
Bangladesh	1 385	2 024	8.48	3.37	9.3	12.0	7.25	2.23
China[3]	14 730	19 872	3.79	2.98	11.0	14.3	3.27	2.69
India	23 472	31 813	2.85	2.67	19.2	22.7	1.40	1.47
Indonesia	5 448	8 699	4.86	4.25	22.7	32.9	3.75	3.39
Iran, Islamic Republic of	2 427	3 132	2.47	2.52	32.8	38.4	1.28	1.68
Korea	1 322	1 626	0.76	1.86	27.4	32.6	0.30	1.56
Malaysia	1 405	1 738	2.43	2.33	49.5	52.0	0.64	0.87
Pakistan	4 822	7 385	2.92	3.68	27.8	35.4	1.12	2.05
Saudi Arabia	952	1 511	4.16	4.27	34.7	44.3	1.21	2.35
Thailand	2 677	3 488	2.60	2.50	38.7	48.3	1.80	2.13
Turkey	2 438	2 853	3.75	2.07	33.5	35.0	2.44	1.08
LEAST DEVELOPED COUNTRIES (LDC)	**8 132**	**12 313**	**6.25**	**3.79**	**9.8**	**11.6**	**4.00**	**1.63**
OECD[4]	**43 156**	**46 674**	**0.20**	**0.75**	**34.0**	**34.7**	**-0.48**	**0.24**
NON-OECD	**117 409**	**155 530**	**2.58**	**2.52**	**20.9**	**24.4**	**1.29**	**1.39**

Note: Crop year: Beginning crop marketing year - see Glossary of Terms for definitions.

1. Least-squares growth rate (see glossary).
2. Includes Israel and also transition economies: Kazakhstan, Kyrgyzstan, Tajikistan, Turkmenistan, Uzbekistan, Armenia, Azerbaijan and Georgia.
3. Refers to mainland only. The economies of Chinese Taipei, Hong Kong (China) and Macau (China) are included in the Other Asia Pacific aggregate.
4. Excludes Iceland but includes all EU27 member countries.

Source: OECD and FAO Secretariats.

StatLink http://dx.doi.org/10.1787/888932642972

Table A.18. Main policy assumptions for sugar markets

Crop year

		Avg. 09/10-11/12est.	12/13	13/14	14/15	15/16	16/17	17/18	18/19	19/20	20/21	21/22
ARGENTINA												
Tariff, sugar	ARS/t	35.0	35.0	35.0	35.0	35.0	35.0	35.0	35.0	35.0	35.0	35.0
BANGLADESH												
Tariff, white sugar	%	32.5	32.5	32.5	32.5	32.5	32.5	32.5	32.5	32.5	32.5	32.5
BRAZIL												
Tariff, raw sugar	%	22.3	16.0	16.0	16.0	16.0	16.0	16.0	16.0	16.0	16.0	16.0
Tariff, white sugar	%	22.3	16.0	16.0	16.0	16.0	16.0	16.0	16.0	16.0	16.0	16.0
CANADA												
Tariff, raw sugar	CAD/t	24.7	24.7	24.7	24.7	24.7	24.7	24.7	24.7	24.7	24.7	24.7
Tariff, white sugar	CAD/t	30.9	30.9	30.9	30.9	30.9	30.9	30.9	30.9	30.9	30.9	30.9
CHINA[1]												
TRQ sugar	kt	1 954	1 954	1 954	1 954	1 954	1 954	1 954	1 954	1 954	1 954	1 954
Tariff, in-quota, raw sugar	%	15.0	15.0	15.0	15.0	15.0	15.0	15.0	15.0	15.0	15.0	15.0
Tariff, in-quota, white sugar	%	20.0	20.0	20.0	20.0	20.0	20.0	20.0	20.0	20.0	20.0	20.0
Tariff, over-quota	%	50.0	50.0	50.0	50.0	50.0	50.0	50.0	50.0	50.0	50.0	50.0
EUROPEAN UNION[2]												
Reference price, white sugar[3]	EUR/t	404.4	404.4	404.4	404.4	0.0	0.0	0.0	0.0	0.0	0.0	0.0
Production quota[4]	mt wse	13.5	13.5	13.5	13.5	0.0	0.0	0.0	0.0	0.0	0.0	0.0
Subsidised export limits												
Quantity Limit	kt rse	1 431	1 431	1 431	1 431	1 431	1 431	1 431	1 431	1 431	1 431	1 431
Value Limit	'000 EUR	499 100	499 100	499 100	499 100	499 100	499 100	499 100	499 100	499 100	499 100	499 100
Tariff, raw sugar	EUR/t	339.0	339.0	339.0	339.0	339.0	339.0	339.0	339.0	339.0	339.0	339.0
Tariff, white sugar	EUR/t	419.0	419.0	419.0	419.0	419.0	419.0	419.0	419.0	419.0	419.0	419.0
INDIA												
Intervention price, sugarcane	INR/t	1 366.1	1 400.0	1 400.0	1 400.0	1 400.0	1 400.0	1 400.0	1 400.0	1 400.0	1 400.0	1 400.0
Applied tariff, raw sugar	%	60.0	60.0	60.0	60.0	60.0	60.0	60.0	60.0	60.0	60.0	60.0
INDONESIA												
Tariff, white sugar	%	25.0	25.0	25.0	25.0	25.0	25.0	25.0	25.0	25.0	25.0	25.0
JAPAN												
Minimum stabilisation price, raw sugar	JPY/kg	152.9	153.2	153.2	153.2	153.2	153.2	153.2	153.2	153.2	153.2	153.2
Tariff, raw sugar	JPY/kg	71.8	71.8	71.8	71.8	71.8	71.8	71.8	71.8	71.8	71.8	71.8
Tariff, white sugar	JPY/kg	103.1	103.1	103.1	103.1	103.1	103.1	103.1	103.1	103.1	103.1	103.1
KOREA												
Tariff, raw sugar	%	18.0	18.0	18.0	18.0	18.0	18.0	18.0	18.0	18.0	18.0	18.0
MEXICO												
Ad valorem tariff,raw sugar	%	156.0	156.0	156.0	156.0	156.0	156.0	156.0	156.0	156.0	156.0	157.0
RUSSIAN FEDERATION												
Minimum tariff, raw sugar	USD/t	140.0	140.0	140.0	140.0	140.0	140.0	140.0	140.0	140.0	140.0	140.0
Minimum tariff, white sugar	USD/t	250.0	243.0	236.3	229.7	223.0	223.0	223.0	223.0	223.0	223.0	223.0
UNITED STATES[2]												
Loan rate, raw sugar	USD/t	407.9	413.4	413.4	413.4	413.4	413.4	413.4	413.4	413.4	413.4	413.4
Loan rate, white sugar	USD/t	524.1	531.2	531.2	531.2	531.2	531.2	531.2	531.2	531.2	531.2	531.2
TRQ, raw sugar	kt rse	1 532	1 704	1 569	1 560	1 358	1 353	1 511	1 654	1 720	1 772	1 841
TRQ, refined sugar	kt rse	49.0	49.0	49.0	49.0	49.0	49.0	49.0	49.0	49.0	49.0	49.0
Raw sugar 2nd tier WTO tariff	USD/t	338.6	338.6	338.6	338.6	338.6	338.6	338.6	338.6	338.6	338.6	338.6
White sugar 2nd tier WTO tariff	USD/t	357.4	357.4	357.4	357.4	357.4	357.4	357.4	357.4	357.4	357.4	357.4
SOUTH AFRICA												
Tariff, raw sugar	%	105.0	105.0	105.0	105.0	105.0	105.0	105.0	105.0	105.0	105.0	105.0
TANZANIA												
Applied tariff, white sugar	%	25.0	25.0	25.0	25.0	25.0	25.0	25.0	25.0	25.0	25.0	25.0
VIETNAM												
Applied tariff, white sugar	%	40.0	40.0	40.0	40.0	40.0	40.0	40.0	40.0	40.0	40.0	40.0

Note: Crop year: Beginning crop marketing year - see the Glossary of Terms for definitions.

The source for tariffs (except United States and Russia) is AMAD. The source for the Russian Federation and United States tariffs is ERS, USDA.

1. Refers to mainland only.
2. In addition, price based special safeguard actions may apply.
3. Reference price for consumers.
4. Production that receives official support.

Source: OECD and FAO Secretariats.

StatLink http://dx.doi.org/10.1787/888932642991

Table A.19. World meat projections

Calendar year

		Avg 2009-11	2012	2013	2014	2015	2016	2017	2018	2019	2020	2021
OECD[1]												
BEEF AND VEAL												
Production	kt cwe	27 300	26 881	26 096	26 220	27 057	27 635	28 254	28 505	28 763	28 939	29 023
Consumption	kt cwe	26 747	26 268	25 511	25 792	26 627	27 215	27 861	28 146	28 438	28 614	28 705
PIGMEAT												
Production	kt cwe	39 535	39 921	40 271	40 626	41 066	41 151	41 330	41 526	41 837	42 133	42 243
Consumption	kt cwe	36 959	37 201	37 492	37 920	38 312	38 422	38 584	38 757	39 040	39 244	39 356
POULTRY MEAT												
Production	kt rtc	40 613	41 449	42 032	42 840	43 432	44 127	44 676	45 306	45 941	46 486	47 159
Consumption	kt rtc	38 038	38 617	39 265	39 926	40 411	41 131	41 662	42 291	42 880	43 283	43 905
SHEEPMEAT												
Production	kt cwe	2 537	2 506	2 518	2 575	2 601	2 646	2 654	2 694	2 688	2 737	2 745
Consumption	kt cwe	2 095	2 081	2 082	2 113	2 101	2 125	2 111	2 141	2 110	2 134	2 137
TOTAL MEAT												
Per capita consumption[2]	kg rwt	65.2	64.8	64.6	65.1	65.8	66.3	66.7	67.1	67.5	67.6	67.9
NON-OECD												
BEEF AND VEAL												
Production	kt cwe	38 331	39 223	40 096	41 158	42 218	42 990	43 911	44 746	45 636	46 599	47 480
Consumption	kt cwe	38 097	39 224	40 089	41 001	42 018	42 820	43 732	44 501	45 354	46 330	47 190
PIGMEAT												
Production	kt cwe	69 096	71 809	72 925	74 468	75 655	77 100	78 399	79 913	81 055	82 744	83 999
Consumption	kt cwe	71 131	73 697	74 882	76 334	77 554	78 981	80 307	81 850	82 999	84 774	86 048
POULTRY MEAT												
Production	kt rtc	58 057	62 708	64 703	66 460	68 326	70 096	72 156	74 070	76 146	77 882	80 041
Consumption	kt rtc	60 444	65 340	67 233	69 148	71 131	72 892	74 956	76 874	78 989	80 868	83 098
SHEEPMEAT												
Production	kt cwe	10 574	10 973	11 216	11 495	11 677	11 933	12 068	12 423	12 576	12 898	13 114
Consumption	kt cwe	10 961	11 346	11 595	11 897	12 117	12 396	12 555	12 926	13 103	13 449	13 669
TOTAL MEAT												
Per capita consumption[2]	kg rwt	25.8	26.4	26.7	27.0	27.3	27.6	27.8	28.1	28.4	28.7	29.0
WORLD												
BEEF AND VEAL												
Production	kt cwe	65 631	66 104	66 192	67 379	69 275	70 625	72 165	73 251	74 400	75 538	76 503
Consumption	kt cwe	64 844	65 492	65 600	66 793	68 646	70 035	71 593	72 646	73 792	74 944	75 895
PIGMEAT												
Production	kt cwe	108 631	111 730	113 196	115 094	116 721	118 251	119 729	121 439	122 892	124 877	126 242
Consumption	kt cwe	108 090	110 898	112 374	114 254	115 866	117 403	118 891	120 608	122 039	124 019	125 404
POULTRY MEAT												
Production	kt rtc	98 670	104 158	106 736	109 300	111 758	114 223	116 832	119 376	122 087	124 368	127 201
Consumption	kt rtc	98 482	103 957	106 498	109 074	111 543	114 023	116 619	119 165	121 869	124 150	127 002
SHEEPMEAT												
Production	kt cwe	13 111	13 479	13 734	14 070	14 278	14 578	14 722	15 117	15 264	15 635	15 859
Consumption	kt cwe	13 056	13 427	13 678	14 010	14 218	14 522	14 667	15 067	15 213	15 583	15 807
TOTAL MEAT												
Per capita consumption[2]	kg rwt	33.0	33.4	33.6	33.9	34.2	34.5	34.7	35.0	35.3	35.5	35.8

Note: Calendar Year: Year ending 30 September for New Zealand.

1. Excludes Iceland but includes all EU27 member countries.
2. Per capita consumption expressed in retail weight. Carcass weight to retail weight conversion factors of 0.7 for beef and veal, 0.78 for pigmeat and 0.88 for both sheepmeat and poultry meat.

Source: OECD and FAO Secretariats.

StatLink http://dx.doi.org/10.1787/888932643010

Table A.20. Beef and veal projections

Calendar year

	PRODUCTION (Kt cwe)[4]		Growth (%)[5]		IMPORTS (Kt cwe)[6]		Growth (%)[5]		EXPORTS (Kt cwe)[6]		Growth (%)[5]	
	Avg 2009-11est	2021	2002-2011	2012-2021	Avg 2009-11est	2021	2002-2011	2012-2021	Avg 2009-11est	2021	2002-2011	2012-2021
WORLD	**65 631**	**76 503**	**1.18**	**1.77**	**6 914**	**8 603**	**2.36**	**1.74**	**7 545**	**9 098**	**2.46**	**1.57**
DEVELOPED	**29 726**	**32 396**	**0.03**	**1.45**	**3 419**	**3 707**	**-1.37**	**0.70**	**3 586**	**4 375**	**0.33**	**1.24**
NORTH AMERICA	**13 019**	**14 071**	**0.03**	**2.10**	**1 296**	**1 530**	**-4.42**	**1.21**	**1 512**	**1 805**	**2.10**	**0.62**
Canada	1 515	1 768	-2.04	1.50	222	197	0.20	-2.20	462	554	-2.32	1.00
United States	11 505	12 303	0.34	2.19	1 074	1 333	-5.03	1.79	1 050	1 251	6.50	0.55
EUROPE	**11 142**	**12 041**	**-0.58**	**0.83**	**1 185**	**1 066**	**1.25**	**-0.71**	**377**	**668**	**-3.93**	**6.06**
EU(27)	8 197	8 166	-0.18	-0.01	322	309	-6.84	-0.30	232	271	-9.48	1.11
Russian Federation	1 726	2 068	-1.79	2.02	751	698	5.63	-0.18	0	0	..	..
Ukraine	420	489	-6.86	1.92	13	8	20.25	-6.55	0	0	0.00	0.47
OCEANIA DEVELOPED	**2 754**	**3 003**	**0.18**	**0.85**	**10**	**11**	**-1.88**	**0.00**	**1 685**	**1 891**	**0.22**	**0.52**
Australia	2 118	2 346	0.25	1.06	4	3	2.71	0.00	1 163	1 362	0.42	0.76
New Zealand	636	657	-0.02	0.13	6	8	-2.62	0.00	522	529	-0.23	-0.08
OTHER DEVELOPED[1]	**2 811**	**3 281**	**2.65**	**1.58**	**929**	**1 100**	**0.38**	**1.52**	**11**	**11**	**2.78**	**-0.12**
Japan	510	495	-0.14	0.03	715	769	0.18	0.42	1	1	36.82	1.86
South Africa	779	859	3.36	1.05	37	199	-7.85	14.68	7	5	-0.29	-2.35
DEVELOPING	**35 905**	**44 107**	**2.20**	**2.02**	**3 495**	**4 896**	**7.18**	**2.61**	**3 959**	**4 723**	**5.24**	**1.90**
AFRICA	**4 754**	**6 251**	**2.85**	**2.63**	**524**	**609**	**6.08**	**3.22**	**85**	**53**	**9.39**	**-2.51**
NORTH AFRICA	**1 090**	**1 393**	**3.86**	**2.23**	**357**	**509**	**7.12**	**4.42**	**1**	**1**	**6.35**	**-0.30**
Algeria	128	152	1.39	1.21	86	105	5.78	3.11	0	0	-19.59	-0.22
Egypt	707	916	5.11	2.37	226	377	6.41	4.84	0	0	7.02	-0.37
SUB-SAHARAN AFRICA	**3 664**	**4 858**	**2.57**	**2.75**	**167**	**100**	**4.04**	**-1.50**	**84**	**53**	**9.08**	**-2.53**
LATIN AMERICA and CARIBBEAN	**18 091**	**21 400**	**2.60**	**1.52**	**693**	**898**	**2.46**	**2.24**	**2 956**	**3 438**	**3.97**	**2.19**
Argentina	2 834	3 152	0.23	2.06	4	17	-8.46	11.60	412	555	-3.50	6.06
Brazil	9 685	11 399	3.15	1.15	36	83	-6.35	1.35	1 643	1 818	2.85	1.54
Chile	211	286	0.93	2.62	173	184	1.23	0.31	7	7	4.75	-0.20
Mexico	1 757	2 049	2.42	1.18	214	335	-3.25	4.56	68	30	41.05	-11.93
Uruguay	649	759	5.02	1.99	1	0	-14.54	-2.94	316	399	2.57	2.94
ASIA and PACIFIC	**13 059**	**16 456**	**1.46**	**2.48**	**2 277**	**3 389**	**9.33**	**2.60**	**919**	**1 232**	**9.54**	**1.34**
Bangladesh	195	263	0.87	2.67	0	0	17.50	-13.04	0	0	-17.77	0.93
China[2]	5 615	6 440	0.05	2.00	49	113	16.73	4.47	82	91	7.88	-0.50
India	2 554	3 459	1.78	2.36	1	0	51.14	-7.78	696	961	9.43	1.98
Indonesia	343	428	-0.73	2.15	110	183	17.65	4.03	1	0	8.95	-0.33
Iran, Islamic Republic of	410	480	3.63	1.47	229	346	24.79	1.55	0	0	34.62	-0.12
Korea	259	307	3.97	2.14	354	425	1.71	0.72	2	1	3.16	0.00
Malaysia	17	27	6.44	2.86	155	198	1.56	1.79	7	7	15.40	-1.79
Pakistan	1 464	2 177	6.06	4.09	4	0	49.10	-14.70	22	38	32.61	3.72
Saudi Arabia	23	24	3.94	-1.18	135	247	9.42	1.54	10	78	-12.84	-1.54
Turkey	296	295	-1.34	2.19	44	161	79.44	6.81	1	1	7.63	-0.62
LEAST DEVELOPED COUNTRIES (LDC)	**3 574**	**4 846**	**2.26**	**2.72**	**111**	**25**	**3.73**	**-12.35**	**3**	**15**	**5.33**	**12.82**
OECD[3]	**27 300**	**29 023**	**0.15**	**1.25**	**3 260**	**3 861**	**-2.36**	**1.19**	**3 509**	**4 008**	**0.24**	**0.43**
NON-OECD	**38 331**	**47 480**	**1.96**	**2.11**	**3 654**	**4 742**	**8.40**	**2.23**	**4 036**	**5 091**	**5.24**	**2.58**

Note: Calendar year: Year ending 30 September for New Zealand.

1. Includes Israel and also transition economies: Kazakhstan, Kyrgyzstan, Tajikistan, Turkmenistan, Uzbekistan, Armenia, Azerbaijan and Georgia.
2. Refers to mainland only. The economies of Chinese Taipei, Hong Kong (China) and Macau (China) are included in the Other Asia Pacific aggregate.
3. Excludes Iceland but includes all EU27 member countries.
4. Gross indigenous production.
5. Least-squares growth rate (see glossary).
6. Excludes trade of live animals.

Source: OECD and FAO Secretariats.

Table A.20. Beef and veal projections *(cont.)*

Calendar year

	CONSUMPTION (Kt cwe)		Growth (%)[1]		PER CAPITA (Kg rwt)[5]		Growth (%)[1]	
	Avg 2009-11est	2021	2002-2011	2012-2021	Avg 2009-11est	2021	2002-2011	2012-2021
WORLD	**64 844**	**75 895**	**1.11**	**1.79**	**6.6**	**6.9**	**-0.06**	**0.77**
DEVELOPED	**29 590**	**31 754**	**-0.20**	**1.40**	**15.1**	**15.6**	**-0.66**	**1.04**
NORTH AMERICA	**13 000**	**14 044**	**-0.67**	**2.19**	**26.4**	**26.1**	**-1.59**	**1.38**
Canada	984	1 021	-0.52	0.53	20.3	19.1	-1.56	-0.33
United States	12 015	13 023	-0.68	2.33	27.1	26.8	-1.59	1.52
EUROPE	**11 984**	**12 391**	**-0.29**	**0.51**	**11.3**	**11.6**	**-0.48**	**0.40**
EU(27)	8 178	8 137	-0.21	0.05	11.4	11.1	-0.61	-0.19
Russian Federation	2 607	2 807	0.63	1.25	12.8	14.0	0.82	1.41
Ukraine	433	494	-6.80	1.51	6.7	8.1	-6.13	2.05
OCEANIA DEVELOPED	**864**	**930**	**-0.11**	**1.31**	**22.7**	**21.4**	**-1.67**	**0.15**
Australia	748	802	-0.26	1.35	23.5	22.0	-1.90	0.15
New Zealand	116	128	1.03	1.07	18.5	18.5	-0.16	0.10
OTHER DEVELOPED[2]	**3 743**	**4 389**	**1.85**	**1.56**	**10.0**	**11.3**	**1.32**	**1.21**
Japan	1 218	1 263	-0.48	0.24	6.7	7.1	-0.52	0.42
South Africa	803	1 040	2.68	2.68	11.2	13.8	1.65	2.24
DEVELOPING	**35 254**	**44 141**	**2.31**	**2.08**	**4.5**	**4.9**	**0.95**	**0.90**
AFRICA	**5 236**	**6 825**	**3.01**	**2.73**	**3.8**	**3.8**	**0.63**	**0.44**
NORTH AFRICA	**1 494**	**1 935**	**4.73**	**2.73**	**6.3**	**7.1**	**3.18**	**1.46**
Algeria	217	261	2.59	1.92	4.3	4.5	1.08	0.74
Egypt	968	1 314	5.57	2.99	8.4	9.6	3.76	1.49
SUB-SAHARAN AFRICA	**3 742**	**4 890**	**2.38**	**2.73**	**3.2**	**3.2**	**-0.17**	**0.25**
LATIN AMERICA and CARIBBEAN	**15 417**	**18 433**	**2.33**	**1.41**	**18.3**	**19.6**	**1.14**	**0.45**
Argentina	2 426	2 614	0.89	1.41	42.0	41.4	0.01	0.61
Brazil	7 830	9 459	2.77	1.06	28.1	31.3	1.75	0.34
Chile	377	463	1.18	1.69	15.4	17.4	0.18	0.92
Mexico	1 696	2 078	1.68	1.93	10.5	11.5	0.43	0.94
Uruguay	261	298	4.81	1.24	54.2	59.5	4.62	0.86
ASIA and PACIFIC	**14 602**	**18 883**	**2.04**	**2.53**	**2.6**	**3.0**	**0.91**	**1.62**
Bangladesh	196	263	0.89	2.66	0.9	1.1	-0.35	1.52
China[3]	5 588	6 472	0.06	2.07	2.9	3.3	-0.46	1.78
India	1 859	2 498	-0.24	2.51	1.1	1.2	-1.70	1.31
Indonesia	555	708	3.10	2.28	1.6	1.9	1.98	1.41
Iran, Islamic Republic of	639	826	8.06	1.50	6.0	7.1	6.87	0.66
Korea	579	732	1.70	1.30	8.4	10.3	1.23	1.00
Malaysia	177	228	1.38	1.95	4.4	4.8	-0.41	0.49
Pakistan	1 442	2 140	5.93	4.09	5.8	7.2	4.14	2.45
Saudi Arabia	155	197	8.08	2.57	3.9	4.1	5.13	0.65
Turkey	375	532	2.18	3.09	3.6	4.6	0.86	2.10
LEAST DEVELOPED COUNTRIES (LDC)	**3 637**	**4 809**	**2.47**	**2.55**	**3.1**	**3.2**	**0.22**	**0.39**
OECD[4]	**26 747**	**28 705**	**-0.21**	**1.38**	**14.7**	**14.9**	**-0.89**	**0.87**
NON-OECD	**38 097**	**47 190**	**2.12**	**2.05**	**4.7**	**5.2**	**0.83**	**0.91**

Note: Calendar year: Year ending 30 September New Zealand.

1. Least-squares growth rate (see glossary).
2. Includes Israel and also transition economies: Kazakhstan, Kyrgyzstan, Tajikistan, Turkmenistan, Uzbekistan, Armenia, Azerbaijan and Georgia.
3. Refers to mainland only. The economies of Chinese Taipei, Hong Kong (China) and Macau (China) are included in the Other Asia Pacific aggregate.
4. Excludes Iceland but includes all EU27 member countries.
5. Per capita consumption expressed in retail weight. Carcass weight to retail weight conversion factors of 0.7 for beef and veal, 0.78 for pigmeat and 0.88 for both sheepmeat and poultry meat.

Source: OECD and FAO Secretariats.

StatLink http://dx.doi.org/10.1787/888932643048

Table A.21. Pigmeat projections

Calendar year

	PRODUCTION (Kt cwe)[4]		Growth (%)[5]		IMPORTS (Kt cwe)[6]		Growth (%)[5]		EXPORTS (Kt cwe)[6]		Growth (%)[5]	
	Avg 2009-11est	2021	2002-2011	2012-2021	Avg 2009-11est	2021	2002-2011	2012-2021	Avg 2009-11est	2021	2002-2011	2012-2021
WORLD	**108 631**	**126 242**	**1.80**	**1.36**	**5 898**	**6 669**	**5.99**	**1.12**	**6 536**	**7 484**	**6.10**	**0.99**
DEVELOPED	**41 417**	**45 189**	**1.11**	**0.73**	**3 195**	**3 055**	**2.97**	**0.00**	**5 203**	**5 978**	**7.04**	**0.86**
NORTH AMERICA	**12 262**	**13 637**	**1.62**	**1.11**	**578**	**695**	**-0.42**	**1.64**	**3 196**	**3 749**	**8.51**	**1.29**
Canada	2 121	2 370	-0.29	1.52	186	217	9.70	0.64	1 178	1 359	3.01	1.58
United States	10 140	11 266	2.06	1.02	392	478	-3.54	2.09	2 018	2 391	12.98	1.13
EUROPE	**26 881**	**29 168**	**0.89**	**0.59**	**1 109**	**649**	**6.37**	**-3.42**	**1 952**	**2 162**	**5.47**	**0.11**
EU(27)	22 551	23 537	0.56	0.26	31	26	2.68	-0.34	1 857	2 071	5.43	0.08
Russian Federation	2 320	3 191	4.87	2.23	683	290	4.15	-5.04	0	0	0.39	0.60
Ukraine	592	779	0.51	2.29	149	101	30.10	-5.37	6	20	-23.14	3.92
OCEANIA DEVELOPED	**381**	**419**	**-2.40**	**0.78**	**326**	**421**	**12.59**	**1.76**	**49**	**62**	**-7.18**	**1.74**
Australia	333	367	-2.73	0.78	287	371	13.57	1.72	49	62	-7.22	1.74
New Zealand	47	52	0.23	0.78	39	51	6.97	2.08	0	0	12.61	0.00
OTHER DEVELOPED[1]	**1 894**	**1 965**	**1.79**	**0.27**	**1 182**	**1 290**	**0.27**	**0.56**	**5**	**5**	**9.32**	**3.57**
Japan	1 288	1 195	0.32	-0.52	1 068	1 160	-0.64	0.19	1	0	40.68	1.81
South Africa	319	400	13.40	2.07	35	39	9.84	1.26	4	4	14.14	4.09
DEVELOPING	**67 214**	**81 053**	**2.26**	**1.73**	**2 704**	**3 614**	**10.75**	**2.16**	**1 334**	**1 506**	**2.89**	**1.55**
AFRICA	**890**	**1 238**	**3.00**	**3.20**	**166**	**275**	**14.07**	**4.15**	**5**	**3**	**-0.13**	**-3.22**
NORTH AFRICA	**2**	**2**	**-8.42**	**1.34**	**2**	**2**	**17.45**	**1.61**	**0**	**0**	**36.30**	**-2.70**
Algeria	0	0	0.00	4.05	1	2	26.72	1.46	0	0	0.00	-1.46
Egypt	0	0	-18.26	1.03	0	0	-18.04	2.38	0	0	52.14	-2.38
SUB-SAHARAN AFRICA	**888**	**1 236**	**3.03**	**3.21**	**165**	**273**	**14.05**	**4.17**	**5**	**3**	**-0.98**	**-3.27**
LATIN AMERICA and CARIBBEAN	**6 713**	**8 063**	**3.59**	**1.76**	**790**	**987**	**9.38**	**1.98**	**789**	**936**	**3.95**	**2.00**
Argentina	291	404	7.40	2.97	36	35	4.02	-0.37	7	27	23.78	12.02
Brazil	3 237	3 714	3.15	1.30	10	12	62.08	2.02	576	618	2.41	1.01
Chile	504	652	4.70	2.65	11	9	30.65	-4.86	130	211	6.44	4.86
Mexico	1 158	1 332	1.81	1.42	504	608	8.56	1.61	58	70	13.94	1.37
Uruguay	22	29	2.79	1.03	17	25	9.79	4.17	0	0	4.39	-0.36
ASIA and PACIFIC	**59 611**	**71 752**	**2.10**	**1.70**	**1 748**	**2 352**	**11.13**	**2.02**	**539**	**567**	**1.60**	**0.89**
Bangladesh	0	0	0.00	0.00	0	0	46.50	0.00	0	0	0.00	0.00
China[2]	50 564	60 095	2.04	1.56	280	542	12.25	3.93	287	369	-4.26	0.85
India	485	577	0.02	1.39	1	1	38.02	-3.80	1	1	4.02	-8.42
Indonesia	668	826	4.05	2.08	3	35	-0.82	26.41	0	0	-61.17	-2.01
Iran, Islamic Republic of	0	0	0.00	0.00	1	1	46.38	0.00	0	0	-33.30	0.00
Korea	1 002	1 065	-0.54	1.58	429	616	12.59	2.50	3	0	-101.07	0.00
Malaysia	227	284	1.84	1.57	16	34	-5.34	15.97	6	11	25.18	6.33
Pakistan	0	0	0.00	0.00	0	0	34.88	0.00	0	0	33.36	0.00
Saudi Arabia	0	0	0.00	0.00	7	7	9.03	0.00	0	0	23.55	0.00
Turkey	0	0	-30.70	0.00	1	1	29.98	0.00	1	1	18.44	0.00
LEAST DEVELOPED COUNTRIES (LDC)	**1 219**	**1 817**	**4.86**	**3.80**	**141**	**199**	**17.33**	**2.56**	**1**	**1**	**29.45**	**0.27**
OECD[3]	**39 535**	**42 243**	**0.90**	**0.61**	**2 964**	**3 554**	**3.27**	**1.25**	**5 300**	**6 170**	**7.00**	**0.97**
NON-OECD	**69 096**	**83 999**	**2.35**	**1.76**	**2 935**	**3 115**	**9.28**	**0.96**	**1 236**	**1 315**	**2.76**	**1.08**

Note: Calendar year: Year ending 30 September New Zealand.

1. Includes Israel and also transition economies: Kazakhstan, Kyrgyzstan, Tajikistan, Turkmenistan, Uzbekistan, Armenia, Azerbaijan and Georgia.
2. Refers to mainland only. The economies of Chinese Taipei, Hong Kong (China) and Macau (China) are included in the Other Asia Pacific aggregate.
3. Excludes Iceland but includes all EU27 member countries.
4. Gross indigenous production.
5. Least-squares growth rate (see glossary).
6. Excludes trade of live animals.

Source: OECD and FAO Secretariats.

Table A.21. Pigmeat projections *(cont.)*

Calendar year

	CONSUMPTION (Kt cwe)		Growth (%)[1]		PER CAPITA (Kg rwe)[5]		Growth (%)[1]	
	Avg 2009-11est	2021	2002-2011	2012-2021	Avg 2009-11est	2021	2002-2011	2012-2021
WORLD	**108 090**	**125 404**	**1.82**	**1.37**	**12.2**	**12.7**	**0.65**	**0.35**
DEVELOPED	**39 477**	**42 253**	**0.72**	**0.65**	**22.4**	**23.1**	**0.26**	**0.30**
NORTH AMERICA	**9 451**	**10 364**	**-0.23**	**1.08**	**21.4**	**21.4**	**-1.15**	**0.27**
Canada	747	755	-1.57	0.55	17.1	15.7	-2.62	-0.31
United States	8 705	9 609	-0.11	1.13	21.9	22.1	-1.01	0.31
EUROPE	**26 289**	**27 862**	**0.93**	**0.51**	**27.7**	**29.0**	**0.74**	**0.41**
EU(27)	20 632	21 427	0.13	0.28	32.1	32.5	-0.27	0.03
Russian Federation	3 276	3 682	6.03	1.38	17.9	20.4	6.22	1.53
Ukraine	746	872	3.21	0.95	12.8	15.9	3.89	1.48
OCEANIA DEVELOPED	**660**	**778**	**3.90**	**1.23**	**19.3**	**20.0**	**2.34**	**0.07**
Australia	573	675	4.06	1.20	20.1	20.6	2.43	0.00
New Zealand	87	103	2.87	1.41	15.5	16.4	1.68	0.44
OTHER DEVELOPED[2]	**3 077**	**3 250**	**1.29**	**0.40**	**9.2**	**9.3**	**0.76**	**0.05**
Japan	2 362	2 355	0.01	-0.15	14.6	14.8	-0.03	0.02
South Africa	350	434	12.90	1.98	5.4	6.4	11.87	1.54
DEVELOPING	**68 613**	**83 151**	**2.50**	**1.75**	**9.7**	**10.3**	**1.15**	**0.57**
AFRICA	**1 051**	**1 510**	**4.28**	**3.38**	**0.8**	**0.9**	**1.91**	**1.09**
NORTH AFRICA	**3**	**3**	**-2.52**	**1.89**	**0.0**	**0.0**	**-4.06**	**0.62**
Algeria	1	2	18.28	1.70	0.0	0.0	16.77	0.52
Egypt	1	0	-19.58	2.88	0.0	0.0	-21.39	1.39
SUB-SAHARAN AFRICA	**1 048**	**1 507**	**4.31**	**3.39**	**1.0**	**1.1**	**1.76**	**0.90**
LATIN AMERICA and CARIBBEAN	**6 722**	**8 125**	**4.01**	**1.76**	**8.9**	**9.6**	**2.82**	**0.80**
Argentina	321	411	6.60	2.26	6.2	7.3	5.72	1.46
Brazil	2 670	3 109	3.38	1.37	10.7	11.5	2.36	0.65
Chile	385	450	4.54	1.54	17.5	18.8	3.55	0.78
Mexico	1 612	1 881	2.83	1.47	11.1	11.6	1.58	0.48
Uruguay	39	54	5.45	2.33	9.1	12.0	5.26	1.96
ASIA and PACIFIC	**60 840**	**73 515**	**2.32**	**1.72**	**12.0**	**13.1**	**1.18**	**0.81**
Bangladesh	0	0	24.75	0.00	0.0	0.0	23.51	-1.14
China[3]	50 428	60 140	2.15	1.59	29.3	33.7	1.63	1.29
India	486	578	0.04	1.39	0.3	0.3	-1.42	0.19
Indonesia	669	859	4.61	2.46	2.2	2.5	3.50	1.60
Iran, Islamic Republic of	1	1	46.56	0.00	0.0	0.0	45.37	-0.84
Korea	1 455	1 681	3.42	1.87	23.6	26.3	2.95	1.57
Malaysia	237	306	1.05	2.32	6.5	7.1	-0.74	0.86
Pakistan	0	0	20.81	0.00	0.0	0.0	19.02	-1.63
Saudi Arabia	7	7	9.27	0.00	0.2	0.2	6.32	-1.92
Turkey	0	0	9.20	0.00	0.0	0.0	7.88	-0.99
LEAST DEVELOPED COUNTRIES (LDC)	**1 362**	**2 018**	**5.76**	**3.67**	**1.3**	**1.5**	**3.51**	**1.51**
OECD[4]	**36 959**	**39 356**	**0.36**	**0.61**	**22.7**	**22.8**	**-0.31**	**0.10**
NON-OECD	**71 131**	**86 048**	**2.65**	**1.73**	**9.9**	**10.5**	**1.36**	**0.60**

Note: Calendar year: Year ending 30 September for New Zealand.

1. Least-squares growth rate (see glossary).
2. Includes Israel and also transition economies: Kazakhstan, Kyrgyzstan, Tajikistan, Turkmenistan, Uzbekistan, Armenia, Azerbaijan and Georgia.
3. Refers to mainland only. The economies of Chinese Taipei, Hong Kong (China) and Macau (China) are included in the Other Asia Pacific aggregate.
4. Excludes Iceland but includes all EU27 member countries.
5. Per capita consumption expressed in retail weight. Carcass weight to retail weight conversion factors of 0.7 for beef and veal, 0.78 for pigmeat and 0.88 for both sheepmeat and poultry meat.

Source: OECD and FAO Secretariats.

StatLink http://dx.doi.org/10.1787/888932643086

Table A.22. Poultry meat projections

Calendar year

	PRODUCTION (Kt rtc)		Growth (%)[4]		IMPORTS (Kt rtc)		Growth (%)[4]		EXPORTS (Kt rtc)		Growth (%)[4]	
	Avg 2009-11est	2021	2002-2011	2012-2021	Avg 2009-11est	2021	2002-2011	2012-2021	Avg 2009-11est	2021	2002-2011	2012-2021
WORLD	98 670	127 201	3.42	2.21	11 128	13 461	5.83	1.55	11 288	13 656	5.53	1.52
DEVELOPED	40 595	47 438	2.03	1.36	3 456	2 938	0.74	-1.06	4 901	5 633	4.05	0.95
NORTH AMERICA	20 229	23 841	1.21	1.66	289	297	7.03	0.80	3 543	4 270	4.62	1.72
Canada	1 205	1 364	1.22	0.94	235	238	4.87	0.86	178	187	6.56	0.78
United States	19 024	22 477	1.21	1.70	54	59	24.88	0.58	3 365	4 084	4.53	1.76
EUROPE	16 087	18 597	3.01	0.98	2 089	1 295	-1.56	-3.28	1 263	1 265	2.12	-1.24
EU(27)	11 879	12 534	0.73	0.44	805	827	2.56	0.68	1 166	1 093	1.52	-2.01
Russian Federation	2 860	4 113	14.06	1.93	745	20	-7.83	-26.57	18	100	42.93	10.88
Ukraine	832	1 309	13.45	3.22	268	182	9.72	-1.77	16	21	9.65	0.59
OCEANIA DEVELOPED	1 117	1 388	3.36	1.37	0	0	..	..	39	38	5.54	1.29
Australia	968	1 206	3.79	1.32	0	0	..	..	39	38	5.54	1.29
New Zealand	149	182	0.85	1.77	0	0	..	..	0	0	..	..
OTHER DEVELOPED[1]	3 161	3 613	2.10	1.35	1 078	1 346	4.31	1.29	55	59	11.13	-0.37
Japan	1 392	1 422	1.63	0.43	405	358	-1.33	-1.50	10	11	23.57	0.00
South Africa	1 019	1 281	1.45	2.13	281	378	10.13	1.17	29	32	12.30	-0.66
DEVELOPING	58 075	79 763	4.48	2.75	7 672	10 523	8.90	2.42	6 387	8 023	6.79	1.95
AFRICA	3 119	4 349	3.50	3.17	951	1 463	10.43	3.13	15	24	-9.89	7.50
NORTH AFRICA	1 891	2 456	3.83	2.41	142	296	23.27	4.85	5	6	-4.32	-0.61
Algeria	259	288	0.37	1.12	11	26	11.25	7.24	0	0	43.62	-0.54
Egypt	787	1 019	2.55	2.53	111	263	40.42	5.76	5	6	-4.73	-0.69
SUB-SAHARAN AFRICA	1 228	1 894	3.01	4.24	809	1 166	9.04	2.73	10	18	-10.07	12.07
LATIN AMERICA and CARIBBEAN	22 215	29 454	5.54	2.32	1 648	1 846	9.59	-0.08	4 187	5 525	9.20	1.92
Argentina	1 649	2 180	11.21	2.22	17	21	4.84	0.08	233	281	28.16	3.25
Brazil	12 544	16 446	6.23	1.90	1	2	12.49	1.88	3 743	5 060	8.38	1.89
Chile	594	756	3.14	2.64	67	81	46.33	-0.92	112	115	13.29	0.92
Mexico	2 741	3 760	3.13	3.01	586	518	6.39	-2.67	2	1	50.34	2.09
Uruguay	90	117	11.14	1.92	2	1	17.45	-4.51	8	16	73.04	4.49
ASIA and PACIFIC	32 741	45 959	3.89	2.99	5 072	7 214	8.46	3.01	2 185	2 474	4.04	1.98
Bangladesh	170	268	3.92	4.88	2	0	2.68	-34.15	0	0	2.10	2.43
China[2]	16 832	23 801	3.03	2.91	668	1 045	6.11	4.70	530	618	-0.16	0.56
India	2 658	4 403	7.14	4.71	1	0	46.73	-6.55	2	3	-16.02	0.50
Indonesia	1 436	1 808	3.33	2.28	7	139	4.65	25.43	0	0	-77.34	-1.81
Iran, Islamic Republic of	1 664	2 519	6.26	3.79	45	2	28.93	-37.01	30	105	8.24	11.55
Korea	650	846	5.31	2.05	106	96	0.72	-2.64	13	15	20.83	0.87
Malaysia	1 500	1 675	6.75	0.88	58	247	4.06	11.95	139	39	4.97	-11.93
Pakistan	704	883	9.08	1.84	3	36	14.34	18.95	2	1	-15.27	-1.66
Saudi Arabia	575	733	2.95	2.67	682	1 002	6.11	2.52	3	3	-10.65	-0.29
Turkey	1 348	1 795	6.88	2.49	100	173	1.76	6.08	156	100	22.12	-6.08
LEAST DEVELOPED COUNTRIES (LDC)	1 905	3 081	6.06	4.72	775	1 041	9.14	1.81	2	154	-16.71	71.35
OECD[3]	40 613	47 159	1.52	1.43	2 415	2 407	3.12	-0.36	5 057	5 661	4.34	0.68
NON-OECD	58 057	80 041	4.91	2.69	8 713	11 055	6.68	2.01	6 231	7 995	6.57	2.16

Note: Calendar year: Year ending 30 September for New Zealand.

1. Includes Israel and also transition economies: Kazakhstan, Kyrgyzstan, Tajikistan, Turkmenistan, Uzbekistan, Armenia, Azerbaijan and Georgia.
2. Refers to mainland only. The economies of Chinese Taipei, Hong Kong (China) and Macau (China) are included in the Other Asia Pacific aggregate.
3. Excludes Iceland but includes all EU27 member countries.
4. Least-squares growth rate (see glossary).

Source: OECD and FAO Secretariats.

Table A.22. Poultry meat projections *(cont.)*

Calendar year

	CONSUMPTION (Kt rtc)		Growth (%)[1]		PER CAPITA (Kg rwt)[5]		Growth (%)[1]	
	Avg 2009-11est	2021	2002-2011	2012-2021	Avg 2009-11est	2021	2002-2011	2012-2021
WORLD	**98 482**	**127 002**	**3.44**	**2.21**	**12.6**	**14.5**	**2.27**	**1.19**
DEVELOPED	**39 216**	**44 743**	**1.69**	**1.23**	**25.1**	**27.6**	**1.23**	**0.87**
NORTH AMERICA	**17 033**	**19 867**	**0.71**	**1.63**	**43.5**	**46.4**	**-0.21**	**0.82**
Canada	1 268	1 416	1.22	0.96	32.8	33.3	0.17	0.10
United States	15 764	18 451	0.67	1.68	44.7	47.8	-0.24	0.87
EUROPE	**16 912**	**18 627**	**2.42**	**0.77**	**20.1**	**21.9**	**2.23**	**0.67**
EU(27)	11 518	12 268	0.76	0.70	20.2	21.0	0.36	0.45
Russian Federation	3 585	4 032	6.11	0.27	22.1	25.2	6.30	0.43
Ukraine	1 084	1 469	12.37	2.51	21.0	30.2	13.04	3.05
OCEANIA DEVELOPED	**1 078**	**1 350**	**3.28**	**1.38**	**35.6**	**39.1**	**1.72**	**0.22**
Australia	929	1 168	3.72	1.32	36.7	40.3	2.08	0.12
New Zealand	149	182	0.85	1.77	30.0	32.8	-0.33	0.80
OTHER DEVELOPED[2]	**4 194**	**4 900**	**2.56**	**1.35**	**14.1**	**15.9**	**2.03**	**1.01**
Japan	1 795	1 769	0.78	0.02	12.5	12.5	0.74	0.20
South Africa	1 271	1 627	2.70	1.96	22.3	27.1	1.67	1.52
DEVELOPING	**59 266**	**82 259**	**4.72**	**2.79**	**9.4**	**11.5**	**3.37**	**1.61**
AFRICA	**4 055**	**5 787**	**4.92**	**3.14**	**3.7**	**4.1**	**2.55**	**0.85**
NORTH AFRICA	**2 027**	**2 746**	**4.66**	**2.65**	**10.8**	**12.7**	**3.12**	**1.38**
Algeria	269	314	0.64	1.51	6.7	6.8	-0.87	0.32
Egypt	893	1 277	4.35	3.13	9.7	11.7	2.55	1.64
SUB-SAHARAN AFRICA	**2 028**	**3 041**	**5.19**	**3.60**	**2.2**	**2.5**	**2.64**	**1.11**
LATIN AMERICA and CARIBBEAN	**19 581**	**25 773**	**5.12**	**2.23**	**29.2**	**34.5**	**3.93**	**1.26**
Argentina	1 337	1 919	8.69	2.12	29.1	38.2	7.81	1.32
Brazil	8 803	11 389	5.46	1.91	39.7	47.3	4.45	1.19
Chile	550	722	3.43	2.47	28.3	34.1	2.43	1.71
Mexico	3 325	4 277	3.63	2.14	25.8	29.6	2.38	1.15
Uruguay	83	103	10.00	1.47	21.8	25.7	9.81	1.10
ASIA and PACIFIC	**35 630**	**50 699**	**4.49**	**3.04**	**7.9**	**10.2**	**3.35**	**2.13**
Bangladesh	173	268	3.83	4.87	1.0	1.4	2.59	3.73
China[3]	16 971	24 228	3.23	3.04	11.1	15.3	2.71	2.75
India	2 657	4 401	7.19	4.71	1.9	2.8	5.73	3.51
Indonesia	1 442	1 947	3.37	2.97	5.3	6.5	2.25	2.11
Iran, Islamic Republic of	1 679	2 416	6.53	3.32	20.0	26.1	5.34	2.48
Korea	742	926	4.08	1.49	13.6	16.3	3.61	1.19
Malaysia	1 419	1 883	6.77	2.39	43.9	49.6	4.97	0.93
Pakistan	705	918	9.16	2.18	3.6	3.9	7.37	0.55
Saudi Arabia	1 254	1 732	4.74	2.59	40.2	44.7	1.79	0.67
Turkey	1 292	1 868	5.39	3.48	15.6	20.2	4.08	2.49
LEAST DEVELOPED COUNTRIES (LDC)	**2 678**	**3 968**	**6.94**	**3.47**	**2.8**	**3.3**	**4.70**	**1.31**
OECD[4]	**38 038**	**43 905**	**1.29**	**1.42**	**26.3**	**28.7**	**0.62**	**0.91**
NON-OECD	**60 444**	**83 098**	**4.97**	**2.66**	**9.5**	**11.5**	**3.68**	**1.52**

Note: Calendar year: Year ending 30 September for New Zealand.

1. Least-squares growth rate (see glossary).
2. Includes Israel and also transition economies: Kazakhstan, Kyrgyzstan, Tajikistan, Turkmenistan, Uzbekistan, Armenia, Azerbaijan and Georgia.
3. Refers to mainland only. The economies of Chinese Taipei, Hong Kong (China) and Macau (China) are included in the Other Asia Pacific aggregate.
4. Excludes Iceland but includes all EU27 member countries.
5. Per capita consumption expressed in retail weight. Carcass weight to retail weight conversion factors of 0.7 for beef and veal, 0.78 for pigmeat and 0.88 for both sheepmeat and poultry meat.

Source: OECD and FAO Secretariats.

StatLink http://dx.doi.org/10.1787/888932643124

Table A.23. Sheepmeat projections

Calendar year

	PRODUCTION (Kt cwe)[4]		Growth (%)[5]		IMPORTS (Kt cwe)[6]		Growth (%)[5]		EXPORTS (Kt cwe)[6]		Growth (%)[5]	
	Avg 2009-11est	2021	2002-2011	2012-2021	Avg 2009-11est	2021	2002-2011	2012-2021	Avg 2009-11est	2021	2002-2011	2012-2021
WORLD	**13 111**	**15 859**	**1.54**	**1.79**	**871**	**1 122**	**2.24**	**2.39**	**996**	**1 201**	**2.30**	**2.35**
DEVELOPED	**3 115**	**3 462**	**-0.37**	**1.20**	**411**	**440**	**-0.18**	**0.34**	**831**	**978**	**0.91**	**2.80**
NORTH AMERICA	**96**	**86**	**-3.07**	**-0.20**	**99**	**106**	**0.83**	**0.58**	**8**	**8**	**9.53**	**0.23**
Canada	15	17	-2.39	1.97	22	20	2.57	-0.11	0	0	3.61	0.00
United States	81	69	-3.19	-0.68	77	86	0.38	0.75	7	8	9.86	0.24
EUROPE	**1 199**	**1 197**	**-1.70**	**0.08**	**261**	**286**	**0.86**	**0.27**	**19**	**23**	**10.64**	**1.62**
EU(27)	896	818	-2.92	-0.65	243	262	0.63	0.25	12	15	..	0.54
Russian Federation	186	242	4.14	2.18	8	15	8.46	1.77	0	0	..	..
Ukraine	20	22	2.60	-0.12	0	0	35.38	5.91	0	0	-78.63	-0.42
OCEANIA DEVELOPED	**1 135**	**1 357**	**-0.75**	**2.08**	**5**	**3**	**-1.32**	**0.00**	**804**	**947**	**0.67**	**2.87**
Australia	656	793	-0.42	2.07	0	0	..	..	345	447	0.79	3.49
New Zealand	479	564	-1.14	2.10	5	3	-1.32	0.00	459	499	0.62	2.34
OTHER DEVELOPED[1]	**685**	**822**	**3.66**	**1.67**	**47**	**45**	**-6.01**	**0.23**	**1**	**1**	**24.10**	**-3.08**
Japan	0	0	..	..	35	33	-3.33	-0.05	0	0	..	..
South Africa	131	130	-0.86	-0.65	10	10	-13.52	1.71	0	0	11.43	-7.00
DEVELOPING	**9 996**	**12 397**	**2.19**	**1.97**	**460**	**682**	**4.81**	**3.95**	**165**	**223**	**13.33**	**0.37**
AFRICA	**2 442**	**3 332**	**2.50**	**2.98**	**37**	**115**	**6.91**	**11.02**	**19**	**58**	**1.97**	**3.25**
NORTH AFRICA	**524**	**612**	**1.45**	**1.30**	**16**	**30**	**3.04**	**8.86**	**0**	**0**	**2.31**	**0.39**
Algeria	194	244	1.04	2.05	1	1	-25.23	-10.47	0	0	44.43	0.76
Egypt	68	85	0.11	1.86	9	11	36.97	-1.30	0	0	-5.12	0.10
SUB-SAHARAN AFRICA	**1 918**	**2 720**	**2.81**	**3.39**	**21**	**86**	**10.98**	**11.96**	**19**	**58**	**1.97**	**3.25**
LATIN AMERICA and CARIBBEAN	**365**	**432**	**0.81**	**1.63**	**42**	**53**	**-4.95**	**1.21**	**35**	**20**	**10.26**	**-2.79**
Argentina	58	52	-0.76	-1.35	0	0	-9.39	0.00	10	11	12.99	-0.37
Brazil	80	83	1.70	0.58	7	6	9.62	1.44	0	0	-20.19	-1.82
Chile	16	17	1.35	0.67	0	0	43.53	0.18	7	7	4.26	-1.24
Mexico	55	73	3.96	2.52	22	26	-10.06	-0.96	0	0	..	..
Uruguay	26	16	-4.13	0.31	0	0	50.92	-50.35	18	1	12.24	-23.05
ASIA and PACIFIC	**7 189**	**8 633**	**2.16**	**1.61**	**382**	**513**	**6.26**	**3.16**	**112**	**145**	**18.80**	**-0.01**
Bangladesh	242	313	6.78	2.41	0	0	-17.23	11.43	0	0	-15.76	-0.81
China[2]	3 973	4 281	3.04	0.62	65	78	8.99	2.39	13	19	1.98	1.12
India	724	931	0.23	2.67	0	0	22.72	23.55	70	41	28.59	-5.78
Indonesia	132	163	-0.04	1.60	1	15	9.01	21.48	0	0	17.13	-1.61
Iran, Islamic Republic of	504	689	1.34	2.60	7	0	118.28	-35.17	0	0	61.34	-9.36
Korea	1	1	-9.67	0.00	5	4	6.45	0.00	0	0	-23.06	0.36
Malaysia	1	2	-1.34	7.01	21	30	6.56	2.33	0	0	24.09	-0.33
Pakistan	433	705	-2.31	3.82	0	0	31.96	-4.80	12	66	22.40	9.94
Saudi Arabia	9	12	15.13	1.83	57	107	1.24	3.87	2	16	0.29	-0.55
Turkey	300	351	-0.96	1.51	1	2	77.77	43.35	0	0	11.67	-3.11
LEAST DEVELOPED COUNTRIES (LDC)	**1 806**	**2 616**	**3.65**	**3.48**	**11**	**23**	**13.15**	**0.68**	**9**	**58**	**0.28**	**4.22**
OECD[3]	**2 537**	**2 745**	**-1.58**	**1.05**	**417**	**445**	**-0.42**	**0.26**	**831**	**977**	**0.92**	**2.77**
NON-OECD	**10 574**	**13 114**	**2.40**	**1.96**	**454**	**677**	**5.24**	**4.06**	**166**	**224**	**13.16**	**0.52**

Note: Calendar year: Year ending 30 September for New Zealand.

1. Includes Israel and also transition economies: Kazakhstan, Kyrgyzstan, Tajikistan, Turkmenistan, Uzbekistan, Armenia, Azerbaijan and Georgia.
2. Refers to mainland only. The economies of Chinese Taipei, Hong Kong (China) and Macau (China) are included in the Other Asia Pacific aggregate.
3. Excludes Iceland but includes all EU27 member countries.
4. Gross indigenous production.
5. Least-squares growth rate (see glossary).
6. Excludes trade of live animals.

Source: OECD and FAO Secretariats.

Table A.23. Sheepmeat projections *(cont.)*

Calendar year

	CONSUMPTION (Kt cwe)		Growth (%)[4]		PER CAPITA (Kg rwt)[5]		Growth (%)[4]	
	Avg 2009-11est	2021	2002-2011	2012-2021	Avg 2009-11est	2021	2002-2011	2012-2021
WORLD	**13 056**	**15 807**	**1.60**	**1.80**	**1.7**	**1.8**	**0.43**	**0.78**
DEVELOPED	**2 653**	**2 828**	**-0.53**	**0.65**	**1.7**	**1.7**	**-0.99**	**0.29**
NORTH AMERICA	**182**	**178**	**-1.45**	**0.23**	**0.5**	**0.4**	**-2.37**	**-0.58**
Canada	37	38	1.25	0.78	1.0	0.9	0.20	-0.07
United States	144	141	-2.06	0.09	0.4	0.4	-2.97	-0.72
EUROPE	**1 424**	**1 436**	**-1.50**	**0.14**	**1.7**	**1.7**	**-1.69**	**0.03**
EU(27)	1 115	1 055	-2.54	-0.40	2.0	1.8	-2.94	-0.65
Russian Federation	187	243	4.89	2.30	1.1	1.5	5.09	2.45
Ukraine	20	22	2.85	-0.08	0.4	0.5	3.52	0.45
OCEANIA DEVELOPED	**319**	**348**	**-1.54**	**0.72**	**10.5**	**10.1**	**-3.09**	**-0.44**
Australia	252	278	-0.64	0.83	10.0	9.6	-2.28	-0.37
New Zealand	67	70	-4.36	0.31	13.4	12.6	-5.54	-0.66
OTHER DEVELOPED[1]	**729**	**866**	**2.45**	**1.60**	**2.5**	**2.8**	**1.92**	**1.26**
Japan	35	33	-3.33	-0.05	0.2	0.2	-3.37	0.12
South Africa	140	141	-3.36	-0.47	2.5	2.3	-4.39	-0.91
DEVELOPING	**10 403**	**12 979**	**2.20**	**2.07**	**1.7**	**1.8**	**0.85**	**0.90**
AFRICA	**2 426**	**3 370**	**2.72**	**3.18**	**2.2**	**2.4**	**0.34**	**0.88**
NORTH AFRICA	**539**	**641**	**1.46**	**1.54**	**2.9**	**3.0**	**-0.08**	**0.27**
Algeria	195	245	0.66	2.03	4.8	5.3	-0.84	0.84
Egypt	77	96	1.51	1.46	0.8	0.9	-0.30	-0.04
SUB-SAHARAN AFRICA	**1 887**	**2 728**	**3.10**	**3.60**	**2.1**	**2.3**	**0.54**	**1.11**
LATIN AMERICA and CARIBBEAN	**379**	**470**	**-0.62**	**1.98**	**0.6**	**0.6**	**-1.80**	**1.01**
Argentina	48	40	-2.31	-1.63	1.0	0.8	-3.19	-2.43
Brazil	92	89	2.86	0.71	0.4	0.4	1.85	-0.01
Chile	9	11	-0.52	2.14	0.5	0.5	-1.52	1.37
Mexico	80	102	-2.87	1.43	0.6	0.7	-4.13	0.44
Uruguay	7	17	-15.54	9.91	1.9	4.2	-15.74	9.54
ASIA and PACIFIC	**7 598**	**9 139**	**2.20**	**1.70**	**1.7**	**1.8**	**1.06**	**0.79**
Bangladesh	242	314	6.75	2.42	1.4	1.6	5.52	1.28
China[2]	4 026	4 341	3.14	0.65	2.6	2.7	2.62	0.35
India	650	887	-0.99	3.26	0.5	0.6	-2.45	2.06
Indonesia	132	176	-0.09	2.37	0.5	0.6	-1.20	1.51
Iran, Islamic Republic of	505	680	1.57	2.53	6.0	7.3	0.38	1.69
Korea	6	6	0.58	0.00	0.1	0.1	0.11	-0.30
Malaysia	23	33	6.60	2.50	0.7	0.9	4.81	1.05
Pakistan	420	637	-2.64	3.38	2.1	2.7	-4.43	1.75
Saudi Arabia	137	180	-0.48	2.33	4.4	4.6	-3.43	0.42
Turkey	301	352	-0.81	1.57	3.6	3.8	-2.13	0.58
LEAST DEVELOPED COUNTRIES (LDC)	**1 777**	**2 555**	**3.88**	**3.47**	**1.9**	**2.1**	**1.64**	**1.31**
OECD[3]	**2 095**	**2 137**	**-2.02**	**0.27**	**1.5**	**1.4**	**-2.69**	**-0.24**
NON-OECD	**10 961**	**13 669**	**2.42**	**2.06**	**1.7**	**1.9**	**1.13**	**0.93**

Note: Calendar year: Year ending 30 September for New Zealand.

1. Includes Israel and also transition economies: Kazakhstan, Kyrgyzstan, Tajikistan, Turkmenistan, Uzbekistan, Armenia, Azerbaijan and Georgia.
2. Refers to mainland only. The economies of Chinese Taipei, Hong Kong (China) and Macau (China) are included in the Other Asia Pacific aggregate.
3. Excludes Iceland but includes all EU27 member countries.
4. Least-squares growth rate (see glossary).
5. Per capita consumption expressed in retail weight. Carcass weight to retail weight conversion factors of 0.7 for beef and veal, 0.78 for pigmeat and 0.88 for both sheepmeat and poultry meat.

Source: OECD and FAO Secretariats.

StatLink http://dx.doi.org/10.1787/888932643162

Table A.24. Main policy assumptions for meat markets

		Avg. 2009-11est	2012	2013	2014	2015	2016	2017	2018	2019	2020	2021
ARGENTINA												
Beef export tax	%	15.0	15.0	15.0	15.0	15.0	15.0	15.0	15.0	15.0	15.0	15.0
CANADA												
Beef tariff-quota	kt pw	76.4	76.4	76.4	76.4	76.4	76.4	76.4	76.4	76.4	76.4	76.4
In-quota tariff	%	0.0	0.0	0.0	0.0	0.0	0.0	0.0	0.0	0.0	0.0	0.0
Out-of-quota tariff	%	26.5	26.5	26.5	26.5	26.5	26.5	26.5	26.5	26.5	26.5	26.5
Poultry meat tariff-quota	kt pw	45.4	45.4	45.4	45.4	45.4	45.4	45.4	45.4	45.4	45.4	45.4
In-quota tariff	%	2.5	2.5	2.5	2.5	2.5	2.5	2.5	2.5	2.5	2.5	2.5
Out-of-quota tariff	%	196.6	196.6	196.6	196.6	196.6	196.6	196.6	196.6	196.6	196.6	196.6
EUROPEAN UNION[1]												
Beef budget ceiling[2]	'000 EUR	1 475 134	1 135 000	1 135 000	1 135 000	1 135 000	1 135 000	1 135 000	1 135 000	1 135 000	1 135 000	1 135 000
Beef basic price[3,4,5]	EUR/kg dw	2.2	2.2	2.2	2.2	2.2	2.2	2.2	2.2	2.2	2.2	2.2
Beef buy-in price[4]	EUR/kg dw	1.6	1.6	1.6	1.6	1.6	1.6	1.6	1.6	1.6	1.6	1.6
Pigmeat basic price[5]	EUR/kg dw	1.5	1.5	1.5	1.5	1.5	1.5	1.5	1.5	1.5	1.5	1.5
Sheep basic rate[6]	EUR/head	21.0	21.0	21.0	21.0	21.0	21.0	21.0	21.0	21.0	21.0	21.0
Beef tariff-quota	kt pw	215.7	215.7	215.7	215.7	215.7	215.7	215.7	215.7	215.7	215.7	215.7
Pig tariff-quota	kt pw	167.4	167.4	167.4	167.4	167.4	167.4	167.4	167.4	167.4	167.4	167.4
Poultry tariff-quota	kt pw	96.2	96.2	96.2	96.2	96.2	96.2	96.2	96.2	96.2	96.2	96.2
Sheepmeat tariff-quota	kt cwe	285.2	285.2	285.2	285.2	285.2	285.2	285.2	285.2	285.2	285.2	285.2
Subsidised export limits[5]												
Beef[7]	kt cwe	821.7	821.7	821.7	821.7	821.7	821.7	821.7	821.7	821.7	821.7	821.7
Pigmeat[7]	kt cwe	588.4	588.4	588.4	588.4	588.4	588.4	588.4	588.4	588.4	588.4	588.4
Poultry meat	kt cwe	430.8	430.8	430.8	430.8	430.8	430.8	430.8	430.8	430.8	430.8	430.8
JAPAN[8]												
Beef stabilisation prices												
Upper price	JPY/kg dw	1 060.0	1 060.0	1 060.0	1 060.0	1 060.0	1 060.0	1 060.0	1 060.0	1 060.0	1 060.0	1 060.0
Lower price	JPY/kg dw	815.0	815.0	815.0	815.0	815.0	815.0	815.0	815.0	815.0	815.0	815.0
Beef tariff	%	38.5	38.5	38.5	38.5	38.5	38.5	38.5	38.5	38.5	38.5	38.5
Pigmeat stabilisation prices												
Upper price	JPY/kg dw	545.0	545.0	545.0	545.0	545.0	545.0	545.0	545.0	545.0	545.0	545.0
Lower price	JPY/kg dw	400.0	400.0	400.0	400.0	400.0	400.0	400.0	400.0	400.0	400.0	400.0
Pigmeat import system[9]												
Tariff	%	4.3	4.3	4.3	4.3	4.3	4.3	4.3	4.3	4.3	4.3	4.3
Standard import price	JPY/kg dw	409.9	409.9	409.9	409.9	409.9	409.9	409.9	409.9	409.9	409.9	409.9
Poultry meat tariff	%	7.4	7.4	7.4	7.4	7.4	7.4	7.4	7.4	7.4	7.4	7.4
KOREA												
Beef tariff	%	40.0	37.3	34.7	32.0	29.3	26.7	18.0	16.0	14.0	12.0	10.0
Beef mark-up	%	0.0	0.0	0.0	0.0	0.0	0.0	0.0	0.0	0.0	0.0	0.0
Pigmeat tariff	%	22.5	22.5	22.5	22.5	22.5	22.5	15.8	14.6	13.5	12.4	11.3
Poultry meat tariff	%	21.0	21.0	21.0	21.0	21.0	21.0	21.0	21.0	21.0	21.0	21.0
MEXICO												
Pigmeat tariff	%	45.0	45.0	45.0	45.0	45.0	45.0	45.0	45.0	45.0	45.0	45.0
Pigmeat NAFTA tariff	%	0.0	0.0	0.0	0.0	0.0	0.0	0.0	0.0	0.0	0.0	0.0
Poultry meat tariff-quota	kt pw	40.5	40.5	40.5	40.5	40.5	40.5	40.5	40.5	40.5	40.5	40.5
In-quota tariff	%	50.0	50.0	50.0	50.0	50.0	50.0	50.0	50.0	50.0	50.0	50.0
Out-of-quota tariff	%	227.7	227.7	227.7	227.7	227.7	227.7	227.7	227.7	227.7	227.7	227.7
RUSSIAN FEDERATION												
Beef tariff-quota	kt pw	533.2	570.0	570.0	570.0	570.0	570.0	570.0	570.0	570.0	0.0	0.0
In-quota tariff	%	15.0	15.0	15.0	15.0	15.0	15.0	15.0	15.0	15.0	0.0	0.0
Out-of-quota tariff	%	43.3	50.0	50.0	55.0	55.0	55.0	55.0	55.0	55.0	27.5	27.5
Pigmeat tariff-quota	kt pw	510.6	430.0	430.0	430.0	430.0	430.0	430.0	430.0	430.0	0.0	0.0
In-quota tariff	%	15.0	0.0	0.0	0.0	0.0	0.0	0.0	0.0	0.0	0.0	0.0
Out-of-quota tariff	%	75.0	65.0	65.0	65.0	65.0	65.0	65.0	65.0	65.0	25.0	25.0
Poultry tariff-quota	kt pw	694.0	364.0	364.0	364.0	364.0	364.0	364.0	364.0	364.0	0.0	0.0
In-quota tariff	%	25.0	25.0	25.0	25.0	25.0	25.0	25.0	25.0	25.0	0.0	0.0
Out-of-quota tariff	%	85.0	80.0	80.0	80.0	80.0	80.0	80.0	80.0	80.0	37.5	37.5
UNITED STATES												
Beef tariff-quota	kt pw	696.6	696.6	696.6	696.6	696.6	696.6	696.6	696.6	696.6	696.6	696.6
In-quota tariff	%	4.8	4.8	4.8	4.8	4.8	4.8	4.8	4.8	4.8	4.8	4.8
Out-of-quota tariff	%	26.4	26.4	26.4	26.4	26.4	26.4	26.4	26.4	26.4	26.4	26.4

Table A.24. Main policy assumptions for meat markets *(cont.)*

		Avg. 2009-11est	2012	2013	2014	2015	2016	2017	2018	2019	2020	2021
CHINA												
Beef tariff	%	15.5	15.5	15.5	15.5	15.5	15.5	15.5	15.5	15.5	15.5	16.5
Pigmeat tariff	%	16.0	16.0	16.0	16.0	16.0	16.0	16.0	16.0	16.0	16.0	16.0
Sheepmeat tariff	%	15.0	15.0	15.0	15.0	15.0	15.0	15.0	15.0	15.0	15.0	15.0
Poultry meat tariff	%	19.0	19.0	19.0	19.0	19.0	19.0	19.0	19.0	19.0	19.0	19.0
INDIA												
Beef tariff	%	100.0	100.0	100.0	100.0	100.0	100.0	100.0	100.0	100.0	100.0	100.0
Pigmeat tariff	%	100.0	100.0	100.0	100.0	100.0	100.0	100.0	100.0	100.0	100.0	100.0
Sheepmeat tariff	%	91.9	91.9	91.9	91.9	91.9	91.9	91.9	91.9	91.9	91.9	91.9
Poultry meat tariff	%	87.0	87.0	87.0	87.0	87.0	87.0	87.0	87.0	87.0	87.0	87.0
Eggs tariff	%	150.0	150.0	150.0	150.0	150.0	150.0	150.0	150.0	150.0	150.0	150.0
SOUTH AFRICA												
Sheepmeat tariff-quota	kt pw	6.0	6.0	6.0	6.0	6.0	6.0	6.0	6.0	6.0	6.0	6.0
In-quota tariff	%	20.0	20.0	20.0	20.0	20.0	20.0	20.0	20.0	20.0	20.0	20.0
Out-of-quota tariff	%	96.0	96.0	96.0	96.0	96.0	96.0	96.0	96.0	96.0	96.0	96.0

Note: est.: estimate.

1. EU farmers also benefit from the Single Farm Payment (SFP) Scheme, which provides flat-rate payments independent from current production decisions and market developments. For the new member states, payments are phased in with the assumption of maximum top-ups from national budgets up to 2013 through the Single Area Payment (SAP), and through the (SFP) from 2014. Due to modulation, an increasing share of the total SFP will go to rural development spending rather than directly to farmers.
2. Coupled payment EU budget ceiling for suckler cow premia and other beef payments.
3. Price for R3 grade male cattle.
4. Basic price for storage.
5. Year beginning 1 July. Indicate WTO ceiling, not actual expenditure.
6. A supplementary payment of 7 euro per head is provided for Less Favoured Areas. Member states decide whether the payment remains coupled or is part of the Single Farm Payment; health check limit premia to 50%.
7. Includes live trade.
8. Year beginning 1 April.
9. Pig carcass imports. Emergency import procedures triggered from November 1995 to March 1996, from July 1996 to June 1997, from August 2001 to March 2002, from August 2002 to March 2003, from August 2003 to March 2004 and from August 2004 to March 2005.

Source: OECD and FAO Secretariats.

StatLink http://dx.doi.org/10.1787/888932643181

Table A.25. World fish and seafood projections

Calendar year

		Avg 2009-2011	2012	2013	2014	2015	2016	2017	2018	2019	2020	2021
FISH												
OECD												
Production	kt	31 519	31 551	31 904	32 233	31 725	32 211	32 547	32 697	32 773	32 064	32 573
of which aquaculture	kt	5 460	5 610	5 797	5 957	6 116	6 168	6 292	6 440	6 545	6 637	6 653
Consumption	kt	38 834	39 651	40 174	40 439	40 134	40 632	40 931	41 233	41 361	40 887	41 538
of which for food	kt	32 225	33 048	33 580	33 902	34 175	34 353	34 640	34 981	35 166	35 225	35 533
of which for reduction	kt	6 140	5 934	5 954	5 967	5 449	5 780	5 801	5 763	5 706	5 173	5 516
Non-OECD												
Production	kt	117 768	123 252	124 957	126 778	127 129	129 851	130 937	132 781	134 960	135 493	138 938
of which aquaculture	kt	53 519	57 266	58 628	60 252	62 081	63 168	64 074	65 906	68 089	69 791	71 860
Consumption	kt	109 564	114 741	116 527	118 561	118 710	121 419	122 543	124 234	126 362	126 660	129 963
of which for food	kt	95 036	100 388	102 255	104 227	105 614	107 203	108 480	110 224	112 421	113 785	116 019
of which for reduction	kt	10 459	10 708	10 711	10 699	9 511	10 681	10 577	10 575	10 556	9 539	10 658
World												
Production	kt	149 287	154 803	156 861	159 011	158 854	162 062	163 484	165 478	167 734	167 557	171 511
of which aquaculture	kt	58 979	62 876	64 425	66 209	68 198	69 336	70 367	72 347	74 634	76 428	78 514
Consumption	kt	148 397	154 392	156 700	159 000	158 844	162 052	163 473	165 467	167 723	167 547	171 501
of which for food	kt	127 260	133 435	135 835	138 129	139 789	141 556	143 120	145 204	147 587	149 010	151 552
of which for reduction	kt	16 599	16 642	16 665	16 667	14 959	16 461	16 378	16 338	16 262	14 712	16 174
Price												
Aquaculture[1]	USD/t	1 904.7	1 985.8	2 001.5	2 048.8	2 175.4	2 233.5	2 361.1	2 471.5	2 578.2	2 763.3	2 818.0
Capture[2]	USD/t	1 288.7	1 362.3	1 391.2	1 437.7	1 510.2	1 550.5	1 611.2	1 665.0	1 730.9	1 792.6	1 843.1
Product traded[3]	USD/t	2 500.1	2 684.0	2 689.8	2 759.3	2 914.9	2 964.3	3 082.6	3 169.1	3 295.8	3 400.8	3 474.8
FISH MEAL												
OECD												
Production	kt	1 857.7	1 890.3	1 909.5	1 933.2	1 827.7	1 923.4	1 945.9	1 953.2	1 954.5	1 839.5	1 940.8
from whole fish	kt	1 351.9	1 326.1	1 336.8	1 345.9	1 227.7	1 309.6	1 318.9	1 313.8	1 304.0	1 179.7	1 270.5
Consumption	kt	2 157.0	2 182.3	2 192.3	2 120.6	2 076.4	2 006.4	1 992.3	1 969.4	1 929.7	1 903.2	1 865.8
Variation in stocks	kt	-11.2	0.1	13.9	9.4	-73.7	73.7	-0.1	-0.2	-0.2	-63.1	63.0
Non-OECD												
Production	kt	4 102.2	4 371.6	4 413.9	4 442.0	4 195.2	4 547.5	4 574.9	4 628.4	4 696.9	4 501.2	4 885.5
from whole fish	kt	2 424.1	2 500.2	2 518.2	2 533.4	2 265.7	2 565.8	2 551.1	2 561.3	2 567.4	2 325.0	2 614.9
Consumption	kt	3 896.8	4 104.4	4 173.3	4 201.3	4 281.3	4 290.0	4 482.8	4 566.5	4 676.0	4 709.8	4 801.8
Variation in stocks	kt	-82.9	-25.7	-57.0	43.0	-262.0	100.0	45.0	45.0	45.0	-210.0	95.0
World												
Production	kt	5 960.0	6 262.0	6 323.3	6 375.2	6 022.9	6 470.9	6 520.8	6 581.5	6 651.4	6 340.7	6 826.3
from whole fish	kt	3 776.0	3 826.3	3 855.0	3 879.2	3 493.4	3 875.4	3 869.9	3 875.1	3 871.3	3 504.8	3 885.4
Consumption	kt	6 053.8	6 286.7	6 365.6	6 321.9	6 357.7	6 296.4	6 475.1	6 535.9	6 605.7	6 613.0	6 667.5
Variation in stocks	kt	-94.1	-25.6	-43.1	52.4	-335.7	173.7	44.9	44.8	44.8	-273.1	158.0
Price[4]	USD/t	1 484.8	1 565.1	1 702.4	1 868.1	1 995.5	2 016.7	2 047.8	2 146.6	2 246.5	2 347.2	2 360.8
FISH OIL												
OECD												
Production	kt	583.8	629.4	637.9	647.5	615.1	653.6	660.2	662.5	664.4	624.5	664.7
from whole fish	kt	293.3	288.5	289.6	290.4	265.3	281.2	282.6	280.8	278.1	252.3	269.4
Consumption	kt	859.1	868.7	868.6	861.0	820.5	807.3	823.7	822.0	820.1	795.8	821.4
Variation in stocks	kt	-30.9	16.3	4.0	3.6	-14.1	7.0	3.7	4.2	4.2	-18.3	5.1
Non-OECD												
Production	kt	422.8	456.8	456.8	456.3	397.3	456.7	454.2	454.3	454.0	402.6	456.7
from whole fish	kt	367.8	390.7	389.3	387.5	327.1	385.0	381.4	380.2	378.7	326.0	378.8
Consumption	kt	165.9	184.2	206.1	234.5	230.6	274.4	286.8	292.2	297.1	270.9	301.6
Variation in stocks	kt	1.2	-6.9	1.9	0.8	-24.5	21.6	0.2	-1.6	-2.9	-21.2	-6.7
World												
Production	kt	1 006.7	1 086.2	1 094.6	1 103.8	1 012.4	1 110.3	1 114.4	1 116.8	1 118.4	1 027.2	1 121.4
from whole fish	kt	661.1	679.1	678.9	677.9	592.4	666.2	664.0	661.0	656.8	578.3	648.1
Consumption	kt	1 024.9	1 052.9	1 074.7	1 095.5	1 051.0	1 081.8	1 110.5	1 114.2	1 117.1	1 066.8	1 123.0
Variation in stocks	kt	-29.6	9.3	5.9	4.3	-38.6	28.5	3.9	2.6	1.3	-39.6	-1.6
Price[5]	USD/t	1 148.2	1 378.4	1 432.1	1 464.5	1 789.9	1 615.5	1 622.5	1 702.6	1 769.1	2 176.5	1 785.5

Note: The term "fish" indicates fish, crustaceans, molluscs and other aquatic animals, but excludes aquatic mammals, crocodiles, caimans, alligators and aquatic plants.

1. World unit value of aquaculture fisheries production (live weight basis).
2. FAO estimated value of world ex vessel value of capture fisheries production excluding for reduction.
3. World unit value of trade (sum of exports and imports).
4. Fish meal, 64-65% protein, Hamburg, Germany.
5. Fish oil, any origin, N.W. Europe.

Source: OECD and FAO Secretariats.

StatLink http://dx.doi.org/10.1787/888932643200

Table A.26. Fish and seafood projections

Calendar year

	PRODUCTION (Kt)		Growth (%)[3]		IMPORTS (Kt)		Growth (%)[3]		EXPORTS (Kt)		Growth (%)[3]	
	Avg 2009-11est	2021	2002-2011	2012-2021	Avg 2009-11est	2021	2002-2011	2012-2021	Avg 2009-11est	2021	2002-2011	2012-2021
WORLD	**149 287**	**171 511**	**2.00**	**0.57**	**34 742**	**44 683**	**3.11**	**2.01**	**35 652**	**44 683**	**3.62**	**1.94**
DEVELOPED	**28 081**	**29 397**	**-1.25**	**0.57**	**20 350**	**24 841**	**1.81**	**1.70**	**12 048**	**14 934**	**1.52**	**2.26**
NORTH AMERICA	**5 865**	**5 931**	**-2.00**	**-0.79**	**5 530**	**7 240**	**2.66**	**1.95**	**2 727**	**3 529**	**1.06**	**1.79**
Canada	1 092	1 115	-2.20	1.55	647	805	1.70	1.58	878	1 032	-0.84	1.64
United States	4 773	4 817	-1.96	2.35	4 882	6 435	2.79	2.00	1 849	2 496	2.03	1.85
EUROPE	**16 032**	**17 023**	**-0.65**	**-0.82**	**10 207**	**12 746**	**2.92**	**2.04**	**8 035**	**9 678**	**1.81**	**2.26**
EU(27)	6 596	6 754	-1.53	1.34	7 469	9 483	2.41	2.25	2 270	2 988	-0.19	3.00
Norway	3 443	3 824	0.94	-0.54	252	162	-1.50	-3.79	2 864	3 231	4.05	2.06
Russian Federation	4 116	4 451	3.09	-0.83	1 291	1 773	6.72	2.81	1 459	1 937	0.66	3.20
OCEANIA DEVELOPED	**787**	**896**	**-1.98**	**-2.27**	**491**	**703**	**3.27**	**3.10**	**480**	**473**	**-2.33**	**0.03**
Australia	241	281	-0.74	-0.80	442	655	3.06	3.36	62	28	-4.27	-6.82
New Zealand	546	614	-2.48	0.00	48	48	5.43	0.00	419	444	-2.02	0.64
OTHER DEVELOPED[1]	**5 397**	**5 547**	**-2.05**	**1.00**	**4 123**	**4 153**	**-1.63**	**0.11**	**806**	**1 254**	**2.88**	**4.79**
Japan	4 711	4 579	-1.79	-3.32	3 680	3 575	-2.48	-0.26	570	975	4.22	5.94
South Africa	560	794	-5.18	-6.34	164	213	14.73	2.76	164	202	-2.16	1.56
DEVELOPING	**121 206**	**142 115**	**2.87**	**-1.18**	**14 392**	**19 842**	**5.17**	**2.41**	**23 604**	**29 750**	**4.81**	**1.78**
AFRICA	**8 128**	**9 142**	**2.20**	**0.34**	**2 519**	**3 192**	**2.78**	**1.71**	**1 661**	**1 431**	**0.43**	**-1.09**
NORTH AFRICA	**2 627**	**3 000**	**3.31**	**0.17**	**513**	**776**	**7.23**	**3.29**	**479**	**438**	**4.29**	**-0.07**
Egypt	1 227	1 459	5.33	0.37	322	514	5.22	3.65	9	0	9.09	..
SUB-SAHARAN AFRICA	**5 501**	**6 143**	**1.70**	**-0.37**	**2 007**	**2 415**	**1.86**	**1.26**	**1 182**	**993**	**-0.87**	**-1.51**
Ghana	353	371	-1.28	-2.93	243	314	2.73	2.54	19	13	-19.53	-2.02
Nigeria	793	910	6.15	0.06	488	754	-4.41	3.75	6	6	4.45	0.00
LATIN AMERICA and CARIBBEAN	**16 429**	**17 216**	**-1.60**	**3.80**	**1 916**	**2 500**	**8.64**	**1.97**	**3 652**	**4 467**	**1.27**	**2.09**
Argentina	844	1 024	-1.31	2.08	50	50	12.07	0.00	646	815	-0.87	1.55
Brazil	1 258	1 605	3.20	-1.03	527	597	9.39	-0.84	46	93	-13.07	7.13
Chile	3 962	4 231	-2.71	-0.23	39	40	2.99	0.00	1 147	1 477	-0.48	2.24
Mexico	1 691	1 603	2.17	-1.38	268	500	11.20	4.79	183	212	1.57	1.88
Peru	6 518	6 554	-3.17	-1.43	114	70	11.94	-1.26	533	732	10.93	4.64
ASIA and PACIFIC	**96 649**	**115 757**	**3.86**	**0.91**	**9 957**	**14 151**	**5.24**	**2.66**	**18 291**	**23 852**	**6.14**	**1.92**
China[2]	51 717	62 899	3.78	-0.48	3 070	4 060	6.87	1.24	8 038	11 317	8.78	2.32
India	8 869	11 007	5.42	-3.94	22	330	19.06	17.81	784	540	5.07	-3.84
Indonesia	7 404	8 694	4.30	0.24	248	480	25.70	5.20	1 072	1 057	3.63	-3.30
Korea	2 266	2 232	1.84	-0.60	1 529	1 611	0.39	0.13	829	614	7.56	-1.51
Philippines	3 370	3 669	3.75	5.88	272	426	10.30	2.22	327	234	6.80	-1.53
Thailand	3 159	3 323	-3.30	0.36	1 700	2 924	5.98	5.07	2 492	3 929	3.63	4.22
Viet Nam	5 050	6 666	8.47	2.42	118	199	20.28	4.88	1 550	2 398	10.80	3.92
LEAST DEVELOPED COUNTRIES (LDC)	**11 262**	**13 447**	**5.87**	**1.30**	**537**	**657**	**8.26**	**0.84**	**1 300**	**1 591**	**3.11**	**1.58**
OECD	**31 519**	**32 573**	**-1.47**	**0.31**	**19 690**	**23 904**	**1.31**	**1.62**	**12 376**	**14 939**	**1.71**	**2.02**
NON-OECD	**117 768**	**138 938**	**3.09**	**1.27**	**15 052**	**20 779**	**5.86**	**2.47**	**23 276**	**29 744**	**4.74**	**1.90**

Table A.26. Fish and seafood projections *(cont.)*

Calendar year

	REDUCTION (Kt)		Growth (%)[3]		FOOD USE (Kt)		Growth (%)[3]		PER CAPITA (Kg)		Growth (%)[3]	
	Avg 2009-11est	2021	2002-2011	2012-2021	Avg 2009-11est	2021	2002-2011	2012-2021	Avg 2009-11est	2021	2002-2011	2012-2021
WORLD	**16 599**	**16 174**	**-3.98**	**-0.60**	**127 260**	**151 552**	**2.87**	**1.36**	**18.5**	**19.6**	**1.70**	**0.34**
DEVELOPED	**3 522**	**3 137**	**-7.72**	**-0.83**	**32 448**	**35 718**	**0.66**	**0.71**	**23.6**	**25.0**	**0.20**	**0.35**
NORTH AMERICA	**523**	**411**	**-6.26**	**-2.93**	**7 888**	**9 002**	**0.35**	**1.18**	**22.9**	**23.9**	**-0.57**	**0.36**
Canada	52	83	-8.08	2.35	777	774	0.36	-0.36	22.8	20.7	-0.69	-1.22
United States	472	328	-6.05	-3.94	7 111	8 227	0.35	1.33	22.9	24.2	-0.55	0.52
EUROPE	**1 885**	**1 606**	**-10.34**	**-0.80**	**16 163**	**18 265**	**2.19**	**0.91**	**21.8**	**24.4**	**2.00**	**0.81**
EU(27)	691	555	-7.47	-2.27	11 038	12 544	1.27	1.06	22.0	24.4	0.87	0.81
Norway	530	429	-10.71	0.06	281	325	3.29	0.96	57.6	61.7	2.33	0.29
Russian Federation	334	302	7.04	-1.43	3 553	3 926	5.21	0.49	24.9	27.9	5.40	0.64
OCEANIA DEVELOPED	**99**	**120**	**-5.07**	**2.08**	**698**	**1 006**	**2.49**	**3.17**	**26.2**	**33.1**	**0.93**	**2.01**
Australia	43	39	0.16	-0.79	578	869	2.55	3.47	26.0	34.1	0.92	2.27
New Zealand	56	81	-7.75	3.80	120	137	2.18	1.43	27.5	28.1	1.00	0.46
OTHER DEVELOPED[1]	**1 014**	**999**	**-2.46**	**-0.23**	**7 699**	**7 446**	**-1.96**	**-0.57**	**29.5**	**27.4**	**-2.49**	**-0.91**
Japan	750	564	-2.22	-3.32	7 071	6 616	-2.21	-0.83	55.9	53.2	-2.25	-0.65
South Africa	265	435	-3.26	5.88	295	369	-2.46	0.57	5.9	7.0	-3.50	0.13
DEVELOPING	**13 077**	**13 037**	**-2.79**	**-0.54**	**94 812**	**115 834**	**3.71**	**1.57**	**17.2**	**18.4**	**2.36**	**0.40**
AFRICA	**439**	**447**	**-5.59**	**0.57**	**8 530**	**10 438**	**3.49**	**1.42**	**8.8**	**8.3**	**1.11**	**-0.87**
NORTH AFRICA	**297**	**281**	**-2.49**	**0.37**	**2 355**	**3 050**	**4.70**	**1.85**	**14.2**	**16.0**	**3.16**	**0.58**
Egypt	0	0	..	..	1 541	1 973	5.27	1.73	19.0	20.5	3.46	0.24
SUB-SAHARAN AFRICA	**142**	**166**	**-10.24**	**0.91**	**6 175**	**7 387**	**3.06**	**1.25**	**7.7**	**7.0**	**0.50**	**-1.24**
Ghana	0	0	..	..	576	672	1.39	1.23	23.6	21.7	-1.01	-0.90
Nigeria	0	0	..	..	1 275	1 658	1.11	2.27	8.0	7.9	-1.38	-0.24
LATIN AMERICA and CARIBBEAN	**8 668**	**8 283**	**-3.58**	**-1.18**	**5 891**	**6 915**	**2.95**	**1.19**	**10.0**	**10.5**	**1.76**	**0.22**
Argentina	0	0	..	..	248	258	-1.14	-0.56	6.1	5.8	-2.02	-1.36
Brazil	95	109	5.14	1.55	1 643	2 000	5.81	1.41	8.4	9.4	4.80	0.69
Chile	2 502	2 427	-3.26	-0.82	336	367	-1.55	-0.29	19.6	19.7	-2.55	-1.05
Mexico	385	360	2.83	0.17	1 390	1 531	3.44	0.51	12.3	12.1	2.19	-0.48
Peru	5 405	5 200	-4.64	-1.38	693	693	4.24	1.13	23.8	21.1	3.13	0.05
ASIA and PACIFIC	**3 970**	**4 307**	**-0.62**	**0.57**	**80 391**	**98 481**	**3.80**	**1.62**	**20.3**	**22.4**	**2.66**	**0.71**
China[2]	1 390	1 583	-4.88	1.34	42 932	51 558	3.59	1.49	32.0	37.1	3.07	1.20
India	465	536	7.71	1.00	7 443	10 161	5.21	2.34	6.1	7.2	3.75	1.14
Indonesia	73	65	10.65	0.00	6 457	8 052	4.82	1.92	26.9	30.4	3.70	1.06
Korea	175	72	2.36	-6.34	2 694	3 058	1.07	0.96	55.9	61.3	0.60	0.66
Philippines	0	0	..	..	3 315	3 860	3.88	0.84	35.5	34.6	2.09	-0.76
Thailand	727	729	-3.98	-0.48	1 667	1 590	-3.43	-0.20	24.1	22.0	-4.23	-0.57
Viet Nam	353	382	10.69	0.24	3 123	4 059	7.00	2.07	35.5	41.8	5.91	1.20
LEAST DEVELOPED COUNTRIES (LDC)	**240**	**267**	**21.62**	**0.34**	**9 392**	**11 864**	**5.69**	**1.79**	**11.3**	**11.2**	**3.44**	**-0.37**
OECD	**6 140**	**5 516**	**-5.85**	**-1.03**	**32 225**	**35 533**	**0.29**	**0.76**	**25.4**	**26.4**	**-0.39**	**0.24**
NON-OECD	**10 459**	**10 658**	**-2.78**	**-0.37**	**95 036**	**116 019**	**3.86**	**1.56**	**16.9**	**18.2**	**2.58**	**0.43**

Note: Fish: The term "fish" indicates fish, crustaceans, molluscs and other aquatic animals, but excludes aquatic mammals; crocodiles,caimans, alligators and aquatic plants.

1. Includes Israel and also transition economies: Kazakhstan, Kyrgyzstan, Tajikistan, Turkmenistan, Uzbekistan, Armenia, Azerbaijan and Georgia.
2. Refers to mainland only. The economies of Chinese Taipei, Hong Kong (China) and Macau (China) are included in the Other Asia Pacific aggregate.
3. Least-squares growth rate (see glossary).

Source: OECD and FAO Secretariats.

StatLink http://dx.doi.org/10.1787/888932643219

Table A.27. World fish and seafood trade projections

Calendar year

		Avg 2009-11est.	2012	2013	2014	2015	2016	2017	2018	2019	2020	2021
IMPORTS												
Fish												
World	**kt**	**34 742**	**37 126**	**38 095**	**38 959**	**39 856**	**40 679**	**41 376**	**42 100**	**42 967**	**43 788**	**44 683**
OECD	kt	19 690	20 646	21 003	21 235	21 620	21 976	22 295	22 716	23 087	23 481	23 904
Developing	kt	14 392	15 782	16 393	16 907	17 413	17 823	18 154	18 425	18 935	19 370	19 842
Least Developed Countries	kt	537	604	606	605	608	615	616	622	628	640	657
Fishmeal												
World Trade	**kt**	**3 251**	**3 085**	**3 101**	**3 021**	**2 832**	**2 895**	**2 932**	**2 915**	**2 879**	**2 742**	**2 878**
OECD	kt	1 295	1 146	1 168	1 118	1 012	1 073	1 032	1 024	1 008	944	1 019
Developing	kt	1 998	1 978	1 985	1 967	1 894	1 902	1 989	1 989	1 974	1 908	1 967
Least Developed Countries	kt	21	22	25	29	31	31	32	32	33	33	34
Fish oil												
World Trade	**kt**	**818**	**775**	**759**	**747**	**678**	**696**	**718**	**718**	**718**	**670**	**709**
OECD	kt	663	602	596	586	549	544	554	551	548	528	541
Developing	kt	253	300	288	283	246	257	274	275	276	239	271
Least Developed Countries	kt	1	0	0	0	0	0	0	0	0	0	0
EXPORTS												
Fish												
World	**kt**	**35 652**	**37 526**	**38 195**	**38 959**	**39 856**	**40 679**	**41 376**	**42 100**	**42 967**	**43 788**	**44 683**
OECD	kt	12 376	12 546	12 733	13 029	13 212	13 555	13 912	14 180	14 500	14 658	14 939
Developing	kt	23 604	25 292	25 715	26 196	26 767	27 269	27 631	28 108	28 631	29 142	29 750
Least Developed Countries	kt	1 300	1 352	1 410	1 441	1 483	1 469	1 474	1 512	1 530	1 576	1 591
Fishmeal												
World Trade	**kt**	**2 862**	**2 735**	**2 801**	**2 771**	**2 632**	**2 745**	**2 832**	**2 865**	**2 879**	**2 742**	**2 878**
OECD	kt	1 007	854	871	921	837	917	986	1 008	1 033	943	1 031
Developing	kt	2 317	2 192	2 245	2 192	2 032	2 135	2 210	2 226	2 224	2 069	2 195
Least Developed Countries	kt	53	50	49	49	48	48	47	47	46	46	45
Fish oil												
World Trade	**kt**	**739**	**695**	**699**	**707**	**658**	**696**	**718**	**718**	**718**	**670**	**709**
OECD	kt	416	347	361	369	358	383	387	387	388	375	379
Developing	kt	411	419	413	417	361	409	427	429	431	380	431
Least Developed Countries	kt	1	1	2	2	2	2	2	2	2	2	2

Source: OECD and FAO Secretariats.

StatLink http://dx.doi.org/10.1787/888932643238

Table A.28. World dairy projections (butter and cheese)

Calendar year

		Avg 2009-11est.	2012	2013	2014	2015	2016	2017	2018	2019	2020	2021
BUTTER												
OECD[1]												
Production	kt pw	3 868	3 941	3 981	4 035	4 075	4 117	4 142	4 172	4 199	4 234	4 256
Consumption	kt pw	3 296	3 328	3 361	3 387	3 409	3 435	3 445	3 459	3 477	3 501	3 515
Stock changes	kt pw	-13	-1	-11	-4	-5	-2	1	2	2	2	0
Non-OECD												
Production	kt pw	6 517	6 847	6 985	7 186	7 401	7 588	7 765	7 998	8 303	8 627	8 877
Consumption	kt pw	7 063	7 423	7 571	7 792	8 026	8 226	8 416	8 663	8 978	9 312	9 571
WORLD												
Production	kt pw	10 385	10 788	10 966	11 220	11 476	11 704	11 907	12 170	12 502	12 861	13 132
Consumption	kt pw	10 359	10 751	10 932	11 179	11 436	11 661	11 860	12 122	12 455	12 814	13 086
Stock changes	kt pw	-21	-9	-11	-4	-5	-2	1	2	2	2	0
Price[2]	USD/t	3 626	3 860	3 821	3 765	3 774	3 795	3 912	4 032	4 087	4 136	4 214
CHEESE												
OECD[1]												
Production	kt pw	15 310	15 733	15 864	16 112	16 363	16 614	16 813	17 058	17 275	17 547	17 759
Consumption	kt pw	14 607	14 956	15 141	15 340	15 563	15 800	15 995	16 233	16 426	16 661	16 851
Stock changes	kt pw	19	2	-4	-1	-1	-3	-4	-5	-2	-2	-1
Non-OECD												
Production	kt pw	4 624	4 844	5 016	5 107	5 218	5 330	5 446	5 568	5 682	5 783	5 899
Consumption	kt pw	5 331	5 641	5 768	5 899	6 038	6 166	6 286	6 417	6 552	6 690	6 827
WORLD												
Production	kt pw	19 933	20 578	20 881	21 219	21 581	21 944	22 258	22 625	22 957	23 330	23 658
Consumption	kt pw	19 938	20 597	20 909	21 239	21 601	21 966	22 281	22 650	22 978	23 351	23 678
Stock changes	kt pw	25	-1	-9	-1	-1	-3	-4	-5	-2	-2	-1
Price[3]	USD/t	3 761	3 855	3 886	3 824	3 872	3 924	4 025	4 123	4 188	4 252	4 327

Note: Calendar year: Year ending 30 June for Australia and 31 May for New Zealand in OECD aggregate.

1. Excludes Iceland but includes all EU27 member countries.
2. F.o.b. export price, butter, 82% butterfat, Oceania.
3. F.o.b. export price, cheddar cheese, 39% moisture, Oceania.

Source: OECD and FAO Secretariats.

StatLink http://dx.doi.org/10.1787/888932643257

Table A.29. World dairy projections (powders and casein)

Calendar year

		Avg 2009-11est.	2012	2013	2014	2015	2016	2017	2018	2019	2020	2021
SKIM MILK POWDER												
OECD[1]												
Production	kt pw	2 789	2 877	2 965	3 017	3 066	3 119	3 159	3 199	3 255	3 317	3 374
Consumption	kt pw	1 706	1 635	1 635	1 665	1 687	1 703	1 712	1 724	1 747	1 777	1 798
Stock changes	kt pw	-38	-96	8	-7	-6	4	8	4	4	2	8
Non-OECD												
Production	kt pw	652	675	711	722	742	760	781	801	820	836	859
Consumption	kt pw	1 718	1 975	1 995	2 042	2 088	2 134	2 181	2 234	2 285	2 335	2 388
WORLD												
Production	kt pw	3 441	3 552	3 676	3 739	3 807	3 880	3 939	4 000	4 075	4 153	4 233
Consumption	kt pw	3 424	3 609	3 630	3 707	3 774	3 837	3 893	3 957	4 032	4 112	4 186
Stock changes	kt pw	-8	-96	8	-7	-6	4	8	4	4	2	8
Price[2]	USD/t	3 018	3 188	3 348	3 338	3 428	3 480	3 567	3 622	3 689	3 761	3 841
WHOLE MILK POWDER												
OECD[1]												
Production	kt pw	2 187	2 486	2 490	2 545	2 588	2 633	2 666	2 710	2 757	2 800	2 851
Consumption	kt pw	765	838	836	842	849	857	865	875	883	892	902
Stock changes	kt pw	1	0	0	0	0	0	0	0	0	0	0
Non-OECD												
Production	kt pw	2 081	2 239	2 302	2 342	2 398	2 454	2 523	2 586	2 643	2 703	2 764
Consumption	kt pw	3 620	3 816	3 886	3 974	4 067	4 160	4 255	4 351	4 448	4 541	4 643
WORLD												
Production	kt pw	4 267	4 724	4 792	4 886	4 986	5 086	5 190	5 296	5 400	5 503	5 614
Consumption	kt pw	4 385	4 655	4 722	4 817	4 916	5 017	5 120	5 226	5 330	5 433	5 545
Stock changes	kt pw	40	0	0	0	0	0	0	0	0	0	0
Price[3]	USD/t	3 252	3 423	3 428	3 455	3 528	3 574	3 667	3 750	3 826	3 913	4 007
WHEY POWDER												
Wholesale price, United States[4]	USD/t	897	1 094	1 104	1 120	1 129	1 144	1 164	1 179	1 209	1 233	1 247
CASEIN												
Price[5]	USD/t	7 917	7 850	8 434	8 554	8 668	8 674	8 931	9 036	9 253	9 433	9 665

Note: Calendar year: Year ending 30 June for Australia and 31 May for New Zealand in OECD aggregate.

1. Excludes Iceland but includes all EU27 member countries.
2. F.o.b. export price, non-fat dry milk, 1.25% butterfat,Oceania.
3. F.o.b. export price, WMP 26% butterfat, Oceania.
4. Dry whey, West Region, United States.
5. Export price, New Zealand.

Source: OECD and FAO Secretariats.

StatLink http://dx.doi.org/10.1787/888932643276

Table A.30. Butter projections

Calendar year

	PRODUCTION (Kt)		Growth (%)[4]		IMPORTS (Kt)		Growth (%)[4]		EXPORTS (Kt)		Growth (%)[4]	
	Avg 2009-11est	2021	2002-2011	2012-2021	Avg 2009-11est	2021	2002-2011	2012-2021	Avg 2009-11est	2021	2002-2011	2012-2021
WORLD	**10 385**	**13 132**	**3.33**	**2.19**	**855**	**1 001**	**-2.74**	**1.73**	**858**	**1 025**	**-0.23**	**1.68**
DEVELOPED	**4 290**	**4 730**	**0.40**	**0.84**	**286**	**250**	**-8.39**	**-0.20**	**772**	**928**	**-0.90**	**1.69**
NORTH AMERICA	**825**	**1 034**	**3.19**	**1.82**	**19**	**14**	**-6.57**	**0.00**	**45**	**84**	**27.74**	**3.78**
Canada	82	79	0.28	-0.21	6	6	-6.82	0.00	1	1	-1.35	0.00
United States	743	955	3.55	2.00	13	8	-6.79	0.01	44	83	30.40	3.83
EUROPE	**2 705**	**2 783**	**-0.46**	**0.19**	**194**	**154**	**-11.36**	**-0.80**	**257**	**224**	**-4.85**	**-0.53**
EU(27)	2 106	2 104	-0.74	-0.08	44	31	-12.13	1.27	153	119	-9.23	-1.14
Russian Federation	319	374	1.45	1.34	133	116	-12.46	-1.04	4	3	-4.44	0.00
Ukraine	74	90	-8.09	1.58	9	0	43.58	-27.07	1	2	-41.47	5.41
OCEANIA DEVELOPED	**608**	**734**	**0.46**	**1.99**	**18**	**20**	**11.71**	**0.57**	**467**	**618**	**0.21**	**2.36**
Australia	133	125	-3.40	0.32	17	19	11.78	0.59	67	60	-6.93	0.31
New Zealand	475	609	1.78	2.36	1	1	..	0.00	401	557	1.84	2.60
OTHER DEVELOPED[1]	**152**	**179**	**2.28**	**1.51**	**56**	**62**	**1.53**	**1.14**	**3**	**3**	**2.59**	**-0.74**
Japan	72	66	-2.50	-0.24	5	6	-6.61	-2.16	0	0	..	..
South Africa	12	12	0.55	-0.15	3	5	0.20	4.46	1	1	-1.69	-3.66
DEVELOPING	**6 096**	**8 403**	**5.83**	**3.04**	**569**	**751**	**1.31**	**2.45**	**86**	**97**	**8.03**	**1.60**
AFRICA	**279**	**364**	**4.04**	**2.31**	**137**	**177**	**-0.14**	**2.02**	**2**	**2**	**10.25**	**0.15**
NORTH AFRICA	**174**	**206**	**3.56**	**1.39**	**118**	**146**	**-0.08**	**1.77**	**0**	**0**	**-1.33**	**-1.41**
Algeria	2	3	4.46	2.08	11	16	-3.46	1.52	0	0	-6.51	-0.22
Egypt	138	158	3.50	1.02	75	96	2.89	2.08	0	0	4.63	-3.58
SUB-SAHARAN AFRICA	**106**	**158**	**4.89**	**3.65**	**19**	**31**	**-0.53**	**3.24**	**2**	**2**	**13.20**	**0.27**
LATIN AMERICA and CARIBBEAN	**244**	**309**	**2.86**	**2.01**	**60**	**51**	**-3.54**	**0.13**	**47**	**62**	**12.64**	**2.85**
Argentina	57	67	5.33	1.55	0	0	..	..	23	37	34.27	3.06
Brazil	78	94	1.29	1.53	3	3	-4.21	0.77	2	1	27.61	-5.19
Chile	20	24	7.00	1.92	1	1	-11.15	-1.14	3	3	36.61	1.14
Mexico	14	15	-2.58	0.31	33	30	-4.86	1.43	1	1	17.91	1.97
Uruguay	21	24	2.87	1.97	0	0	15.18	-0.39	13	14	0.67	2.72
ASIA and PACIFIC	**5 572**	**7 730**	**6.08**	**3.12**	**372**	**523**	**2.85**	**2.86**	**37**	**33**	**4.14**	**-0.39**
Bangladesh	24	41	3.53	4.68	1	0	-9.20	-54.11	0	0	-9.50	3.89
China[2]	112	136	2.75	1.50	33	49	16.65	2.07	3	3	61.88	0.00
India	4 269	6 076	7.14	3.36	21	5	25.87	-6.88	7	10	16.35	1.45
Indonesia	0	0	0.00	-35.94	16	19	2.43	2.11	0	0	9.59	-0.30
Iran, Islamic Republic of	236	302	5.27	2.06	58	107	4.86	5.33	1	1	39.46	-0.59
Korea	4	7	-5.06	2.58	5	5	15.69	2.07	0	0	..	..
Malaysia	0	0	0.00	1.23	13	17	2.57	1.84	3	3	24.01	-0.26
Pakistan	644	778	2.41	1.95	0	20	14.53	15.59	0	0	32.54	-1.13
Saudi Arabia	5	7	-0.15	8.09	50	80	3.22	3.47	2	2	-9.82	-0.50
Turkey	151	201	4.01	2.88	12	8	10.48	-3.74	0	0	14.73	0.31
LEAST DEVELOPED COUNTRIES (LDC)	**181**	**268**	**2.03**	**3.54**	**1 218**	**177**	**-0.60**	**-15.04**	**5**	**3**	**33.77**	**-2.46**
OECD[3]	**3 868**	**4 256**	**0.40**	**0.85**	**137**	**116**	**-5.11**	**0.41**	**678**	**835**	**-1.49**	**1.86**
NON-OECD	**6 517**	**8 877**	**5.41**	**2.91**	**717**	**884**	**-2.31**	**1.91**	**180**	**190**	**6.28**	**0.93**

Note: Calendar year: Year ending 30 June for Australia and 31 May for New Zealand.

1. Includes Israel and also transition economies: Kazakhstan, Kyrgyzstan, Tajikistan, Turkmenistan, Uzbekistan, Armenia, Azerbaijan and Georgia.
2. Refers to mainland only. The economies of Chinese Taipei, Hong Kong (China) and Macau (China) are included in the Other Asia Pacific aggregate.
3. Excludes Iceland but includes all EU27 member countries.
4. Least-squares growth rate (see glossary).

Source: OECD and FAO Secretariats.

Table A.30. Butter projections *(cont.)*

Calendar year

	CONSUMPTION (Kt)		Growth (%)[1]		PER CAPITA (Kg)		Growth (%)[1]	
	Avg 2009-11est	2021	2002-2011	2012-2021	Avg 2009-11est	2021	2002-2011	2012-2021
WORLD	**10 359**	**13 086**	**3.14**	**2.19**	**1.5**	**1.7**	**1.97**	**1.17**
DEVELOPED	**3 780**	**4 030**	**-0.05**	**0.56**	**2.8**	**2.8**	**-0.51**	**0.20**
NORTH AMERICA	**801**	**964**	**2.18**	**1.58**	**2.3**	**2.6**	**1.26**	**0.77**
Canada	89	85	-0.46	-0.15	2.6	2.3	-1.51	-1.01
United States	712	880	2.55	1.76	2.3	2.6	1.64	0.95
EUROPE	**2 658**	**2 713**	**-0.89**	**0.15**	**3.6**	**3.6**	**-1.08**	**0.05**
EU(27)	2 005	2 016	0.07	-0.03	4.0	3.9	-0.33	-0.27
Russian Federation	455	487	-3.86	0.65	3.2	3.5	-3.67	0.81
Ukraine	82	88	-4.04	1.00	1.8	2.1	-3.36	1.54
OCEANIA DEVELOPED	**116**	**115**	**2.14**	**0.29**	**4.4**	**3.8**	**0.58**	**-0.87**
Australia	85	84	3.04	0.39	3.8	3.3	1.40	-0.81
New Zealand	31	32	0.01	0.04	7.2	6.5	-1.18	-0.93
OTHER DEVELOPED[2]	**206**	**238**	**2.19**	**1.46**	**0.8**	**0.9**	**1.65**	**1.12**
Japan	79	72	-1.82	-0.34	0.6	0.6	-1.86	-0.16
South Africa	14	16	0.59	1.26	0.3	0.3	-0.44	0.82
DEVELOPING	**6 579**	**9 056**	**5.36**	**3.00**	**1.2**	**1.4**	**4.01**	**1.83**
AFRICA	**414**	**539**	**2.52**	**2.22**	**0.4**	**0.4**	**0.15**	**-0.07**
NORTH AFRICA	**291**	**352**	**2.00**	**1.55**	**1.8**	**1.8**	**0.46**	**0.28**
Algeria	13	19	-2.33	1.61	0.4	0.5	-3.83	0.43
Egypt	213	254	3.30	1.41	2.6	2.6	1.50	-0.09
SUB-SAHARAN AFRICA	**123**	**187**	**3.81**	**3.61**	**0.2**	**0.2**	**1.25**	**1.12**
LATIN AMERICA and CARIBBEAN	**258**	**298**	**0.06**	**1.49**	**0.4**	**0.5**	**-1.12**	**0.52**
Argentina	35	30	-1.33	-0.09	0.9	0.7	-2.21	-0.89
Brazil	79	95	0.57	1.62	0.4	0.5	-0.44	0.90
Chile	18	22	3.24	1.87	1.1	1.2	2.24	1.11
Mexico	46	44	-4.27	1.04	0.4	0.3	-5.52	0.04
Uruguay	7	10	4.78	1.05	2.2	3.0	4.59	0.67
ASIA and PACIFIC	**5 907**	**8 219**	**5.87**	**3.12**	**1.5**	**1.9**	**4.74**	**2.21**
Bangladesh	25	41	2.86	4.33	0.2	0.2	1.63	3.19
China[3]	142	182	4.45	1.68	0.1	0.1	3.93	1.38
India	4 283	6 071	7.16	3.35	3.5	4.3	5.70	2.15
Indonesia	16	19	2.37	2.13	0.1	0.1	1.26	1.27
Iran, Islamic Republic of	293	408	5.08	2.84	4.0	5.0	3.89	2.00
Korea	9	12	1.84	2.36	0.2	0.2	1.37	2.06
Malaysia	10	14	-0.66	2.26	0.3	0.4	-2.46	0.80
Pakistan	644	799	2.42	2.16	3.7	3.8	0.62	0.52
Saudi Arabia	52	84	5.34	3.92	1.9	2.5	2.39	2.00
Turkey	162	209	4.36	2.52	2.2	2.6	3.05	1.53
LEAST DEVELOPED COUNTRIES (LDC)	**189**	**276**	**1.29**	**3.49**	**0.2**	**0.3**	**-0.96**	**1.33**
OECD[4]	**3 296**	**3 515**	**0.71**	**0.58**	**2.6**	**2.6**	**0.04**	**0.07**
NON-OECD	**7 063**	**9 571**	**4.42**	**2.85**	**1.3**	**1.5**	**3.14**	**1.72**

Note: Calendar year: Year ending 30 June for Australia and 31 May for New Zealand.

1. Least-squares growth rate (see glossary).
2. Includes Israel and also transition economies: Kazakhstan, Kyrgyzstan, Tajikistan, Turkmenistan, Uzbekistan, Armenia, Azerbaijan and Georgia.
3. Refers to mainland only. The economies of Chinese Taipei, Hong Kong (China) and Macau (China) are included in the Other Asia Pacific aggregate.
4. Excludes Iceland but includes all EU27 member countries.

Source: OECD and FAO Secretariats.

StatLink http://dx.doi.org/10.1787/888932643314

Table A.31. Cheese projections

Calendar year

	PRODUCTION (Kt)		Growth (%)[4]		IMPORTS (Kt)		Growth (%)[4]		EXPORTS (Kt)		Growth (%)[4]	
	Avg 2009-11est	2021	2002-2011	2012-2021	Avg 2009-11est	2021	2002-2011	2012-2021	Avg 2009-11est	2021	2002-2011	2012-2021
WORLD	**19 933**	**23 658**	**2.08**	**1.56**	**2 254**	**2 853**	**5.78**	**2.05**	**2 225**	**2 834**	**5.98**	**2.07**
DEVELOPED	**16 007**	**18 683**	**1.68**	**1.48**	**1 065**	**1 188**	**2.31**	**1.27**	**1 620**	**2 156**	**3.06**	**2.61**
NORTH AMERICA	**5 082**	**6 438**	**2.48**	**2.24**	**158**	**177**	**-5.94**	**3.34**	**189**	**409**	**15.15**	**6.91**
Canada	377	395	1.49	0.48	22	20	0.55	0.00	9	9	-3.78	-0.67
United States	4 704	6 043	2.56	2.36	137	156	-6.80	3.85	180	400	17.32	7.15
EUROPE	**9 927**	**11 011**	**1.40**	**1.01**	**585**	**631**	**5.35**	**0.55**	**956**	**1 158**	**3.55**	**1.51**
EU(27)	8 731	9 589	1.12	0.88	82	79	-5.34	-0.06	650	734	1.97	0.61
Russian Federation	444	479	3.77	1.24	373	453	8.26	1.16	15	25	15.00	0.00
Ukraine	235	292	3.19	3.04	11	8	22.77	-6.04	77	127	3.82	6.04
OCEANIA DEVELOPED	**668**	**812**	**-0.51**	**2.04**	**73**	**83**	**5.91**	**1.39**	**467**	**583**	**-0.81**	**2.47**
Australia	337	392	-2.08	1.31	68	79	5.37	1.48	159	191	-3.79	1.65
New Zealand	331	419	1.29	2.77	5	5	..	0.00	307	392	1.05	2.89
OTHER DEVELOPED[1]	**329**	**421**	**3.10**	**1.81**	**249**	**297**	**1.86**	**1.67**	**8**	**6**	**1.09**	**-3.97**
Japan	47	53	4.19	1.75	191	235	-1.11	0.71	0	0	..	..
South Africa	45	46	2.93	-0.40	6	14	4.66	10.05	2	1	-4.25	-8.01
DEVELOPING	**3 927**	**4 976**	**3.83**	**1.90**	**1 189**	**1 665**	**9.82**	**2.66**	**605**	**678**	**20.07**	**0.51**
AFRICA	**871**	**1 038**	**0.63**	**1.81**	**226**	**321**	**15.35**	**1.51**	**184**	**168**	**27.23**	**-0.17**
NORTH AFRICA	**655**	**803**	**0.57**	**2.11**	**162**	**257**	**15.35**	**0.37**	**184**	**168**	**27.44**	**-0.17**
Algeria	2	2	0.00	0.67	21	27	-2.05	1.91	0	0	-27.95	-0.27
Egypt	611	742	0.38	2.04	97	207	29.28	0.16	158	164	42.20	-0.16
SUB-SAHARAN AFRICA	**216**	**235**	**0.85**	**0.84**	**64**	**64**	**11.45**	**7.82**	**0**	**0**	**-9.11**	**-3.10**
LATIN AMERICA and CARIBBEAN	**1 847**	**2 301**	**4.80**	**1.87**	**260**	**290**	**6.23**	**1.32**	**147**	**209**	**8.47**	**4.17**
Argentina	517	638	4.87	1.69	3	0	..	..	49	105	7.42	6.20
Brazil	646	802	4.67	1.78	23	26	15.68	0.86	4	5	2.42	3.91
Chile	71	90	1.90	1.92	10	18	10.72	2.65	10	7	9.75	-2.65
Mexico	160	176	2.55	0.98	93	129	2.89	2.41	5	14	22.30	8.57
Uruguay	80	100	14.27	2.04	1	0	9.46	-2.60	39	50	9.96	2.60
ASIA and PACIFIC	**1 209**	**1 637**	**5.05**	**2.02**	**703**	**1 054**	**9.79**	**3.44**	**274**	**301**	**28.04**	**-1.18**
Bangladesh	1	2	0.00	4.99	0	0	6.49	-3.61	0	0	0.00	3.61
China[2]	293	373	2.95	1.65	23	61	23.86	6.88	0	1	-15.28	1.29
India	1	0	7.58	-5.64	2	3	17.08	0.82	3	3	29.05	-0.82
Indonesia	0	0	0.00	-10.99	17	28	9.54	3.21	1	1	25.18	-0.46
Iran, Islamic Republic of	293	437	3.35	2.71	0	0	30.47	-6.19	18	72	52.71	6.00
Korea	22	28	0.00	1.66	63	107	7.84	2.33	0	0	..	..
Malaysia	0	0	0.00	6.19	11	13	8.79	2.17	0	0	20.26	-2.17
Pakistan	0	0	0.00	-4.64	2	3	17.24	2.68	0	0	-7.37	-2.68
Saudi Arabia	128	157	121.35	1.40	146	160	5.09	0.62	209	207	37.93	-0.62
Turkey	159	178	3.32	-1.45	10	26	8.71	19.90	27	11	15.87	-19.90
LEAST DEVELOPED COUNTRIES (LDC)	**310**	**384**	**2.25**	**2.03**	**1 261**	**270**	**-0.15**	**-12.01**	**0**	**0**	**5.08**	**-1.76**
OECD[3]	**15 310**	**17 759**	**1.51**	**1.38**	**738**	**920**	**-0.63**	**1.92**	**1 421**	**1 830**	**2.36**	**1.99**
NON-OECD	**4 624**	**5 899**	**4.15**	**2.13**	**1 516**	**1 932**	**10.48**	**2.12**	**804**	**1 004**	**16.93**	**2.21**

Note: Calendar year: Year ending 30 June for Australia and 31 May for New Zealand.

1. Includes Israel and also transition economies: Kazakhstan, Kyrgyzstan, Tajikistan, Turkmenistan, Uzbekistan, Armenia, Azerbaijan and Georgia.
2. Refers to mainland only. The economies of Chinese Taipei, Hong Kong (China) and Macau (China) are included in the Other Asia Pacific aggregate.
3. Excludes Iceland but includes all EU27 member countries.
4. Least-squares growth rate (see glossary).

Source: OECD and FAO Secretariats.

Table A.31. Cheese projections *(cont.)*

Calendar year

	CONSUMPTION (Kt)		Growth (%)[1]		PER CAPITA (Kg)		Growth (%)[1]	
	Avg 2009-11est	2021	2002-2011	2012-2021	Avg 2009-11est	2021	2002-2011	2012-2021
WORLD	**19 938**	**23 678**	**1.98**	**1.56**	**2.9**	**3.1**	**0.81**	**0.54**
DEVELOPED	**15 428**	**17 715**	**1.48**	**1.33**	**11.2**	**12.4**	**1.02**	**0.97**
NORTH AMERICA	**5 034**	**6 205**	**1.84**	**2.02**	**14.6**	**16.5**	**0.92**	**1.20**
Canada	393	407	1.82	0.48	11.5	10.9	0.77	-0.38
United States	4 642	5 799	1.84	2.13	15.0	17.1	0.93	1.32
EUROPE	**9 550**	**10 484**	**1.25**	**0.92**	**12.9**	**14.0**	**1.07**	**0.81**
EU(27)	8 163	8 935	0.81	0.89	16.3	17.4	0.41	0.65
Russian Federation	796	906	5.42	1.19	5.6	6.4	5.62	1.35
Ukraine	168	173	3.84	0.76	3.7	4.0	4.51	1.30
OCEANIA DEVELOPED	**272**	**313**	**0.72**	**1.17**	**10.2**	**10.3**	**-0.83**	**0.01**
Australia	244	282	0.78	1.20	11.0	11.0	-0.86	0.00
New Zealand	29	32	0.29	0.96	6.5	6.5	-0.89	-0.01
OTHER DEVELOPED[2]	**571**	**712**	**2.59**	**1.80**	**2.2**	**2.6**	**2.06**	**1.46**
Japan	238	288	-0.17	0.90	1.9	2.3	-0.21	1.07
South Africa	48	59	3.64	1.59	1.0	1.1	2.61	1.15
DEVELOPING	**4 510**	**5 963**	**3.83**	**2.28**	**0.8**	**0.9**	**2.47**	**1.10**
AFRICA	**912**	**1 191**	**0.58**	**2.03**	**0.9**	**1.0**	**-1.79**	**-0.26**
NORTH AFRICA	**633**	**892**	**-0.31**	**2.04**	**3.8**	**4.7**	**-1.85**	**0.77**
Algeria	22	28	-1.84	1.84	0.6	0.7	-3.35	0.66
Egypt	549	785	-1.00	2.01	6.8	8.2	-2.80	0.51
SUB-SAHARAN AFRICA	**279**	**299**	**2.72**	**2.01**	**0.3**	**0.3**	**0.16**	**-0.48**
LATIN AMERICA and CARIBBEAN	**1 960**	**2 382**	**4.73**	**1.62**	**3.3**	**3.6**	**3.54**	**0.65**
Argentina	471	533	4.57	0.96	11.7	12.1	3.69	0.16
Brazil	664	823	4.95	1.74	3.4	3.9	3.94	1.02
Chile	71	101	2.14	2.47	4.1	5.4	1.14	1.71
Mexico	247	291	2.44	1.31	2.2	2.3	1.19	0.32
Uruguay	41	50	19.67	1.46	12.3	14.3	19.47	1.08
ASIA and PACIFIC	**1 638**	**2 390**	**4.89**	**3.12**	**0.4**	**0.5**	**3.76**	**2.21**
Bangladesh	1	2	0.79	4.28	0.0	0.0	-0.44	3.14
China[3]	315	433	3.76	2.24	0.2	0.3	3.24	1.95
India	0	0	-22.54	4.58	0.0	0.0	-24.00	3.38
Indonesia	16	27	9.20	3.31	0.1	0.1	8.08	2.45
Iran, Islamic Republic of	275	365	2.50	2.19	3.7	4.5	1.31	1.35
Korea	85	135	5.56	2.19	1.8	2.7	5.09	1.89
Malaysia	11	13	8.65	2.22	0.4	0.4	6.86	0.76
Pakistan	2	3	17.26	2.68	0.0	0.0	15.47	1.05
Saudi Arabia	65	110	-7.35	4.89	2.4	3.2	-10.30	2.97
Turkey	141	192	2.06	3.14	1.9	2.4	0.75	2.16
LEAST DEVELOPED COUNTRIES (LDC)	**364**	**488**	**3.46**	**2.83**	**0.4**	**0.5**	**1.21**	**0.67**
OECD[4]	**14 607**	**16 851**	**1.21**	**1.35**	**11.5**	**12.5**	**0.53**	**0.83**
NON-OECD	**5 331**	**6 827**	**4.34**	**2.11**	**0.9**	**1.1**	**3.05**	**0.98**

Note: Calendar year: Year ending 30 June for Australia and 31 May for New Zealand.

1. Least-squares growth rate (see glossary).
2. Includes Israel and also transition economies: Kazakhstan, Kyrgyzstan, Tajikistan, Turkmenistan, Uzbekistan, Armenia, Azerbaijan and Georgia.
3. Refers to mainland only. The economies of Chinese Taipei, Hong Kong (China) and Macau (China) are included in the Other Asia Pacific aggregate.
4. Excludes Iceland but includes all EU27 member countries.

Source: OECD and FAO Secretariats.

StatLink http://dx.doi.org/10.1787/888932643352

Table A.32. Skim milk powder projections

Calendar year

	PRODUCTION (Kt)		Growth (%)[4]		IMPORTS (Kt)		Growth (%)[4]		EXPORTS (Kt)		Growth (%)[4]	
	Avg 2009-11est	2021	2002-2011	2012-2021	Avg 2009-11est	2021	2002-2011	2012-2021	Avg 2009-11est	2021	2002-2011	2012-2021
WORLD	**3 441**	**4 233**	**0.05**	**1.84**	**1 473**	**2 025**	**2.91**	**1.98**	**1 524**	**2 042**	**3.64**	**1.97**
DEVELOPED	**2 916**	**3 542**	**-0.56**	**1.73**	**139**	**157**	**-1.94**	**0.72**	**1 404**	**1 894**	**3.76**	**1.94**
NORTH AMERICA	**900**	**1 206**	**2.10**	**3.20**	**3**	**4**	**-4.57**	**0.00**	**346**	**579**	**8.88**	**3.76**
Canada	77	72	-1.29	-1.59	3	4	6.90	0.00	8	9	-18.14	-1.02
United States	823	1 135	2.46	3.57	0	0	..	..	338	570	11.51	3.85
EUROPE	**1 152**	**1 292**	**-3.17**	**0.69**	**77**	**86**	**-2.13**	**0.36**	**495**	**563**	**3.60**	**0.54**
EU(27)	943	1 030	-2.92	0.23	4	3	-35.74	3.67	365	434	3.70	-0.04
Russian Federation	55	70	-8.09	3.73	59	71	4.71	1.01	1	1	-10.63	0.00
Ukraine	51	78	-11.14	4.89	5	3	46.81	-8.47	20	42	-12.73	8.47
OCEANIA DEVELOPED	**685**	**863**	**1.49**	**1.84**	**7**	**6**	**8.05**	**-0.21**	**559**	**750**	**1.22**	**1.80**
Australia	208	221	-0.38	0.44	4	4	5.64	-0.34	148	164	-3.37	0.13
New Zealand	476	642	2.45	2.37	3	2	..	0.00	411	586	3.35	2.31
OTHER DEVELOPED[1]	**180**	**181**	**-1.78**	**-0.10**	**52**	**62**	**-2.24**	**1.39**	**4**	**2**	**-0.49**	**-4.46**
Japan	153	153	-2.84	-0.13	26	19	-6.17	-1.18	0	0	..	..
South Africa	15	12	10.90	-2.74	3	9	-3.66	8.08	2	1	11.47	-8.07
DEVELOPING	**525**	**691**	**4.03**	**2.41**	**1 334**	**1 868**	**3.54**	**2.10**	**121**	**148**	**2.41**	**2.36**
AFRICA	**3**	**1**	**0.00**	**-3.08**	**245**	**390**	**5.46**	**2.58**	**3**	**4**	**2.12**	**-0.82**
NORTH AFRICA	**0**	**0**	**0.00**	**-1.86**	**180**	**293**	**6.00**	**2.22**	**1**	**2**	**5.24**	**-0.48**
Algeria	0	0	0.00	-1.00	114	181	5.94	1.70	0	0	-3.08	-0.24
Egypt	0	0	0.00	-25.40	45	89	8.12	3.38	1	2	13.93	-0.48
SUB-SAHARAN AFRICA	**3**	**1**	**0.00**	**-3.12**	**65**	**97**	**4.14**	**3.73**	**2**	**2**	**0.83**	**-1.23**
LATIN AMERICA and CARIBBEAN	**255**	**328**	**3.00**	**2.32**	**204**	**249**	**-0.80**	**0.97**	**39**	**35**	**0.53**	**1.65**
Argentina	36	40	0.72	0.86	0	0	..	..	17	20	-0.55	2.09
Brazil	130	164	2.68	2.10	10	27	1.20	1.71	3	1	-1.31	-6.56
Chile	16	19	6.09	1.40	4	8	-3.72	3.09	2	2	38.09	-3.06
Mexico	32	37	0.72	1.71	137	166	-0.40	1.40	0	0	14.77	-20.45
Uruguay	20	22	3.99	2.58	0	0	30.58	-0.57	14	11	-0.80	3.99
ASIA and PACIFIC	**268**	**361**	**5.15**	**2.51**	**884**	**1 229**	**4.27**	**2.19**	**79**	**109**	**3.49**	**2.73**
Bangladesh	0	0	0.00	9.38	26	37	1.29	2.90	0	0	0.00	-0.41
China[2]	55	61	109.59	0.83	93	166	10.99	3.25	0	0	-21.46	0.00
India	193	271	3.65	2.76	18	12	53.22	-8.46	18	48	7.14	8.44
Indonesia	0	2	0.00	57.23	121	155	6.20	1.80	1	1	-11.17	-0.26
Iran, Islamic Republic of	0	0	0.00	6.39	15	10	15.10	0.71	5	4	36.71	-0.71
Korea	14	20	-10.53	3.74	13	21	8.49	2.36	1	1	58.30	0.00
Malaysia	0	0	0.00	0.84	87	109	5.99	1.80	6	7	-2.99	-0.26
Pakistan	0	0	0.00	-16.57	22	39	26.92	3.16	1	1	25.46	-0.45
Saudi Arabia	0	0	0.00	7.60	36	50	-2.11	1.49	19	16	10.97	-0.21
Turkey	0	0	0.00	-15.11	9	4	6.19	2.97	0	0	17.44	-2.97
LEAST DEVELOPED COUNTRIES (LDC)	**0**	**1**	**0.00**	**12.26**	**1 299**	**302**	**-0.29**	**-11.43**	**2**	**1**	**4.65**	**-1.68**
OECD[3]	**2 789**	**3 374**	**-0.27**	**1.66**	**207**	**235**	**-2.77**	**1.26**	**1 294**	**1 780**	**3.87**	**1.86**
NON-OECD	**652**	**859**	**1.53**	**2.55**	**1 266**	**1 790**	**4.13**	**2.08**	**231**	**261**	**2.50**	**2.71**

Note: Calendar year: Year ending 30 June for Australia and 31 May for New Zealand.

1. Includes Israel and also transition economies: Kazakhstan, Kyrgyzstan, Tajikistan, Turkmenistan, Uzbekistan, Armenia, Azerbaijan and Georgia.
2. Refers to mainland only. The economies of Chinese Taipei, Hong Kong (China) and Macau (China) are included in the Other Asia Pacific aggregate.
3. Excludes Iceland but includes all EU27 member countries.
4. Least-squares growth rate (see glossary).

Source: OECD and FAO Secretariats.

Table A.32. Skim milk powder projections *(cont.)*

Calendar year

	CONSUMPTION (Kt)		Growth (%)[1]		PER CAPITA (Kg)		Growth (%)[1]	
	Avg 2009-11est	2021	2002-2011	2012-2021	Avg 2009-11est	2021	2002-2011	2012-2021
WORLD	**3 424**	**4 186**	**-0.08**	**1.69**	**0.4**	**0.5**	**0.79**	**0.85**
DEVELOPED	**1 685**	**1 775**	**-3.30**	**1.06**	**1.0**	**1.1**	**-0.48**	**1.07**
NORTH AMERICA	**566**	**631**	**-2.47**	**2.78**	**1.4**	**1.6**	**4.58**	**2.16**
Canada	78	67	5.01	-0.76	1.3	1.4	0.76	-0.68
United States	488	564	-3.39	3.27	1.5	1.7	5.47	2.46
EUROPE	**788**	**807**	**-5.01**	**0.27**	**0.8**	**0.8**	**-3.01**	**0.58**
EU(27)	606	592	-5.55	-0.21	0.8	0.7	-3.08	-0.10
Russian Federation	113	140	-2.75	2.30	0.8	1.0	-2.55	2.46
Ukraine	35	39	-7.78	0.72	0.8	0.9	-7.10	1.25
OCEANIA DEVELOPED	**103**	**96**	**6.29**	**-0.53**	**3.8**	**3.2**	**4.73**	**-1.69**
Australia	68	61	11.32	-0.85	3.0	2.4	9.69	-2.05
New Zealand	35	35	0.00	0.05	8.0	7.3	-1.19	-0.92
OTHER DEVELOPED[2]	**228**	**240**	**-1.85**	**0.32**	**0.8**	**0.8**	**-1.75**	**0.03**
Japan	179	171	-3.53	-0.25	1.2	1.2	-3.06	-0.06
South Africa	16	19	3.51	1.52	0.3	0.4	2.48	1.08
DEVELOPING	**1 739**	**2 411**	**3.77**	**2.17**	**0.3**	**0.4**	**2.44**	**1.00**
AFRICA	**246**	**388**	**5.47**	**2.60**	**0.3**	**0.3**	**3.09**	**0.31**
NORTH AFRICA	**179**	**291**	**6.01**	**2.24**	**1.1**	**1.5**	**4.47**	**0.97**
Algeria	114	181	5.94	1.70	3.2	4.5	4.43	0.52
Egypt	45	87	8.15	3.48	0.5	0.9	6.35	1.98
SUB-SAHARAN AFRICA	**66**	**97**	**4.14**	**3.73**	**0.1**	**0.1**	**1.59**	**1.25**
LATIN AMERICA and CARIBBEAN	**419**	**543**	**1.09**	**1.75**	**0.6**	**0.7**	**-0.25**	**0.74**
Argentina	19	19	0.59	-0.18	0.5	0.4	-0.29	-0.97
Brazil	137	190	2.39	2.10	0.5	0.7	1.26	1.36
Chile	18	25	1.61	2.29	1.1	1.4	0.61	1.53
Mexico	167	203	-0.45	1.56	1.5	1.6	-1.70	0.56
Uruguay	7	11	18.72	1.38	2.1	3.3	18.52	1.01
ASIA and PACIFIC	**1 074**	**1 481**	**4.58**	**2.23**	**0.3**	**0.3**	**3.44**	**1.32**
Bangladesh	26	37	1.29	2.90	0.2	0.2	0.06	1.76
China[3]	148	227	16.77	2.54	0.1	0.2	16.25	2.24
India	192	235	4.86	1.00	0.2	0.2	3.41	-0.20
Indonesia	120	156	6.51	1.92	0.5	0.6	5.40	1.06
Iran, Islamic Republic of	10	7	11.47	1.59	0.1	0.1	10.28	0.75
Korea	26	39	-4.17	3.10	0.5	0.8	-4.64	2.80
Malaysia	81	102	7.07	1.95	2.8	3.0	5.27	0.49
Pakistan	21	38	27.84	3.27	0.1	0.2	26.05	1.64
Saudi Arabia	16	34	-8.21	2.36	0.6	1.0	-11.16	0.44
Turkey	8	4	5.94	2.96	0.1	0.1	4.63	1.97
LEAST DEVELOPED COUNTRIES (LDC)	**92**	**136**	**3.16**	**3.50**	**0.1**	**0.1**	**0.91**	**1.34**
OECD[4]	**1 706**	**1 798**	**-3.10**	**1.06**	**1.1**	**1.1**	**-0.42**	**0.91**
NON-OECD	**1 718**	**2 388**	**3.55**	**2.18**	**0.3**	**0.4**	**2.28**	**1.05**

Note: Calendar year: Year ending 30 June for Australia and 31 May for New Zealand.

1. Least-squares growth rate (see glossary).
2. Includes Israel and also transition economies: Kazakhstan, Kyrgyzstan, Tajikistan, Turkmenistan, Uzbekistan, Armenia, Azerbaijan and Georgia.
3. Refers to mainland only. The economies of Chinese Taipei, Hong Kong (China) and Macau (China) are included in the Other Asia Pacific aggregate.
4. Excludes Iceland but includes all EU27 member countries.

Source: OECD and FAO Secretariats.

StatLink http://dx.doi.org/10.1787/888932643390

Table A.33. Whole milk powder projections

Calendar year

	PRODUCTION (Kt)		Growth (%)[4]		IMPORTS (Kt)		Growth (%)[4]		EXPORTS (Kt)		Growth (%)[4]	
	Avg 2009-11est	2021	2002-2011	2012-2021	Avg 2009-11est	2021	2002-2011	2012-2021	Avg 2009-11est	2021	2002-2011	2012-2021
WORLD	**4 267**	**5 614**	**2.85**	**1.95**	**2 170**	**2 640**	**2.94**	**1.91**	**2 091**	**2 710**	**3.73**	**1.86**
DEVELOPED	**2 025**	**2 653**	**1.37**	**1.55**	**120**	**132**	**5.97**	**1.18**	**1 524**	**2 080**	**2.87**	**1.91**
NORTH AMERICA	**42**	**50**	**1.93**	**1.02**	**16**	**16**	**-7.20**	**0.00**	**24**	**28**	**27.66**	**1.39**
Canada	11	13	-6.09	0.70	3	4	-25.47	0.00	0	1	-13.75	0.00
United States	31	37	6.72	1.14	13	12	16.76	0.00	23	28	..	1.42
EUROPE	**893**	**965**	**-1.85**	**0.73**	**78**	**82**	**15.41**	**0.47**	**486**	**503**	**-2.13**	**1.18**
EU(27)	759	783	-1.84	0.43	1	1	-25.64	0.00	440	435	-2.73	1.04
Russian Federation	56	81	-6.09	3.03	69	71	32.01	0.17	3	3	1.57	0.00
Ukraine	14	26	-4.44	3.10	1	0	8.86	-0.76	5	15	-7.83	5.35
OCEANIA DEVELOPED	**1 046**	**1 595**	**4.88**	**2.14**	**11**	**14**	**8.32**	**2.49**	**1 008**	**1 546**	**5.76**	**2.18**
Australia	142	171	-5.74	1.00	9	14	9.07	2.54	105	125	-4.74	1.04
New Zealand	905	1 424	7.66	2.29	1	0	..	..	903	1 420	7.70	2.29
OTHER DEVELOPED[1]	**44**	**44**	**-0.04**	**-0.36**	**15**	**20**	**-2.72**	**4.63**	**6**	**3**	**-1.18**	**-5.41**
Japan	13	12	-2.19	-1.80	0	0	..	..	0	0	..	..
South Africa	15	13	0.65	-1.72	3	5	12.90	6.22	4	3	-4.55	-6.22
DEVELOPING	**2 243**	**2 961**	**4.33**	**2.33**	**2 050**	**2 508**	**2.79**	**1.95**	**568**	**629**	**6.16**	**1.72**
AFRICA	**9**	**11**	**2.92**	**1.88**	**645**	**728**	**5.41**	**2.38**	**16**	**17**	**12.45**	**-0.45**
NORTH AFRICA	**0**	**0**	**0.00**	**0.30**	**246**	**331**	**6.83**	**1.52**	**1**	**1**	**36.95**	**-0.47**
Algeria	0	0	0.00	0.76	200	260	5.12	1.11	0	0	-19.42	-0.16
Egypt	0	0	0.00	-24.61	34	58	24.33	3.44	1	1	48.72	-0.49
SUB-SAHARAN AFRICA	**9**	**11**	**2.93**	**1.88**	**399**	**398**	**4.28**	**3.14**	**15**	**16**	**11.04**	**-0.45**
LATIN AMERICA and CARIBBEAN	**1 157**	**1 499**	**2.88**	**2.30**	**247**	**268**	**-6.14**	**1.84**	**279**	**386**	**3.88**	**3.27**
Argentina	224	319	-0.08	3.49	1	0	..	..	167	289	2.17	3.97
Brazil	494	627	3.94	2.06	44	53	-1.79	1.90	10	11	0.63	1.49
Chile	118	158	8.21	1.99	2	7	-17.01	2.59	13	9	5.36	-2.59
Mexico	188	230	2.98	1.95	34	36	-11.53	0.72	7	10	56.35	1.43
Uruguay	38	57	3.23	2.83	0	0	-2.71	-0.41	56	57	10.36	2.84
ASIA and PACIFIC	**1 077**	**1 451**	**6.10**	**2.36**	**1 158**	**1 511**	**4.15**	**1.78**	**273**	**227**	**8.52**	**-0.38**
Bangladesh	0	0	0.00	13.91	31	30	-1.79	-0.03	0	0	0.00	0.00
China[2]	1 009	1 355	5.40	2.37	284	349	15.17	0.52	5	2	-8.93	2.07
India	3	7	31.20	9.31	7	13	21.25	1.71	2	3	25.14	-0.24
Indonesia	62	79	126.88	1.22	47	58	3.80	2.42	17	6	-3.20	-2.42
Iran, Islamic Republic of	0	1	-6.76	6.03	5	3	28.26	0.09	2	1	36.35	-0.09
Korea	3	4	-13.53	5.74	1	2	-0.68	4.01	0	0	-17.81	0.00
Malaysia	0	0	0.00	1.28	28	35	-11.21	1.57	15	13	4.86	-0.22
Pakistan	0	0	0.00	-8.17	12	16	13.40	1.84	3	3	30.53	-1.84
Saudi Arabia	0	0	0.00	5.86	58	77	-2.33	1.98	10	10	11.38	-0.28
Turkey	0	0	0.00	-6.16	3	2	-4.03	2.33	1	0	25.44	-2.33
LEAST DEVELOPED COUNTRIES (LDC)	**0**	**1**	**0.00**	**17.04**	**1 437**	**492**	**-0.22**	**-8.25**	**6**	**6**	**11.70**	**-0.63**
OECD[3]	**2 187**	**2 851**	**1.97**	**1.58**	**74**	**81**	**-8.59**	**1.10**	**1 495**	**2 030**	**2.88**	**1.89**
NON-OECD	**2 081**	**2 764**	**3.83**	**2.35**	**2 096**	**2 559**	**3.59**	**1.94**	**596**	**680**	**6.00**	**1.79**

Note: Calendar year: Year ending 30 June for Australia and 31 May for New Zealand.

1. Includes Israel and also transition economies: Kazakhstan, Kyrgyzstan, Tajikistan, Turkmenistan, Uzbekistan, Armenia, Azerbaijan and Georgia.
2. Refers to mainland only. The economies of Chinese Taipei, Hong Kong (China) and Macau (China) are included in the Other Asia Pacific aggregate.
3. Excludes Iceland but includes all EU27 member countries.
4. Least-squares growth rate (see glossary).

Source: OECD and FAO Secretariats.

Table A.33. Whole milk powder projections *(cont.)*

Calendar year

	CONSUMPTION (Kt)		Growth (%)[1]		PER CAPITA (Kg)		Growth (%)[1]	
	Avg 2009-11est	2021	2002-2011	2012-2021	Avg 2009-11est	2021	2002-2011	2012-2021
WORLD	**4 385**	**5 545**	**2.61**	**1.98**	**0.6**	**0.7**	**1.44**	**0.96**
DEVELOPED	**620**	**705**	**-1.15**	**0.48**	**0.5**	**0.5**	**-1.58**	**0.13**
NORTH AMERICA	**35**	**37**	**-7.55**	**0.31**	**0.1**	**0.1**	**-8.23**	**-0.50**
Canada	14	15	-13.90	0.55	0.4	0.4	-14.57	-0.31
United States	20	21	0.40	0.14	0.1	0.1	-0.51	-0.67
EUROPE	**484**	**544**	**0.04**	**0.30**	**0.7**	**0.7**	**-0.14**	**0.20**
EU(27)	320	349	-0.84	-0.28	0.6	0.7	-1.24	-0.52
Russian Federation	122	149	3.07	1.66	0.9	1.1	3.27	1.81
Ukraine	10	11	-2.34	0.60	0.2	0.3	-1.66	1.14
OCEANIA DEVELOPED	**48**	**64**	**-6.01**	**1.30**	**1.8**	**2.1**	**-7.57**	**0.14**
Australia	45	60	-6.24	1.21	2.0	2.3	-7.87	0.01
New Zealand	3	4	-2.13	2.69	0.7	0.8	-3.31	1.72
OTHER DEVELOPED[2]	**53**	**61**	**-0.67**	**1.46**	**0.2**	**0.2**	**-1.20**	**1.12**
Japan	13	12	-2.55	-1.80	0.1	0.1	-2.59	-1.63
South Africa	14	15	5.58	1.63	0.3	0.3	4.55	1.19
DEVELOPING	**3 765**	**4 839**	**3.35**	**2.22**	**0.7**	**0.8**	**1.99**	**1.04**
AFRICA	**639**	**722**	**5.23**	**2.44**	**0.7**	**0.6**	**2.85**	**0.15**
NORTH AFRICA	**245**	**330**	**6.75**	**1.53**	**1.5**	**1.7**	**5.21**	**0.26**
Algeria	200	260	5.12	1.11	5.6	6.4	3.61	-0.08
Egypt	33	57	23.72	3.49	0.4	0.6	21.91	2.00
SUB-SAHARAN AFRICA	**394**	**392**	**4.02**	**3.27**	**0.5**	**0.4**	**1.47**	**0.79**
LATIN AMERICA and CARIBBEAN	**1 145**	**1 382**	**0.30**	**1.96**	**1.9**	**2.1**	**-0.89**	**0.99**
Argentina	58	30	-6.35	-0.36	1.4	0.7	-7.23	-1.16
Brazil	528	669	2.94	2.05	2.7	3.2	1.93	1.33
Chile	108	156	7.17	2.34	6.3	8.4	6.17	1.57
Mexico	216	256	-0.82	1.79	1.9	2.0	-2.08	0.79
Uruguay	0	0	-34.11	1.38	0.1	0.1	-34.30	1.00
ASIA and PACIFIC	**1 982**	**2 735**	**4.78**	**2.29**	**0.5**	**0.6**	**3.64**	**1.38**
Bangladesh	31	30	-1.79	-0.02	0.2	0.2	-3.02	-1.16
China[3]	1 288	1 701	7.17	1.96	1.0	1.2	6.65	1.66
India	7	17	22.89	4.62	0.0	0.0	21.43	3.42
Indonesia	92	131	23.92	1.94	0.4	0.5	22.80	1.08
Iran, Islamic Republic of	3	3	10.70	1.78	0.0	0.0	9.51	0.94
Korea	4	7	-9.82	5.15	0.1	0.1	-10.29	4.85
Malaysia	13	23	-20.80	2.70	0.5	0.7	-22.59	1.25
Pakistan	9	13	10.29	2.82	0.1	0.1	8.50	1.19
Saudi Arabia	48	67	-3.91	2.35	1.7	2.0	-6.85	0.43
Turkey	3	2	-6.14	3.15	0.0	0.0	-7.45	2.17
LEAST DEVELOPED COUNTRIES (LDC)	**224**	**322**	**2.51**	**3.45**	**0.3**	**0.3**	**0.27**	**1.29**
OECD[4]	**765**	**902**	**-0.94**	**0.87**	**0.6**	**0.7**	**-1.59**	**0.36**
NON-OECD	**3 620**	**4 643**	**3.50**	**2.21**	**0.6**	**0.7**	**2.21**	**1.08**

Note: Calendar year: Year ending 30 June for Australia and 31 May for New Zealand.

1. Least-squares growth rate (see glossary).
2. Includes Israel and also transition economies: Kazakhstan, Kyrgyzstan, Tajikistan, Turkmenistan, Uzbekistan, Armenia, Azerbaijan and Georgia.
3. Refers to mainland only. The economies of Chinese Taipei, Hong Kong (China) and Macau (China) are included in the Other Asia Pacific aggregate.
4. Excludes Iceland but includes all EU27 member countries.

Source: OECD and FAO Secretariats.

StatLink http://dx.doi.org/10.1787/888932643428

Table A.34. Milk projections

Calendar year

	PRODUCTION (Kt)		Growth (%)[4]		INVENTORIES ('000 hd)		Growth (%)[4]		YIELD (t/hd)		Growth (%)[4]	
	Avg 2009-11est	2021	2002-2011	2012-2021	Avg 2009-11est	2021	2002-2011	2012-2021	Avg 2009-11est	2021	2002-2011	2012-2021
WORLD	**711 561**	**880 350**	**2.13**	**1.97**	**622 144**	**721 032**	**1.68**	**1.42**	**1.1**	**1.2**	**0.45**	**0.55**
DEVELOPED	**362 668**	**411 426**	**0.54**	**1.18**	**76 275**	**73 930**	**-0.91**	**-0.22**	**4.8**	**5.6**	**1.45**	**1.40**
NORTH AMERICA	**95 865**	**112 319**	**1.62**	**1.48**	**10 126**	**9 806**	**0.01**	**-0.37**	**9.5**	**11.5**	**1.61**	**1.85**
Canada	8 458	8 936	0.45	0.25	980	883	-1.03	-1.06	8.6	10.1	1.48	1.32
United States	87 407	103 383	1.74	1.59	9 146	8 923	0.12	-0.30	9.6	11.6	1.62	1.89
EUROPE	**209 847**	**226 206**	**-0.17**	**0.78**	**43 600**	**38 477**	**-2.55**	**-0.94**	**4.8**	**5.9**	**2.38**	**1.72**
EU(27)	148 991	157 612	-0.10	0.54	23 235	21 251	-1.57	-0.62	6.4	7.4	1.47	1.16
Russian Federation	31 876	35 969	-0.42	1.59	8 765	6 905	-3.26	-2.01	3.6	5.2	2.84	3.60
Ukraine	11 197	12 914	-3.15	1.22	3 741	3 234	-5.74	-1.30	3.0	4.0	2.59	2.51
OCEANIA DEVELOPED	**26 683**	**34 976**	**0.89**	**2.03**	**6 170**	**7 044**	**0.54**	**1.00**	**4.3**	**5.0**	**0.35**	**1.03**
Australia	9 455	10 742	-2.15	1.20	1 625	1 624	-3.52	0.01	5.8	6.6	1.36	1.19
New Zealand	17 228	24 234	2.89	2.41	4 545	5 420	2.32	1.32	3.8	4.5	0.57	1.10
OTHER DEVELOPED[1]	**30 272**	**37 925**	**2.10**	**2.05**	**16 378**	**18 603**	**3.23**	**1.02**	**1.8**	**2.0**	**-1.14**	**1.03**
Japan	7 707	7 393	-1.21	-0.18	937	879	-2.15	-0.39	8.2	8.4	0.94	0.21
South Africa	3 162	3 652	2.39	1.24	965	1 070	2.58	0.58	3.3	3.4	-0.19	0.66
DEVELOPING	**348 893**	**468 925**	**3.99**	**2.71**	**545 870**	**647 102**	**2.08**	**1.62**	**0.6**	**0.7**	**1.91**	**1.09**
AFRICA	**35 717**	**47 130**	**2.91**	**2.79**	**192 373**	**225 108**	**2.21**	**1.57**	**0.2**	**0.2**	**0.71**	**1.22**
NORTH AFRICA	**11 377**	**13 832**	**3.94**	**1.97**	**26 801**	**28 158**	**2.44**	**0.77**	**0.4**	**0.5**	**1.49**	**1.20**
Algeria	2 071	2 434	3.69	2.00	14 142	15 522	6.10	0.95	0.1	0.2	-2.40	1.05
Egypt	6 042	6 989	4.16	1.47	5 491	4 695	-2.22	-0.55	1.1	1.5	6.37	2.02
SUB-SAHARAN AFRICA	**24 340**	**33 298**	**2.46**	**3.14**	**165 571**	**196 950**	**2.17**	**1.69**	**0.1**	**0.2**	**0.29**	**1.46**
LATIN AMERICA and CARIBBEAN	**80 260**	**102 838**	**2.85**	**2.10**	**46 161**	**51 949**	**0.88**	**1.10**	**1.7**	**2.0**	**1.97**	**1.00**
Argentina	10 659	16 745	3.12	3.43	2 420	2 789	2.54	0.96	4.4	6.0	0.58	2.47
Brazil	31 210	38 440	3.40	1.75	22 928	27 390	2.37	1.56	1.4	1.4	1.03	0.19
Chile	2 552	3 194	2.32	2.05	1 269	1 187	-5.36	0.49	2.0	2.7	7.67	1.56
Mexico	10 974	11 635	1.28	0.54	2 362	2 450	1.12	0.31	4.6	4.7	0.16	0.23
Uruguay	1 666	2 079	1.37	2.04	722	733	-0.08	0.49	2.3	2.8	1.45	1.55
ASIA and PACIFIC	**232 916**	**318 956**	**4.59**	**2.90**	**307 336**	**370 044**	**2.19**	**1.73**	**0.8**	**0.9**	**2.40**	**1.17**
Bangladesh	3 242	5 490	4.51	4.68	33 745	51 242	5.32	3.85	0.1	0.1	-0.81	0.83
China[2]	42 773	60 432	10.03	2.46	12 712	15 427	8.33	1.66	3.4	3.9	1.71	0.80
India	116 815	165 632	4.13	3.37	111 018	136 582	1.85	1.90	1.1	1.2	2.28	1.47
Indonesia	1 309	1 724	6.31	2.22	11 963	13 530	4.77	0.49	0.1	0.1	1.54	1.73
Iran, Islamic Republic of	7 985	10 344	3.40	2.23	36 331	37 686	1.04	0.21	0.2	0.3	2.36	2.02
Korea	2 017	2 082	-2.50	0.90	240	244	-3.05	0.54	8.4	8.5	0.55	0.36
Malaysia	52	67	2.41	2.78	106	119	2.23	1.25	0.5	0.6	0.19	1.53
Pakistan	32 658	37 885	2.26	1.95	25 936	27 617	1.51	0.95	1.3	1.4	0.75	0.99
Saudi Arabia	1 552	1 770	5.04	2.35	3 480	3 336	-2.85	0.95	0.4	0.5	7.89	1.40
Turkey	13 192	17 512	4.33	2.22	15 269	15 665	-3.53	0.22	0.9	1.1	7.85	2.01
LEAST DEVELOPED COUNTRIES (LDC)	**27 068**	**38 671**	**2.36**	**3.37**	**207 362**	**262 576**	**2.88**	**2.15**	**0.1**	**0.1**	**-0.53**	**1.22**
OECD[3]	**314 935**	**353 630**	**0.68**	**1.04**	**60 845**	**59 653**	**-1.68**	**-0.11**	**5.2**	**5.9**	**2.36**	**1.16**
NON-OECD	**396 626**	**526 720**	**3.39**	**2.64**	**561 299**	**661 379**	**2.10**	**1.57**	**0.7**	**0.8**	**1.29**	**1.07**

Note: Calendar year: Year ending 30 June for Australia and 31 May for New Zealand.

1. Includes Israel and also transition economies: Kazakhstan, Kyrgyzstan, Tajikistan, Turkmenistan, Uzbekistan, Armenia, Azerbaijan and Georgia.
2. Refers to mainland only. The economies of Chinese Taipei, Hong Kong (China) and Macau (China) are included in the Other Asia Pacific aggregate.
3. Excludes Iceland but includes all EU27 member countries.
4. Least-squares growth rate (see glossary).

Source: OECD and FAO Secretariats.

StatLink http://dx.doi.org/10.1787/888932643447

Table A.35. Whey powder and casein projections

Calendar year

		Average 2009-11est.	2021	Growth (%)[2]	
				2002-2011	2012-21
AUSTRALIA					
Net trade, whey	kt pw	83.7	86.4	4.40	0.01
Exports, casein	kt pw	7.3	5.2	-11.83	1.07
CANADA					
Net trade, whey	kt pw	16.4	26.3	26.32	0.39
EUROPEAN UNION					
Net trade, whey	kt pw	293.9	469.7	2.10	5.20
Casein EU(15)					
Production	kt pw	134.4	141.0	-2.69	-0.11
Consumption	kt pw	110.2	120.7	-3.13	-0.07
Net trade	kt pw	25.5	19.5	2.25	-2.64
JAPAN					
Net trade, whey	kt pw	-52.8	-73.1	0.30	3.37
KOREA					
Net trade, whey	kt pw	-36.8	-33.3	-0.25	1.58
MEXICO					
Net trade, whey	kt pw	-34.1	-30.4	-1.99	-0.77
NEW ZEALAND					
Net trade, whey	kt pw	7.2	10.1	2.94	3.89
Exports, casein	kt pw	157.2	211.7	2.09	2.17
UNITED STATES					
Whey					
Production	kt pw	480.1	500.1	-0.22	0.39
Consumption	kt pw	300.6	222.8	-1.99	-3.01
Exports	kt pw	180.3	277.4	2.77	3.81
Imports, casein	kt pw	68.5	70.1	-0.25	0.04
ARGENTINA					
Net trade, whey	kt pw	47.9	110.6	44.11	4.88
BRAZIL					
Net trade, whey	kt pw	-25.9	-30.0	-0.78	1.99
CHINA[1]					
Net trade, whey	kt pw	-244.9	-460.8	6.25	5.51
RUSSIAN FEDERATION					
Net trade, whey	kt pw	-53.0	-66.7	14.22	2.31

Note: Calendar year: Year ending 30 June for Australia and 31 May for New Zealand.

1. Refers to mainland only. The economies of Chinese Taipei, Hong Kong (China) and Macau (China) are included in the Other Asia Pacific aggregate.
2. Least-squares growth rate (see glossary).

Source: OECD and FAO Secretariats.

StatLink http://dx.doi.org/10.1787/888932643466

Table A.36. Main policy assumptions for dairy markets

Calendar year

		Avg 2009-11est.	2012	2013	2014	2015	2016	2017	2018	2019	2020	2021
CANADA												
Milk target price[1]	CADc/litre	73.0	75.2	76.6	78.9	80.4	81.8	83.0	84.0	85.0	85.8	86.4
Butter support price	CAD/t	7 132.3	7 280.0	7 376.7	7 485.7	7 596.3	7 708.6	7 822.5	7 938.1	8 055.4	8 174.5	8 295.3
SMP support price	CAD/t	6 242.4	6 345.1	6 420.8	6 599.3	6 762.4	6 836.2	6 910.5	6 956.5	7 070.5	7 180.0	7 230.1
Cheese tariff-quota	kt pw	20.4	20.4	20.4	20.4	20.4	20.4	20.4	20.4	20.4	20.4	20.4
In-quota tariff	%	1.2	1.2	1.2	1.2	1.2	1.2	1.2	1.2	1.2	1.2	1.2
Out-of-quota tariff	%	245.6	245.6	245.6	245.6	245.6	245.6	245.6	245.6	245.6	245.6	245.6
Subsidised export limits[2]												
Cheese	kt pw	9.0	9.0	9.0	9.0	9.0	9.0	9.0	9.0	9.0	9.0	9.0
SMP	kt pw	45.0	45.0	45.0	45.0	45.0	45.0	45.0	45.0	45.0	45.0	45.0
EUROPEAN UNION[3]												
Milk quota[4]	kt pw	146 526	149 199	150 446	150 446	0	0	0	0	0	0	0
Butter intervention price	EUR/t	2 217.5	2 217.5	2 217.5	2 217.5	2 217.5	2 217.5	2 217.5	2 217.5	2 217.5	2 217.5	2 217.5
SMP intervention price	EUR/t	1 698.0	1 698.0	1 698.0	1 698.0	1 698.0	1 698.0	1 698.0	1 698.0	1 698.0	1 698.0	1 698.0
Butter tariff-quotas	kt pw	89.8	89.8	89.8	89.8	89.8	89.8	89.8	89.8	89.8	89.8	89.8
Cheese tariff-quotas	kt pw	103.2	103.2	103.2	103.2	103.2	103.2	103.2	103.2	103.2	103.2	103.2
SMP tariff-quota	kt pw	70.7	70.7	70.7	70.7	70.7	70.7	70.7	70.7	70.7	70.7	70.7
Subsidised export limits[2]												
Butter	kt pw	411.6	411.6	411.6	411.6	411.6	411.6	411.6	411.6	411.6	411.6	411.6
Cheese	kt pw	331.7	331.7	331.7	331.7	331.7	331.7	331.7	331.7	331.7	331.7	331.7
SMP	kt pw	323.4	323.4	323.4	323.4	323.4	323.4	323.4	323.4	323.4	323.4	323.4
JAPAN												
Direct payments	JPY/kg	11.4	11.8	11.8	11.8	11.8	11.8	11.8	11.8	11.8	11.8	11.8
Cheese tariff[5]	%	31.2	31.2	31.2	31.2	31.2	31.2	31.2	31.2	31.2	31.2	31.2
Tariff-quotas												
Butter	kt pw	1.9	1.9	1.9	1.9	1.9	1.9	1.9	1.9	1.9	1.9	1.9
In-quota tariff	%	35.0	35.0	35.0	35.0	35.0	35.0	35.0	35.0	35.0	35.0	35.0
Out-of-quota tariff	%	732.6	732.6	732.6	732.6	732.6	732.6	732.6	732.6	732.6	732.6	732.6
SMP	kt pw	115.7	115.7	115.7	115.7	115.7	115.7	115.7	115.7	115.7	115.7	115.7
In-quota tariff	%	15.8	15.8	15.8	15.8	15.8	15.8	15.8	15.8	15.8	15.8	15.8
Out-of-quota tariff	%	210.4	210.4	210.4	210.4	210.4	210.4	210.4	210.4	210.4	210.4	210.4
WMP	kt pw	0.0	0.0	0.0	0.0	0.0	0.0	0.0	0.0	0.0	0.0	0.0
In-quota tariff	%	24.0	24.0	24.0	24.0	24.0	24.0	24.0	24.0	24.0	24.0	24.0
Out-of-quota tariff	%	316.2	316.2	316.2	316.2	316.2	316.2	316.2	316.2	316.2	316.2	316.2
KOREA												
Tariff-quotas												
Butter	kt pw	0.4	0.4	0.4	0.4	0.4	0.4	0.4	0.4	0.4	0.4	0.4
In-quota tariff	%	40.0	40.0	40.0	40.0	40.0	40.0	40.0	40.0	40.0	40.0	40.0
Out-of-quota tariff	%	89.0	89.0	89.0	89.0	89.0	89.0	89.0	89.0	89.0	89.0	89.0
SMP	kt pw	1.0	1.0	1.0	1.0	1.0	1.0	1.0	1.0	1.0	1.0	1.0
In-quota tariff	%	20.0	20.0	20.0	20.0	20.0	20.0	20.0	20.0	20.0	20.0	20.0
Out-of-quota tariff	%	176.0	176.0	176.0	176.0	176.0	176.0	176.0	176.0	176.0	176.0	176.0
WMP	kt pw	0.6	0.6	0.6	0.6	0.6	0.6	0.6	0.6	0.6	0.6	0.6
In-quota tariff	%	40.0	40.0	40.0	40.0	40.0	40.0	40.0	40.0	40.0	40.0	40.0
Out-of-quota tariff	%	176.0	176.0	176.0	176.0	176.0	176.0	176.0	176.0	176.0	176.0	176.0

Table A.36. Main policy assumptions for dairy markets *(cont.)*

Calendar year

		Avg 2009-11est.	2012	2013	2014	2015	2016	2017	2018	2019	2020	2021
MEXICO												
Butter tariff	%	0.0	0.0	0.0	0.0	0.0	0.0	0.0	0.0	0.0	0.0	0.0
Tariff-quotas												
Cheese	kt pw	9.4	9.4	9.4	9.4	9.4	9.4	9.4	9.4	9.4	9.4	9.4
In-quota tariff	%	50.0	50.0	50.0	50.0	50.0	50.0	50.0	50.0	50.0	50.0	50.0
Out-of-quota tariff	%	125.1	125.1	125.1	125.1	125.1	125.1	125.1	125.1	125.1	125.1	125.1
SMP	kt pw	90.0	90.0	90.0	90.0	90.0	90.0	90.0	90.0	90.0	90.0	90.0
In-quota tariff	%	0.0	0.0	0.0	0.0	0.0	0.0	0.0	0.0	0.0	0.0	0.0
Out-of-quota tariff	%	125.1	125.1	125.1	125.1	125.1	125.1	125.1	125.1	125.1	125.1	125.1
Liconsa social program	MXN mn	2 825.0	2 551.5	2 551.5	2 551.5	2 551.5	2 551.5	2 551.5	2 551.5	2 551.5	2 551.5	2 551.5
RUSSIAN FEDERATION												
Butter tariff	%	15.0	15.0	15.0	15.0	15.0	15.0	15.0	15.0	15.0	15.0	15.0
Cheese tariff	%	15.0	15.0	15.0	15.0	15.0	15.0	15.0	15.0	15.0	15.0	15.0
UNITED STATES												
Milk support price[1]	USDc/litre	22.5	22.5	22.5	22.5	22.5	22.5	22.5	22.5	22.5	22.5	22.5
Target price[6]	USDc/litre	37.3	37.3	37.3	37.3	37.3	37.3	37.3	37.3	37.3	37.3	37.3
Butter support price	USD/t	2 315.0	2 315.0	2 315.0	2 315.0	2 315.0	2 315.0	2 315.0	2 315.0	2 315.0	2 315.0	2 315.0
SMP support price	USD/t	1 763.7	1 763.7	1 763.7	1 763.7	1 763.7	1 763.7	1 763.7	1 763.7	1 763.7	1 763.7	1 763.7
Butter tariff-quota	kt pw	13.1	13.1	13.1	13.1	13.1	13.1	13.1	13.1	13.1	13.1	13.1
In-quota tariff	%	10.1	10.1	10.1	10.1	10.1	10.1	10.1	10.1	10.1	10.1	10.1
Out-of-quota tariff	%	112.0	112.0	112.0	112.0	112.0	112.0	112.0	112.0	112.0	112.0	112.0
Cheese tariff-quota	kt pw	135.0	135.0	135.0	135.0	135.0	135.0	135.0	135.0	135.0	135.0	135.0
In-quota tariff	%	12.3	12.3	12.3	12.3	12.3	12.3	12.3	12.3	12.3	12.3	12.3
Out-of-quota tariff	%	87.0	87.0	87.0	87.0	87.0	87.0	87.0	87.0	87.0	87.0	87.0
Subsidised export limits[2]												
Butter	kt pw	21.0	21.0	21.0	21.0	21.0	21.0	21.0	21.0	21.0	21.0	21.0
SMP	kt pw	68.0	68.0	68.0	68.0	68.0	68.0	68.0	68.0	68.0	68.0	68.0
INDIA												
Milk tariff	%	80.0	80.0	80.0	80.0	80.0	80.0	80.0	80.0	80.0	80.0	80.0
Butter tariff	%	40.0	40.0	40.0	40.0	40.0	40.0	40.0	40.0	40.0	40.0	40.0
Cheese tariff	%	40.0	40.0	40.0	40.0	40.0	40.0	40.0	40.0	40.0	40.0	40.0
Whole milk powder tariff	%	20.0	20.0	20.0	20.0	20.0	20.0	20.0	20.0	20.0	20.0	20.0
SOUTH AFRICA												
Milk powder tariff-quota	kt pw	4.5	4.5	4.5	4.5	4.5	4.5	4.5	4.5	4.5	4.5	4.5
In-quota tariff	%	20.0	20.0	20.0	20.0	20.0	20.0	20.0	20.0	20.0	20.0	20.0
Out-of-quota tariff	%	80.8	80.8	80.8	80.8	80.8	80.8	80.8	80.8	80.8	80.8	80.8

Note: The source for tariffs and Tariff Rate Quotas (except Russia) is AMAD (Agricultural market access database). The tariff and TRQ data are based on Most Favoured Nation rates scheduled with the WTO and exclude those under preferential or regional agreements, which may be substantially different. Tariffs are simple averages of several product lines. Specific rates are converted to ad valorem rates using world prices in the Outlook. Import quotas are based on global commitments scheduled in the WTO rather than those allocated to preferential partners under regional or other agreements.

1. For manufacturing milk.
2. The effective volume of cheese and SMP subsidised exports will be lower reflecting the binding nature of subsidised export limits in value terms.
3. EU farmers also benefit from the Single Farm Payment (SFP) Scheme, which provides flat-rate payments independent from current production decisions and market developments. For the new member states, payments are phased in with the assumption of maximum top-ups from national budgets up to 2013 through the Single Area Payment (SAP), and through the (SFP) from 2014. Due to modulation, an increasing share of the total SFP will go to rural development spending rather than directly to farmers.
4. Total quota, EU27 starting in 1999.
5. Excludes processed cheese
6. The counter-cyclical payment for milk is determined as a percentage difference between the target price and the Boston class I price. The difference is set at 34% in 2007 and 2008, at 45% in 2009-2012 and 34% thereafter. The target price is adjusted by 45% of the percentage difference between the National Average Dairy Feed Rations Cost and the target cost of feed rations of 16.20USD/100kg between 2009 and 2012 and 20.94USD/100kg thereafter.

Source: OECD and FAO Secretariats.

StatLink http://dx.doi.org/10.1787/888932643485

Table B.1. Information on food price changes

	Total inflation % change (y to y)		Food inflation % change (y to y)[1]		Expenditure Share of Food		Food contribution to total change in inflation[2]	
	2011	2012	2011	2012	2011	2012	2011	2012
OECD								
Australia	2.7	3.7	2.5	1.8	12.8	12.8	0.3	0.2
Austria	2.5	3.0	2.7	1.5	12.0	12.0	0.3	0.2
Belgium	3.2	3.6	2.1	2.7	17.4	17.4	0.2	0.3
Canada	2.3	2.5	1.9	4.9	11.5	11.5	0.2	0.6
Chile	2.7	4.2	5.5	8.9	18.9	18.9	0.7	1.1
Czech Republic	1.7	3.5	4.3	7.0	17.0	17.0	0.5	0.8
Denmark	2.7	2.8	2.4	5.7	11.5	11.5	0.3	0.7
Estonia	5.2	4.5	11.4	4.0	21.7	21.7	1.4	0.5
Finland	3.0	3.2	4.6	5.0	13.4	13.4	0.6	0.6
France	1.8	2.3	0.1	3.6	14.7	14.7	0.0	0.4
Germany	2.0	2.1	2.8	3.0	10.4	10.4	0.3	0.4
Greece	5.2	2.3	2.4	3.0	17.1	17.1	0.3	0.4
Hungary	3.9	5.4	7.2	5.5	19.6	19.6	0.9	0.7
Iceland	1.9	6.5	1.5	6.2	14.9	14.9	0.2	0.7
Ireland	1.7	2.2	0.3	0.3	11.7	11.7	0.0	0.0
Israel	3.6	2.0	5.8	0.6	14.3	14.3	0.7	0.1
Italy	2.1	3.2	1.6	2.4	16.3	16.3	0.2	0.3
Japan	-0.6	0.1	0.0	1.4	19.0	19.0	0.0	0.2
Korea	3.4	3.4	10.2	5.9	14.4	14.4	1.2	0.7
Luxembourg	3.2	2.9	2.5	1.7	11.1	11.1	0.3	0.2
Mexico	3.8	4.0	3.6	7.7	18.9	18.9	0.4	0.9
Netherlands	2.0	2.5	0.8	2.3	11.3	11.3	0.1	0.3
New Zealand	4.0	4.1	4.8	3.1	17.4	17.4	0.8	0.5
Norway	2.0	0.5	-1.6	2.2	13.3	13.3	-0.2	0.3
Poland	3.7	4.6	4.9	4.7	24.1	24.1	0.6	0.6
Portugal	3.6	3.5	2.2	3.3	18.1	18.1	0.3	0.4
Slovak Republic	3.0	3.9	6.2	3.5	18.4	18.4	0.7	0.4
Slovenia	1.8	2.3	3.7	3.9	17.0	17.0	0.4	0.5
Spain	3.3	2.0	0.9	2.3	18.2	18.2	0.1	0.3
Sweden	2.5	1.9	0.9	1.1	13.9	13.9	0.1	0.1
Switzerland	0.3	-0.8	-2.8	-2.3	10.8	10.8	-0.3	-0.3
Turkey	4.9	10.6	7.1	11.7	26.8	26.8	0.8	1.4
United Kingdom	4.0	3.6	6.3	3.5	11.8	11.8	0.8	0.4
United States	1.6	2.9	2.1	5.3	7.8	7.8	0.2	0.6
OECD Total	2.1	2.8	2.6	4.3	..	..	..	..
Enhanced Engagement								
Brazil	6.0	6.2	10.4	6.9	22.5	22.5	2.3	1.6
China	4.9	4.5	10.3	10.5	33.6	33.6	3.5	3.5
India	9.3	5.3	9.3	7.7	35.4	35.4	3.3	2.7
Indonesia	7.0	3.7	16.2	3.3	19.6	19.6	3.2	0.6
Russian Federation	9.6	4.1	14.2	2.1	32.8	32.8	4.7	0.7
South Africa	3.6	6.4	3.1	10.5	18.3	18.3	0.6	1.9

Table B.1. Information on food price changes *(cont.)*

	Total inflation % change (y to y)		Food inflation % change (y to y)[1]		Expenditure Share of Food		Food contribution to total change in inflation[2]	
	2011	2012	2011	2012	2011	2012	2011	2012
Non OECD								
Algeria	4.1	9.4	3.0	12.3	43.8	43.8	1.3	5.4
Argentina	10.6	9.7	13.1	8.1	20.3	20.3	2.7	1.6
Bangladesh	9.0	9.7	11.9	10.9	52.0	52.0	6.2	5.7
Benin	5.1	5.9	14.9	-2.2	31.9	31.9	4.8	-0.1
Bolivia	8.4	5.8	14.0	4.2	39.3	39.3	5.5	1.6
Botswana	7.9	8.8	4.7	8.9	23.7	23.7	1.1	2.1
Bulgaria	2.3	3.1	4.5	4.6	37.2	37.2	1.7	1.7
Burkina Faso	0.5	2.8	5.4	3.0	36.8	36.8	2.0	1.1
Columbia	3.4	3.5	4.8	4.9	34.7	34.7	1.7	1.7
Costa Rica	4.8	4.2	7.8	3.3	25.7	25.7	2.0	0.8
Côte d'Ivoire	4.6	1.8	12.9	1.2	24.8	24.8	3.2	0.3
Croatia	2.4	1.2	3.7	2.2	36.3	36.3	1.3	0.8
Dominican Republic	6.6	6.5	6.7	8.8	29.2	29.2	2.0	2.6
Ecuador	4.9	5.3	4.9	6.1	31.7	31.7	1.5	1.9
Egypt	11.3	9.2	18.4	11.4	38.1	38.1	7.0	4.3
El Salvador	2.3	4.7	6.7	3.0	26.3	26.3	1.8	0.8
Ethiopia	17.7	32.0	13.6	41.4	57.0	57.0	7.8	7.3
Gambia	5.0	6.8	4.4	5.6	33.8	33.8	1.5	0.3
Ghana	9.1	8.7	4.8	4.5	37.0	37.0	1.8	1.7
Guatelama	4.9	5.4	6.2	11.3	28.6	28.6	1.8	3.2
Haiti	6.2	8.3	7.8	7.7	50.4	50.4	3.9	3.9
Honduras	6.4	5.4	8.4	2.2	31.8	31.8	2.7	0.7
Hong Kong	3.6	6.1	8.2	10.9	16.1	16.1	1.3	1.8
Iraq	2.1	5.4	4.4	3.9	35.0	35.0	1.5	1.4
Jordan	5.0	3.4	5.2	3.9	35.2	35.2	1.8	1.4
Kenya	5.4	18.3	8.6	24.6	36.0	36.0	3.1	8.9
Laos	6.1	6.7	8.9	8.8	41.0	41.0	3.7	3.6
Lebanon	5.8	2.6	6.3	5.6	19.9	19.9	1.3	1.1
Madagascar	11.2	6.0	21.1	6.6	60.0	60.0	3.9	0.3
Malawi	6.5	10.3	2.9	6.1	56.3	56.3	1.6	3.4
Malaysia	2.4	2.7	3.6	4.8	25.0	25.0	0.9	1.2
Mauritania	6.4	5.7	7.3	5.2	53.1	53.1	3.9	0.3
Moldavia	6.7	6.9	7.1	5.1	34.1	34.1	2.4	0.3
Morocco	2.2	0.9	4.4	1.7	40.4	40.4	1.8	0.7
Mozambique	16.1	5.1	20.0	4.9	65.3	65.3	13.1	0.8
New Caledonia	2.2	2.1	3.7	2.8	21.0	21.0	0.8	0.1
Nicaragua	8.2	8.6	10.9	9.6	26.1	26.1	2.8	2.5
Niger	4.2	-0.7	5.9	-1.8	29.0	29.0	1.7	-0.5
Nigeria	12.1	10.9	10.3	13.1	51.8	51.8	5.3	6.8
Pakistan	13.9	10.1	20.2	9.2	45.5	45.5	9.2	4.2
Panama	4.8	6.1	4.9	7.0	33.6	33.6	1.6	0.3
Paraguay	7.8	4.4	12.3	2.0	39.1	39.1	4.8	0.8
Peru	2.2	4.3	2.3	7.5	29.0	29.0	0.7	2.2
Philipinnes	4.0	4.1	3.0	3.2	39.0	39.0	1.2	1.2
Réunion	2.0	1.8	-0.1	4.4	13.2	13.2	0.0	0.6
Romania	7.0	2.7	7.2	0.1	37.4	37.4	2.7	0.0
Rwanda	-1.2	7.9	-1.3	12.8	48.4	48.4	-0.6	6.2
Senegal	2.3	1.9	8.0	2.0	53.4	53.4	4.3	1.1
Singapore	5.1	4.8	2.7	3.8	8.5	8.5	0.2	0.3
Sri Lanka	6.3	3.8	10.3	-0.2	41.0	41.0	4.2	-0.1
Syria	7.1	17.7	12.0	15.7	41.8	41.8	5.0	1.1
Taiwan	1.1	2.4	2.0	4.5	16.6	16.6	0.3	0.8
Tanzania	6.4	19.7	10.0	26.2	47.8	47.8	4.8	12.5
Thailand	3.0	3.4	5.9	7.7	33.0	33.0	1.9	2.5
Tunisia	3.7	5.1	3.7	6.5	33.8	33.8	1.2	2.2
Uganda	5.6	25.7	3.6	27.3	27.2	27.2	1.0	7.4
Uruguay	7.3	8.0	8.4	6.8	19.2	19.2	1.6	0.5
Venezuela	28.9	26.3	37.6	30.1	29.5	29.5	11.1	8.9
Zambia	9.0	6.4	5.2	6.1	52.5	52.5	2.7	0.5

1. CPI food : definition based on national sources
2. Contribution is column 2x3/100

Source: OECD and national sources (for details, see the online version of tables).

StatLink http://dx.doi.org/10.1787/888932643504

ORGANISATION FOR ECONOMIC CO-OPERATION AND DEVELOPMENT

The OECD is a unique forum where governments work together to address the economic, social and environmental challenges of globalisation. The OECD is also at the forefront of efforts to understand and to help governments respond to new developments and concerns, such as corporate governance, the information economy and the challenges of an ageing population. The Organisation provides a setting where governments can compare policy experiences, seek answers to common problems, identify good practice and work to co-ordinate domestic and international policies.

The OECD member countries are: Australia, Austria, Belgium, Canada, Chile, the Czech Republic, Denmark, Estonia, Finland, France, Germany, Greece, Hungary, Iceland, Ireland, Israel, Italy, Japan, Korea, Luxembourg, Mexico, the Netherlands, New Zealand, Norway, Poland, Portugal, the Slovak Republic, Slovenia, Spain, Sweden, Switzerland, Turkey, the United Kingdom and the United States. The European Union takes part in the work of the OECD.

OECD Publishing disseminates widely the results of the Organisation's statistics gathering and research on economic, social and environmental issues, as well as the conventions, guidelines and standards agreed by its members.

THE FOOD AND AGRICULTURE ORGANIZATION (FAO) OF THE UNITED NATIONS

The Food and Agriculture Organization of the United Nations leads international efforts to defeat hunger. FAO's mandate is to raise levels of nutrition, improve agricultural productivity, better the lives of rural populations and contribute to the growth of the world economy. Serving both developed and developing countries, FAO acts as a neutral forum where all nations meet as equals to negotiate agreements and debate policy. FAO is also a source of knowledge providing access to information in print and electronic format. We help developing countries and countries in transition modernize and improve agriculture, forestry and fisheries practices and ensure good nutrition for all. Since our founding in 1945, we have focused special attention on developing rural areas, home to 70 per cent of the world's poor and hungry people. FAO's activities comprise four main areas: putting information within reach; sharing policy expertise; providing a meeting place for nations; bringing knowledge to the field.

OECD PUBLISHING, 2, rue André-Pascal, 75775 PARIS CEDEX 16
(51 2012 05 1 P) ISBN 978-92-64-17302-6 – No. 59953 2012-02